Lasers in Materials Processing

Edited by E. A. Metzbower

Conference Proceedings
American Society for Metals

Published by
American Society for Metals
Metals Park, Ohio 44073

RODMAN PUBLIC LIBRARY

Copyright © 1983
by the
AMERICAN SOCIETY FOR METALS
All rights reserved

No part of this book may be reproduced, stored in a
retrieval system, or transmitted, in any form or by any
means, electronic, mechanical, photocopying, recording, or
otherwise, without the prior written permission of the
publisher.

Nothing contained in this book is to be construed as a grant
of any right of manufacture, sale, or use in connection with
any method, process, apparatus, product, or composition,
whether or not covered by letters patent or registered
trademark, nor as a defense against liability for the
infringement of letters patent or registered trademark.

Library of Congress Catalog Card Number: 83-072954

ISBN: 0-87170-173-1

SAN: 204-7586

Manufactured by Publishers Choice Book Mfg. Co.
Mars, Pennsylvania 16046

PREFACE

The concept for an international conference on the applications of lasers in materials processing evolved from informal discussions and an impromptu meeting in Chicago in February 1981. A tentative date was picked -- early 1983 -- and an organizing committee was formed. After many discussions, the date and location were fixed, a call for papers was sent out and the conference was on its way.

The organizing committee consisted of: Professor Stephen Copley, University of South California (co-chairperson); Professor Jyoti Mazumder, University of Illinois at Urbana-Champaign; Dr. Edward A. Metzbower, Naval Research Laboratory (co-chairperson); and Professor William Steen, Imperial College, London.

Because of the response to the call for papers, a three-day conference, organized into five distinct areas, evolved. The different areas are: fundamental interactions of lasers and materials; marking and cutting; heat treating; cladding and alloying; and welding. Over 140 people attended the conference and judging by the comments received on the survey sheets, the conference was an outstanding success.

On behalf of the organizing committee, and the conference attendees, I would like to express my sincere thanks to the ASM staff.

E. A. Metzbower
Naval Research Laboratory
Washington, DC 20375

10 May 1983

TABLE OF CONTENTS

LASERS FOR MATERIALS PROCESSING .... (paper not available)
  M. Bass

ANALYSIS OF HIGH POWDER CO_2 LASER-MATERIAL INTERACTION RESULTS WITH HIGH SPEED
PHOTOGRAPHY AND 10.6um ABSORPTION MEASUREMENTS 1
  V. Donati, L. Garifo, R. Menin, F. Pandavese, M. Onorato, P. Savorelli

THE EFFECTS OF LASER SHOCK PROCESSING ON FATIGUE PROPERTIES OF 2024-T3 ALUMINUM .. 7
  A. H. Clauer, C. J. Walters, S. C. Ford

NONLINEAR DYNAMIC TEMPERATURE CHARACTERIZATION OF PULSED LASER ANNEALING OF SEMI-
CONDUCTORS 24
  D. L. Kwong, R. Kwor, C. Paz de Araujo

HEAT FLOW-ACOUSTIC EMISSION-MICROSTRUCTURE CORRELATIONS IN RAPID SURFACE
SOLIDIFICATION 37
  R. B. Clough, H. N. G. Wadley, R. Mehrabian

LASER MARKING TECHNIQUES 48
  B. Bernard

LASER MARKING OF COMPONENT PARTS 54
  A. V. Gress

ANALYSIS OF AN IMPROVED MODEL OF REACTIVE GAS ASSISTED LASER CUTTING .... (paper not
available)
  D. Schuocker

INVESTIGATIONS IN OPTIMIZING THE LASER CUTTING PROCESS 64
  F.O. Olsen

AN OVERVIEW OF DRILLING WITH SOLID STATE LASERS .... (paper not available)
  S. Bolin

LASER SHAPING OF MATERIALS 82
  S. M. Copley, R. Wallace, M. Bass

PARAMETRIC EVALUATIONS OF LASER/CLAD INTERACTIONS FOR HARDFACING APPLICATIONS ... 94
  J. I. Nurminen

LASER SURFACE TREATMENT BY RAPID SOLIDIFICATION 108
  A. Tiziani, L. Giordano, E. A. Ramous

LASER SURFACE TREATMENT OF CHROMIUM ELECTROPLATE ON MEDIUM CARBON STEELS 116
  G. Christodoulou, W.M. Steen

LASER SURFACE ALLOYING LOW CARBON STEELS 127
  T. Chande, J. Mazumder

LASER FUSING OF HARDFACING ALLOY POWDERS 138
  S. J. Matthews

A MODEL FOR SURFACE TENSION DRIVEN FLUID FLOW IN LASER SURFACE ALLOYING 150
  C. Chan, J. Mazumder, M. M. Chen

BASIC COMPUTER MODEL OF THE PULSED LASER DRILLING PROCESS WITH NEODYMIUM LASER . 159
  M. G. Jones, G. Georgalas, A. Brutus

LASER CLADDING WITH PNEUMATIC POWDER DELIVERY 166
  V. Weerasinghe, W. M. Steen

LASER PROCESSING OF PLASMA-SPRAYED NiCr COATINGS 176
  H. Bhat, H. Herman, R. J. Coyle, Jr.

A MICROSTRUCTURAL STUDY OF PULSED AND CONTINUOUS LASER WELDED STAINLESS STEEL .. 185
  R.J. Coyle, Jr.

SOLIDIFICATION STRUCTURE AND FATIGUE CRACK PROPOGATION IN LB WELDS 196
  F. W. Fraser, E. A. Metzbower

LASER WELDING OF STEELS AND NICKEL ALLOYS 209
  R. F. Duhamel, C. M. Banas

CRITICAL THERMAL RADIUS IN LASER SOLDERING 218
  U. I. Chang

LASER HARD-SURFACING OF TURBINE BLADE SHROUD INTERLOCKS 230
  R. M. Macintyre

A CORRELATION BETWEEN DENDRITE-ARM-SPACING AND COOLING RATE FOR LASER-MELTED
Ti-15V-3Al-3Sn-3Cr 241
  T. C. Peng, S. M. L. Sastry, J. E. O'Neal

THE EFFECTS OF INCONEL 600 ON THE TOUGHNESS OF HY-STEEL LASER WELDS 248
  D. W. Moon, E. A. Metzbower

POWDER-FEED LAYERGLAZE[sm]/NARROW-GAP LASER WELDING OF TITANIUM 6Al-4V 255
  E. M. Breinan, D. B. Snow

LASER BEAM WELDING AT NIROP, A NAVY MANUFACTURING TECHNOLOGY PROGRAM 266
  E. A. Metzbower, R. A. Hella

ANALYSIS OF HIGH POWER CO₂ LASER-MATERIAL INTERACTION. RESULTS WITH HIGH SPEED PHOTOGRAPHY AND 10.6 μm ABSORPTION MEASUREMENTS.

V. Donati, L. Garifo, R. Menin, F. Pandarese
CISE S.p.A., Segrate, Milano

M. Onorato, P. Savorelli
Politecnico di Torino

ABSTRACT

An extensive experimental research program is being carried out on laser-matter interaction using an AVCO HPL cw CO_2 laser capable of up to 15 kW power, under experimental conditions typical of actual metalworking.

Results from time resolved absorption measurements coupled with high speed photography and monitoring of the high power incident laser beam intensity fluctuations are shown here. Results have been obtained operating the laser at the averaged power of 10 kW, using stainless steel (AISI 304) targets at translational speed of 2.3 m/min.

INTRODUCTION

The interaction of a CO_2 cw high power laser beam with a target material has been subject of intensive study during the past few years [1-14]. This study is of great importance in laser metalworking.

When the intensity of the laser beam is higher than 10^4 W/cm² as required e.g. in welding interaction is dominated by laser sustained absorption waves generated in metal vapors and in the volume above the material. More specifically at laser intensities higher than 10^7 W/cm² laser supported detonation waves can be generated, while for intensities in the range $3 \cdot 10^4 - 10^7$ W/cm² laser combustion waves can be observed.

The plasma production associated with laser combustion waves is in most cases detrimental, resulting in shielding of the beam and therefore attenuation. However, under appropriate conditions, in the case the surface is highly reflecting at the laser wavelenght, the plasma produced on the solid surface can be beneficial for a better energy coupling. As a matter of fact, when the plasma is produced, it absorbs the laser radiation effectively and reradiates absorbed energy at much shorter wavelength resulting in a better energy coupling to the solid. Plasma control appears to be therefore important. For this purpose, knowledge of plasma behaviour in the interaction region appears fundamental. An experimental research program is being carried out under experimental conditions typical of actual metalworking.

As part of this program in ref. 15 some fluidynamical aspects of the development and propagation of laser substained combustion waves are described. Moreover some results of spectroscopic analysis and time integrated absorption measurements are given. In ref. 16 the spatial distribution of backscattered 10.6 m radiation from the metal target and the surrounding vapors are reported. In ref. 17 preliminary results of time resolved absorption experiments are given. In the present work time resolved absorption measurements coupled with high speed photography and monitoring of the high power incident laser beam intensity fluctuations are carried out.

The reason for these further time resolved measurements is to clarify the effect of the observed [17] strong fluctuations of the incident laser beam used in the present research program and also found in other commercial lasers [17, 18]. The influence of these fluctuations on the interaction phenomena is not known. No sufficient attention on the effects of these fluctuations has so far been paid in the available literature, in which the most studied case is the pulsed laser-matter interaction and to a smaller extent the case of constant continuous wave laser.

EXPERIMENTAL SET UP

Experiments have been performed using an AVCO cw CO_2 laser capable of up to 15 kW power, equipped with an f/7 focoussing telescope. A schematic diagram of the experimental set up is shown in Fig. 1. Absorption measurements were carried out using a cw CO_2 10W probe laser. The beam of this source was attenuated down to 3 W and used as a probe crossing the interaction region in a direction parallel to the target surface. A beam-splitter divides the beam into two almost equally parts, one for monitoring and one for probing. The first one impinges on a detector giving the intensity reference signal, the second one after passing through the plume gives a signal proportional to the attenuation due to the plume.

Detectors are the pyroelectric type Molectron P1-72, with a suitable amplifier overall bandwidth of 1÷2000 Hz. A chopper, positioned at the exit of the probe laser, is used to restore dc level of the pyroelectric detectors. A focussing lens reduces the beam diameter to 0.4 mm at the plume position, in order to have good spatial resolution. Measurements were carried out at various height from the target.

A small diffusing element picks up a negligible fraction of the incident high power radiation. A suitable positioned detector reveals this radiation giving a signal proportional to the fluctuations of the main laser beam. Detector type is the same used for the absorption measurements of the probe laser beam.

The target is put on a X-Y moving table, having a maximum velocity of 20 m/min.

The high speed photography camera, not shown in Fig. 1, is located at 45 cm distance normally to the plane defined by the main beam optical axis and the translational motion direction of the target. The camera is a rotating prism device capable up to 10.000 frames per second.

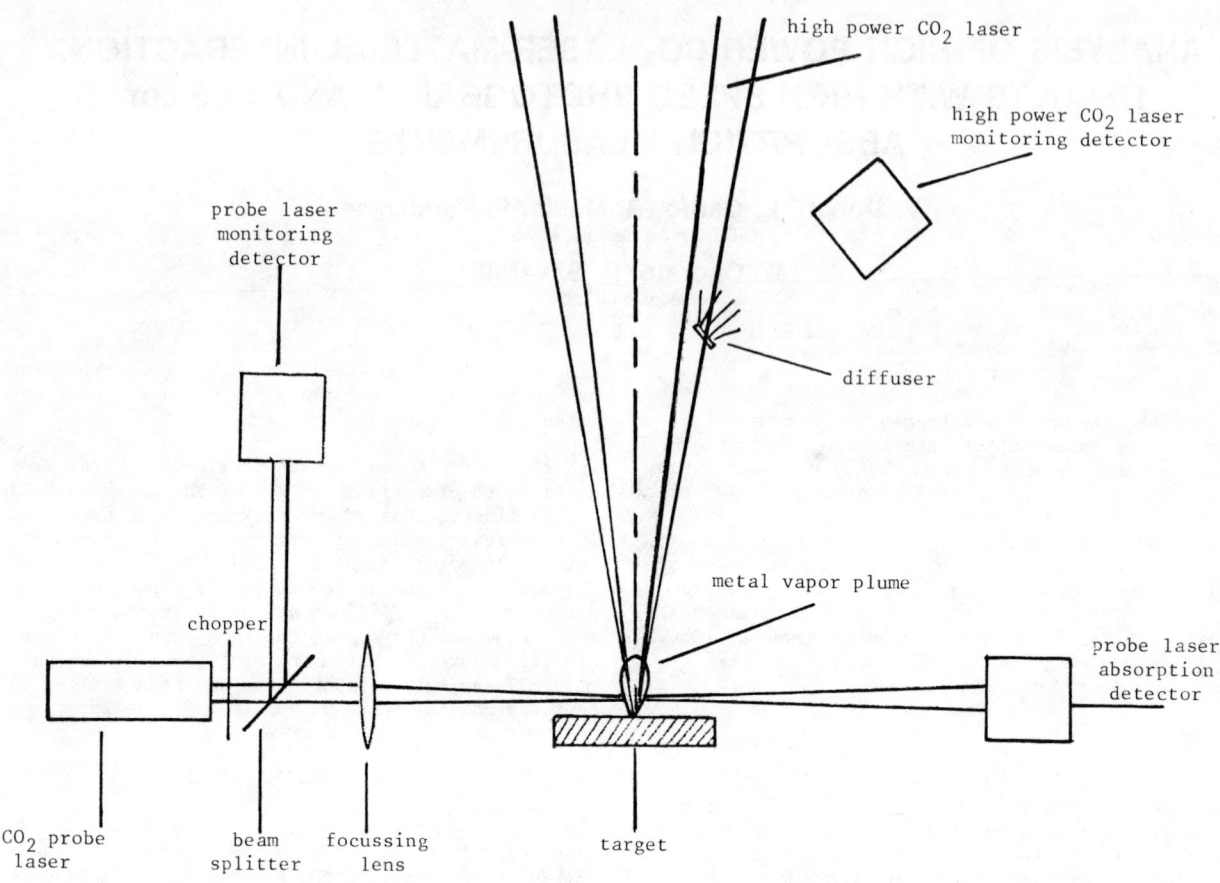

Fig. 1 - Experimental set up

Time correlation between film frames and detectors output is achieved as follows: the signal from a 100 KHz quarz cristal oscillator is downscaled to 1 KHz and simultaneously applied to a marking lamp on the film and to a multichannel fast chart recorder. The signals from the probe laser detectors and the main beam monitoring detector are applied to the same recorder.

RESULTS AND COMMENTS

All results presented here have been obtained operating the laser at an averaged power of 10 kW as measured with an absorbing cone. The estimated beam intensity at the focal spot was about 10^6 W/cm^2.

The amplitude and the shape of the fluctuations of the incident laser beam power are different if the beam is totally absorbed by a calorimetric cone (Fig. 2 upper trace) or if a metal target is present at the focal point (Fig. 2 lower trace). The peak to peak amplitude of the fluctuations is larger in the second case than in the first case, where the structure appears to be also simpler. The difference is almost certainly caused by optical feedback. This phenomenon has already been mentioned in ref. [12]. In the case of the upper trace of Fig. 2, the peak to peak fluctuation amplitude is about 35% of the maximum value. This fluctuations are generated by the ripple of the power supplies. The target material is stainless steel AISI 304, the workpiece surface has not been specially treated. The target translational speed is 2.3 m/min.

From the analysis of high speed photography films the generation and propagation along the direction of the impinging laser beam of the laser sustained combustion wave (LSC) can be observed. The photographic sequence shown in Figs. 3 describes the LSC evolution. It is possible to identify among the less luminescent vapor present above the target a more brillant glowing

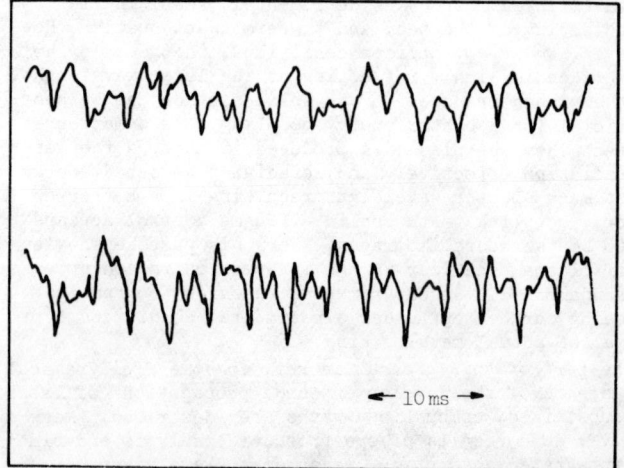

Fig. 2 - Beam power fluctuations of the incident laser at average power of 10 kW.
Upper trace: laser beam totally absorbed
Lower trace: laser beam focussed on metal target

fire ball, whose contour defines the LSC wave. In Fig. 3a the leading edge of a LSC is standing at a distance of 35 mm from the target surface and its thickness is about 14 mm. In Fig. 3b the wave has disappeared and a new one is developing in the vapor emitted from the metal. In next figures the wave propagates in the direction of the impinging laser.

In Figs. 3e and 3f a new wave if starting and developing.

Absorption measurements are shown in Fig. 4. The measuring point is situated 10 mm above the target. Traces 1 and 2 represents the output of the probe laser monitor detector and absorption detector respectively.

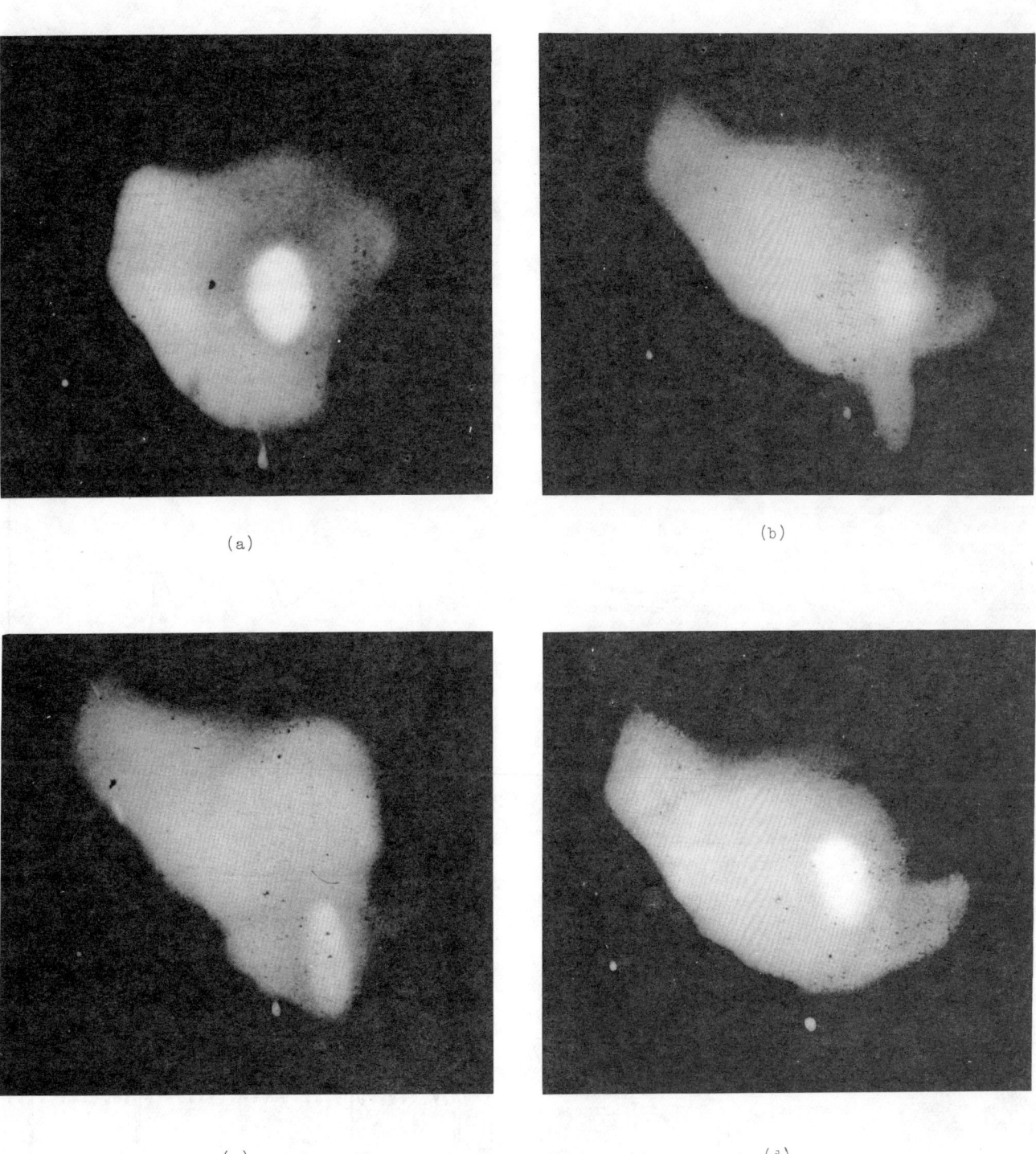

(a) (b) (c) (d)

Figs. 3 - Pictures from the high speed film

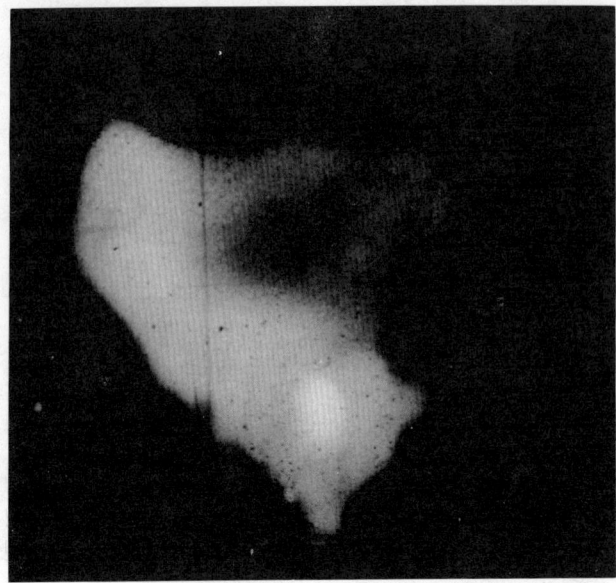 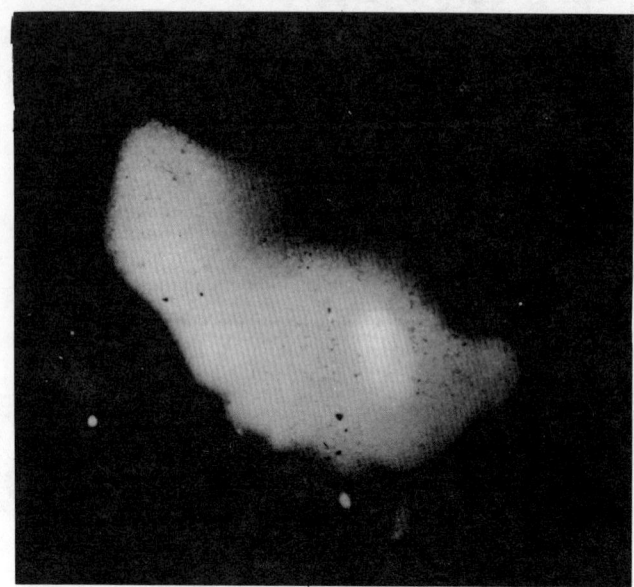

Figs. 3 (cont.) - Pictures from the high speed film

Trace 3 is the output of the high power laser monitor detector. Finally trace 4 is 1 KHz sincronizing reference signal. The ratio between the instantaneous values of trace 2 and 1 is proportional to the absorption due to the plume and reproduces its time dependence. It is clearly shown that time intervals of almost total absorption are followed by intervals of total transmission. In the same figure arrows point to the time instants corresponding to the pictures shown in Figs. 3. It is evident that absorption takes place when the fire ball intercepts the probe laser beam. No significant absorption is detected otherwise, even if vapor is present.

Measurements taken at a distance of 20 mm from the target surface give about the same results. At a distance of 30 mm no relevant absorption is detected.

It is not easy to establish from the present data the instantaneous correlation between the fluctuations of the main laser beam and the absorption measurements. Further investigations are needed, even though at first sight the averaged frequency of occurrence of absorption picks seems to coincide with the dominant 300 Hz component of main beam fluctuations.

Measurements have been repeated under the same conditions but using Helium as shielding gas. A typical nozzle (Linde type) for welding process has been used to generate a flow at 45 degrees from the beam axis. The Helium flux was 5.7 Nm^3/h. From the film appears that the vapor plume is confined very close to the metal surface. Absorption measurements, giving total transmission at a distance of 3 mm from the surface, confirm this observation.

It is planned an extensive investigation on plume behaviour under control by shielding gases.

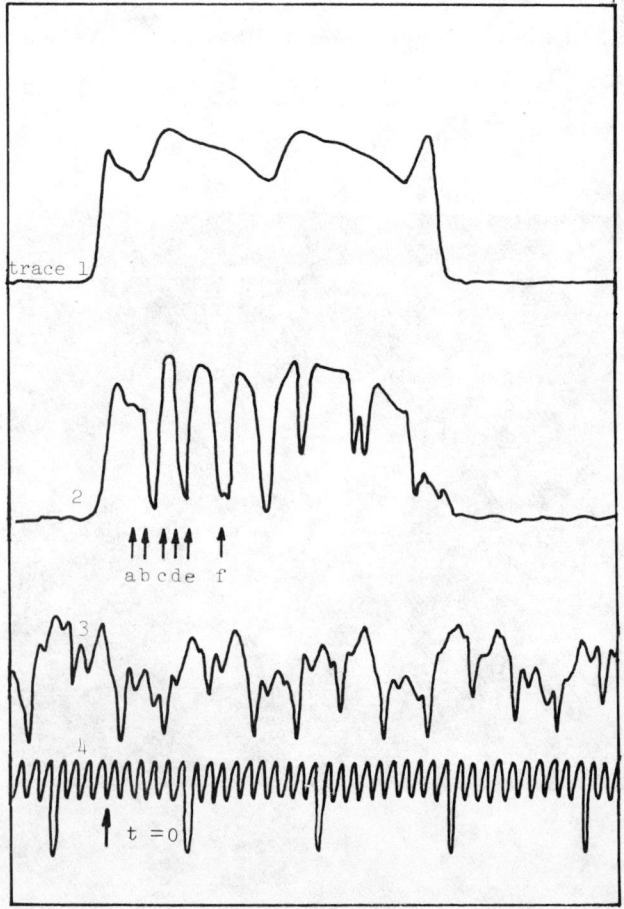

Fig. 4 - Results from absorption measurements

REFERENCES

1. C.J. Knight, AIAA Journal, vol. 17 n. 5 (1979)

2. M. Germano; M.S. Oggiano, VI Congresso Nazionale AIDAA, Roma (1981)

3. P.D. Thomas, AIAA Journal, vol. 13, n. 10 (1975)

4. E.L. Klosterman; S.R. Byron, AFWL-TR-74-003, Math. Sc. NW, Inc., Seattle, Wash (1973)

5. A.N. Pirri, AIAA Journal, vol. 15, n. 1 (1977)

6. E.L. Klosterman, MSNW-75-123-2, Math. Sc. NW Rept. Seattle, Wash (1975)

7. M.C. Fowler; D.C. Smith, Journal of Appl. Phys., vol. 46, n. 1 (1975)

8. Y.P. Raizer, Soviet Phys. JETP, vol. 31, n. 6 (1970)

9. R.G. Root, 3th Int. Symp. on Gas Flow and Chem. Lasers, Marseille (1980)

10. E.L. Klosterman, Journal of Appl. Phys., vol. 45, n. 11 (1974)

11. D.C. Smith, Optical Engineering, vol. 20, n. 6 (1981)

12. G. Herziger, 4th GCL Int. Symp. on Gas Flow and Chem. Lasers, Stresa (1982)

13. J.A. Woodroffe, 4th GCL Int. Symp. on Gas Flow and Chem. Lasers, Stresa (1982)

14. J.F. Ready, Proceed. of the IEEE, vol. 70, n. 6 (1982)

15. M. Cantello; V. Donati; L. Garifo; A.V. La Rocca; R. Menin; M. Onorato, 13th Int. Symp. on Shock Tubes & Waves, Niagara Falls (1981)

16. V. Donati; L. Garifo; A.V. La Rocca, R. Menin; M. Onorato, VI Congresso Nazionale AIMETA, Genova (1982)

17. M. Cantello; V. Donati; L. Garifo; R. Menin; F. Pandarese; A.V. La Rocca; M. Onorato; P. Savorelli, 4th GCL Int. Symp. on Gas Flow and Chem. Lasers, Stresa (1982)

18. G.C. Lim; W.M. Steen, Optics and Laser Technology, (1982).

8301-002

THE EFFECTS OF LASER SHOCK PROCESSING ON THE FATIGUE PROPERTIES OF 2024-T3 ALUMINUM

Allan H. Clauer, Craig T. Walters, and Stephan C. Ford
Battelle's Columbus Laboratories
Columbus, Ohio

INTRODUCTION

The effects of laser shock processing have been investigated in a number of metals and alloys with increases in hardness and tensile and fatigue strengths reported (1-7). A previous study of the fatigue response of laser shocked aluminum alloy plate containing simulated fastener holes showed marked increases in fatigue life in some cases. In addition, the study suggested certain process and geometry changes which would either further enhance the fatigue property improvements or aid in understanding and controlling the laser shock phenomenon influencing the properties (7). This paper describes the effects on fatigue life resulting from several different laser beam geometries and process conditions.

Both solid beam and annular beam geometries were used. The annular beam was added to determine whether a crack could be slowed down by encountering a laser shocked region. In addition, specimens were shocked from both sides simultaneously, from both sides consecutively with a momentum trap on the unirradiated side, and from one side only with a free surface opposite the irradiated side. The purpose of the momentum trap was to minimize the effect of the reflected wave from the surface opposite the irradiated surface. The one-side only shot without the momentum trap was to enable comparison to be made between the full effect of the tensile stress wave reflected from the surface opposite the irradiated side to the effects of a once through passage of the shock wave, i.e., with the momentum trap.

To understand the observed effects on the fatigue life, surface and in-depth residual stress distributions were determined for each of the laser beam geometries and process conditions. Both the residual stress and fatigue results are presented and discussed.

EXPERIMENTAL PROCEDURES

The 2024-T351 material was received as 0.25-inch-thick plate. The T351 condition consists of a solution treatment followed by a light roll-leveling pass which introduces a small amount of deformation (approximately 1 to 2 percent reduction) and natural aging at ambient temperature. The surfaces of the momentum - trapped specimens were given a superficial grinding. All other specimens were treated and tested with the as-received surface intact.

Fatigue Testing

The fatigue specimens were prepared in accordance with Figure 1. Short, narrow notches were electrodischarge machined into the sides of the hole to act as crack initiation sites and provide a consistent crack initiation behavior. The configuration of the hole and the laser shocked zones are shown to scale in Figure 1. The outer diameter of the solid and annular shocked zones are the same. The inner diameter of the annular zone is also indicated.

The fatigue test specimens were instrumented with crack measuring gages to monitor crack initiation and crack growth rates. They were cycled in an electro-hydraulic fatigue test machine in the tension-tension mode at a maximum stress of 15 ksi and a minimum stress of 1.5 ksi, i.e., R=0.1.

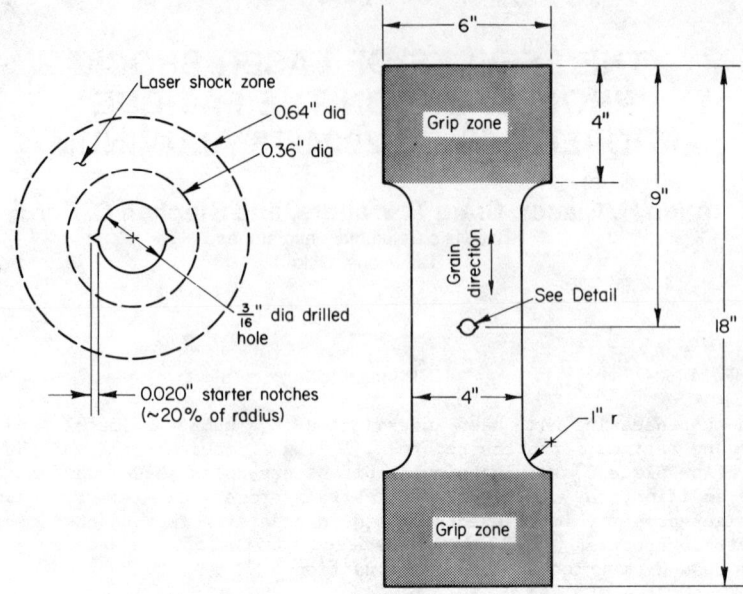

Figure 1. Fatigue Specimen and Detail of the Hole, Starting Notches, and Laser Shocked Zone. Shown here is the Annular Shaped Laser Shocked Zone; the Solid Beam is the same size as the Outside Diameter of the Annular Beam.

Laser Shock Processing

A six amplifier stage CGE high energy neodymium glass pulsed laser was used for the laser shock processing (LSP). Optics were set up to split the 200 J beam from the laser into two beams which were each routed through one additional amplifier. This arrangement permitted irradiations with up to 200 J per side in split beam irradiations. All laser irradiations were conducted with the eight stage laser system operating in a 30 ns pulse length mode. An aluminum coated plastic film (blow-off foil) was employed in the system to suppress superradiance prior to Q-switching. This method provides a sharply rising pulse (risetime less than 10 ns) which is essential to achieving significant shock pressures in the material. Beam diagnostics included a fast photodiode viewing a reflection from the central part of the beam after the sixth amplifier to provide pulse shape records, and two calibrated integrating photodiode assemblies to provide a measurement of total pulse energy per beam delivered to the target area. The energy readings were corrected for reflection losses at the acrylic beam entrance windows to the process chamber. All beams were focused with 1 m focal length antireflection coated lenses and target placement was in front of focus in the geometric region of the beam.

Residual Stress Specimens. Square plate-type specimens (4 X 4 X 0.25 inch) were machined from the as-received plate for the residual stress study. In most specimens, a 0.187-inch-diameter hole was drilled through the center of the plate to simulate a fastener hole. Before laser processing, the work surface of the plates was spray painted with metal primer and flat black paint as an opaque overlay for laser beam absorption. The transparent overlay materials were 1.5-inch-diameter by 0.125-inch-thick disks of either fused quartz or acrylic plastic with optical quality surfaces on both sides. The discs were pressed snugly against the painted surface by a clamping ring secured to the specimen holder with screws. The specimen holder was placed in a wooden enclosure with replaceable acrylic beam entrance ports to confine the debris generated by the explosion of the overlay disc.

The processing parameters employed in the residual stress measurements are summarized in Table 1. The laser shocking geometries are shown in Figure 1. For specimens R1 through R4, an attempt was made to vary fluence in a systematic manner. As a result of uncontrolled energy variation, two specimens ended up with nearly the same fluence. Specimens R5 and R6 were irradiated with a larger beam diameter (0.64 in. as opposed to 0.45 in.) to explore possible beam size effects. Specimen R7 was identical to R5 except that no hole was present so that the effects of the hole on the residual stresses could be assessed. Specimens R8 and R9 are repeats of R1 with acrylic plastic used in place of the fused quartz overlay. All of the above described specimens employed split-beam simultaneous irradiations. Specimen R10 was similar to R7 but irradiated from one side only.

Specimens R11 and R12 were processed using new conditions. Specimen R11 was irradiated simultaneously from both sides with an annular shaped beam (0.64 in. outside diameter, 0.36 in. inside diameter) (Figure 1). The annular beam was formed by mounting a circular aluminum blocking disk on a wire spider located near the target plane with the blocking disk axis coaligned with the axis of both the beam and the drilled hole.

Table 1. Laser Shock Processing Parameters for Aluminum
2024-T351 Residual Stress Samples

Specimen Number	Hole Diameter, in.	Beam Diameter, in.	Symmetry[a]	Overlay	Average Fluence, J/cm^2 Side 1	Side 2	Pulse Width, ns	Peak Power Density, 10^9 W/cm^2 Side 1	Side 2
R1	0.19	0.45	BS	quartz	~160	~160	~13	~12	~12
R2	0.19	0.45	BS	quartz	134	139	15	8.9	9.3
R3	0.19	0.45	BS	quartz	134	129	15	8.9	8.6
R4	0.19	0.45	BS	quartz	98	96	14	7.0	6.9
R5	0.19	0.64	BS	quartz	~77	~78	~13	~5.9	~6.0
R6	0.19	0.64	BS	quartz	69	66	13	5.3	5.1
R7	none	0.64	BS	quartz	~78	~78	13	~6.0	~6.0
R8	0.19	0.45	BS	acrylic	~160	~160	13	~12	~12
R9	0.19	0.45	BS	acrylic	150	146	15	10.0	9.7
R10	none	0.64	OS	quartz	~70	~70	~13	~5.4	~5.4
R11	0.19	0.64/0.36[b]	BS	quartz	80	79	18	4.4	4.4
R12	0.19	0.64	MT	quartz	82/82[c]		23/26[c]	3.6/3.2[c]	

(a) BS = both sides, OS = one side, MT = momentum trap (two separate one-sided irradiations).
(b) Annular beam (outside diameter/inside diameter).
(c) Side 1/Side 2.

Specimen R12 explored the effect of a momentum trap. It was irradiated with sequential one-sided irradiations using a 1.63 cm diameter solid circular beam. In this case, the unirradiated surface was backed up with a spring loaded 2024-T3 aluminum disk (1-inch-diameter X 0.25-inch-thick) during LSP to trap the shock wave and minimize any effects from a reflected tensile wave. Mineral oil was used to couple the momentum trap disk to the specimen, and a tapered cone "disk-catcher" was used to ensure that momentum trapping had occurred, i.e., the momentum trap would not slap back onto the back of the specimen. Before taking the second shot in the sequence, it was necessary to sand flat the plasma-induced "halo" crater effects created on the first surface irradiated so that the momentum trap for the second irradiation could be coupled into this surface. The shocked zone was deformed to a level slightly below the surrounding area so that sanding did not disturb the processed zone. The gap between this surface and the momentum trap surface was not measured, but it was assumed that they were coupled by the mineral oil.

Fatigue Test Specimens. Six fatigue test specimens were prepared and laser shock-processed with the processing conditions in Table 2, and the processing patterns in Figure 1. All specimens were processed with fused quartz overlays. Fatigue Specimens F1 and F2 were duplicate specimens processed with the standard solid split beam geometry. The fluence level was near 80 J/cm^2 with conditions similar to those of Specimens R5 and R6. Specimens F3 and F4 were annular beam processed with conditions similar to those of residual stress Specimen R11. Specimens F5 and F6 were momentum trap specimens with processing sequences similar to those for residual stress Specimen R12 except that Specimen F6 was not sanded between processing the opposite sides of the specimen.

Table 2. Laser Shock Processing Parameters for Aluminum
2024-T351 Fatigue Specimens

Sample Number	Beam Diameter, in.	Symmetry[a]	Overlay	Average Fluence, J/cm^2 Side 1	Side 2	Pulse Width, ns	Peak Power Density, 10^9 W/cm^2 Side 1	Side 2
F1	0.64	BS	quartz	81	80	~15	~5.4	~5.3
F2	0.64	BS	quartz	75	75	15	5.0	5.0
F3	0.64/0.36	BS	quartz	82	78	17	4.8	4.6
F4	0.64/0.36	BS	quartz	84	78	23	3.6	3.4
F5	0.64	MT	quartz	85/86[b]		23/24[b]	3.7/3.6[b]	
F6	0.64	MT	quartz	91/87[b]		20/20[b]	4.6/4.4[b]	

(a) BS = both sides, MT = momentum trap (two separate one-sided irradiations).
(b) Side 1/Side 2.

The residual stresses were measured using the two Inclined Angle X-ray diffraction technique. The conditions used chromium K-alpha radiation diffracted from (311) planes of the 2024 aluminum. Diffraction peak angular positions were determined employing a five-point parabolic regression procedure after correction for the Lorentz-polarization and absorption effects, and for a linearly sloping background intensity. Details of the diffractometer fixturing are outlined below:

- Incident Beam Divergence: 1.0 deg.
- Receiving Slit : 0.5 deg.
- Detector : Si(Li) set for 90 percent acceptance of the chromium K-alpha energy
- Counts per Point : 10,000
- Psi Rotation : 0.0-45.0 deg.
- $E/(1 + v)$ 2024-T351 aluminum : $7.8 \pm 0.07 \times 10^6$ psi
- Irradiated Area : 0.05 X 0.10 in. (long axis aligned in the tangential direction of measurement)

The value of the single crystal elastic constant $E/(1 + v)$ (where E is Young's Modulus and v is Poisson's ratio) for the crystallographic direction normal to the (311) planes of the 2024-T351 aluminum FCC lattice was determined experimentally in the course of the investigation by loading a simple rectangular 2024-T351 beam in four-point bending on the diffractometer and determining the change in the lattice spacing of the (311) planes as a function of applied stress determined by strain gages attached to the beam. The shape of a plot of the change in lattic spacing, d, as a function of applied stress was determined by linear least square regression. The single crystal elastic constant $E/(1 + v)$ in the [311] direction was then determined from the slope of a plot of the change in $d_{(311)}$ as a function of applied stress.

Material was removed for subsurface measurement of stress by electropolishing in nitric acid-methanol electrolytes to minimize alteration of the subsurface residual stress distribution. All data obtained as a function of depth were corrected for the effects of penetration of the radiation employed for residual stress measurement into the subsurface stress gradient, and for the stress relaxation which occurred as a result of the material removal.

RESULTS

The residual stress results are shown in Tables 3 to 8 and Figure 2 to 10. The residual stresses were measured at three sites in some specimens, while less extensive measurements were made in other specimens only to illuminate any similarities or differences due to process changes. The locations on the specimen surfaces at which X-ray diffraction stress measurements were made are shown in Figure 2. The in-depth measurements were made successively at these same locations after electropolishing away a surface layer.

Table 3. Residual Stress Results for Standard Split Beam Laser Shock Processing Around a Drilled Hole with a Fused Quartz Overlay

Specimen Number	Power[a] Density, W/cm²	Depth Below Surface, in.	A	B	C	D Unshocked
R1	12, 12 X 10⁹	0.000	-37.5	-47.5	8.3	-8.0
		0.002	-45.4	-45.1	4.1	-4.5
		0.007	-33.1	-31.3	-3.9	-5.6
		0.012	-27.0	-25.2	2.6	-2.5
		0.022	-16.3	-11.6	0.8	-1.3
R2	8.9, 9.3 X 10⁹	0.000		-52.7		-9.4
		0.002		-50.4		-4.0
		0.006		-45.5		-11.8
		0.015		-34.4		-1.3
		0.025		-16.9		-0.9
		0.035		-14.7		-4.1
		0.043		-6.4		-2.8
R4	7.0, 6.9 X 10⁹	0.000	-43.2	-57.3	-10.7	
		0.004	-43.7	-46.8	-25.8	
		0.008	-31.5	-46.2	-18.9	
		0.012	-33.1	-39.6	-3.0	
		0.023	-26.4	-24.9	-7.4	
R5	5.9, 6.0 X 10⁹	0.000		-58.9		
		0.002		-44.8		
		0.008		-33.9		
		0.012		-32.4		
		0.032		-11.7		
		0.044		-10.5		
		0.052		-9.5		

(a) The power density on each side of the specimen is shown: Side 1, Side 2.

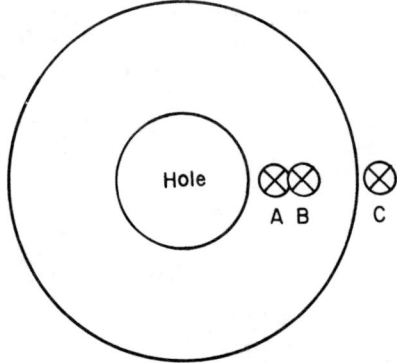

a. Drilled hole, specimens R1, R4, and R8

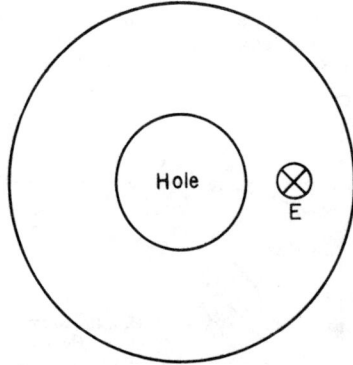

b. Drilled hole, specimens R2, R5, and R9

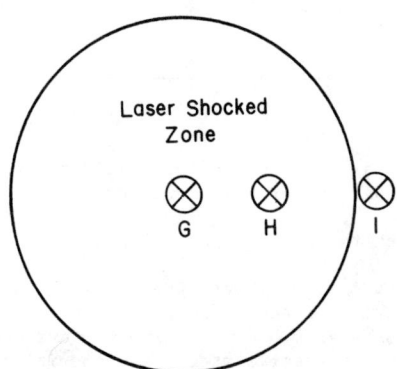

c. No hole, specimens R7 and R10

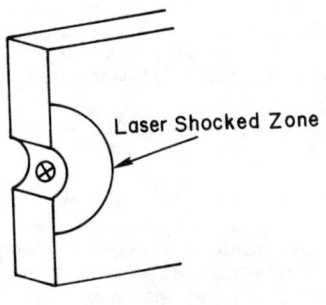

d. One side of drilled hole, specimens R3 and R6

Figure 2. Locations of the Residual Stress Measurements for the Various Specimen and Process Geometries. The Annular Beam Locations are shown in Figure 8.

The variation of the surface residual stresses across the radius of the laser shocked zone can be determined from the two specimens having no drilled hole (Specimens R7, R10). The results are shown in Figure 3 along with previous results on 7075-T6 using a water overlay (7). Each material shows a lower surface residual compressive stress in the center of the laser shocked region, a maximum at about halfway between the center and edge, and a tensile stress near or outside of the laser shocked zone. The lower value at the center of the spot is not understood. It is doubtful that it results from a lower beam intensity in the center of the beam. Rather, it may result from the elastic-plastic constraints of the material response.

<u>Influence of Power Density</u>. The residual stresses for the split solid beam irradiation at different power density levels are shown in Figure 4. These data are taken from Specimens R1, R2, R4, and R5, all of which had the hole drilled before shocking and used fused quartz transparent overlays. The readings all represent the region midway between the edge of the hole and the edge of the laser shocked zone (locations B in Figures 2a and 2b, and H in Figure 2c). Additional results for Specimen R7 without a drilled hole are shown in Table 5. The residual stresses in the as-received condition, unshocked, show a small surface residual compressive stress of about -8 ksi extending to slightly greater than 0.020 in. below the specimen surface. All the laser shocked conditions developed surface stresses of -27 to -59 ksi to projected depths of 0.040 in. or more. The compressive stress profile in Specimen R7 (Table 5) at location H is similar to that of Specimen R5 at the same power density in Figure 4. The trend of the magnitude of the surface stress with power density is shown in Figure 4b. Surprisingly, there is a linear increase in the magnitude of the compressive stress with decreasing power density. Included in Figure 4b is the result from the mid-point of the annular shaped laser shocked zone (Specimen R11, Location D in Figure 8).

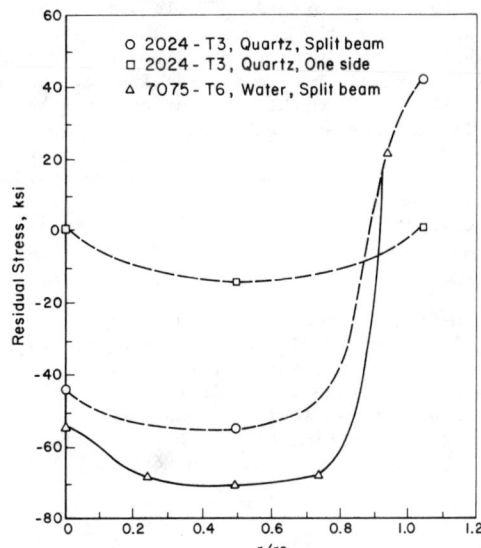

Figure 3. Variation of the Surface Residual Stress Along the Radius of the Laser Shocked Zone. r_o is the Radius of the Laser Shocked zone.

This specimen was processed at the lowest power density and its residual stress level is consistent with the trend in the solid beam data. Eventually, at still lower power densities, the magnitude of the surface compressive stresses must begin to decrease.

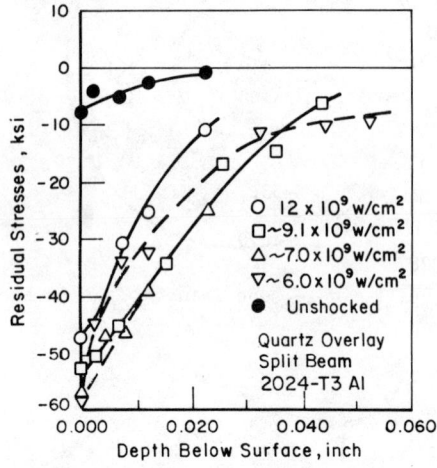 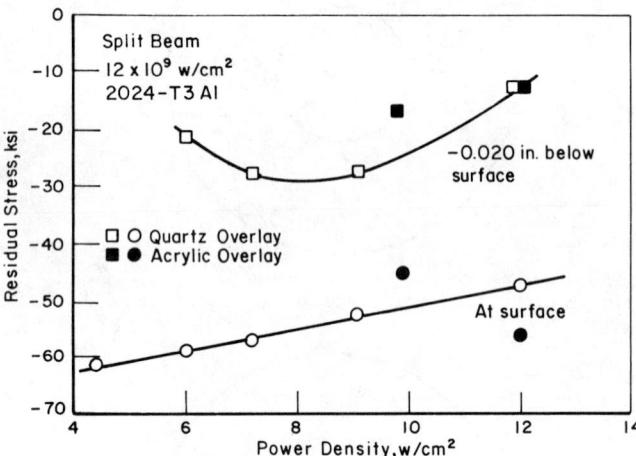

a. In-Depth Residual Stress Profiles

b. Variation of Surface Residual Stress with Laser Beam Power Density

Figure 4. Influence of Laser Beam Power Density on the Residual Stresses.

Not only the magnitude of the surface stresses, but also the depth variation of the stresses will have an effect on the crack initiation and propagation behavior. One indication of the influence of power density on the in-depth stress is shown by a plot of the magnitude of the stresses at 0.020 in. below the surface (Figure 4b). Figure 4b indicates there may exist a maximum level of compressive stress at some intermediate power density, e.g., about 8×10^9 W/cm^2. Higher power densities produce shallower residual stresses.

Influence of a Momentum Trap. For many applications it will not be possible to have line of sight access to both sides of the fatigue critical region for split beam shocking. For this reason, a procedure for irradiating from one side only was investigated, i.e., irradiating from one side with a momentum trap placed against the other side. The momentum trap carries away the tensile wave which would produce some distortion or bulging of the plate when reflected from the free back surface of the target area.

The momentum trap was a disk of the same material as the specimen to minimize any stress wave reflections at the specimen/momentum trap interface. The 0.25 in. thickness of the trap was adequate to carry away all of the shock wave. For example, assuming the entire stress wave is not more than 200 nsec in duration and the sound speed in aluminum is 5×10^5 cm/sec or less, the maximum length of the stress wave would be $(200 \times 10^{-9}$ sec$) \times (5 \times 10^5$ cm/sec$) = 1$ mm $= 0.04$ in.

The results are summarized in Table 4 and plotted in Figure 5. Compressive stresses were produced on both sides of the specimen by the successive shocking procedure. The in-depth stress profiles of the first and second sides shocked are almost identical. The first side shocked has a slightly lower surface stress, which may or may not be a result of the interaction with the second side shock wave. These results illuminate two points. Firstly, they confirm the expectation that the surface residual stresses are created at the irradiated surface by the initial passage of the stress wave into the material. Secondly, the preservation of the residual stresses on the first side shocked after laser shocking the second side demonstrates that passage of a compressive shock wave through a compressive residual stress region will not significantly change the existing residual stresses. However, this may not be true for the interaction between a residual stress field and a stress wave of opposite sign, e.g., a tensile stress wave passing through a residual compressive stress region may degrade or remove the compressive residual stresses as suggested by the one side shock results presented later.

The momentum trap produced residual stress profiles similar to those developed by the split beam, Figure 6. The 6×10^9 W/cm^2 condition is the nearest comparable power density split beam condition. The magnitude of the surfaces stresses for the momentum trap condition are slightly lower than for the split beam shots, possibly because the power density is lower. If the lower surface stresses are a result only of the lower power density and no other interactions, a comparison of the -40 to -44 ksi surface stress at the 3.4×10^9 W/cm power density level of this condition to the residual stresses shown in Figure 4b indicates a sharp decrease in surface stresses below a power density of 4 to 5×10^9 W/cm.

Table 4. Residual Stresses on Both Sides of a Consecutively
Shocked, Momentum Trapped Specimen. Specimen R12.

First Side Shocked		Second Side Shocked	
Depth Below Surface, in.	Residual Stress, ksi	Depth Below Surface, in.	Residual Stress, ksi
0.000	−39.3	0.000	−43.9
0.002	−35.3	0.003	−33.6
0.009	−30.8	0.007	−33.3
0.014	−32.3	0.012	−23.6
0.018	−26.9	0.018	−26.0
0.028	−17.4	0.028	−18.8

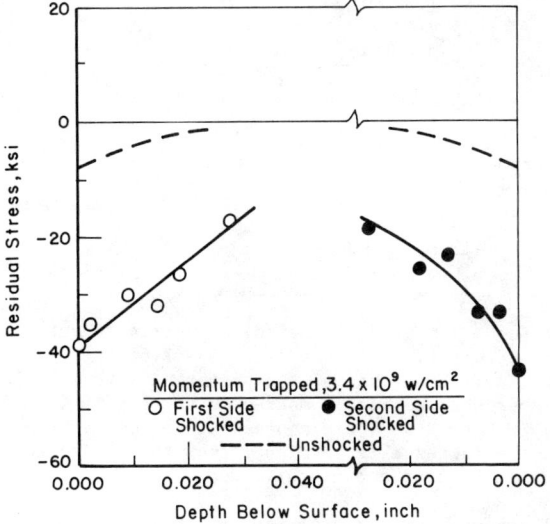

Figure 5. In-Depth Residual Stress Profiles on Both Sides of a Consecutively Shocked, Momentum Trapped Specimen. Specimen 6AB.

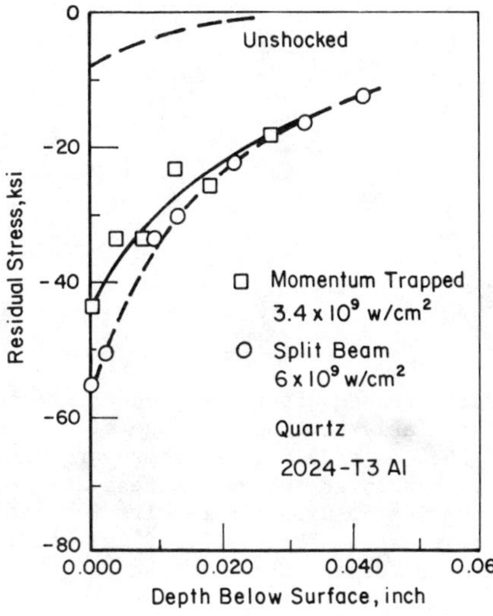

Figure 6. Comparison of the In-Depth Residual Stresses for the Split Beam and Momentum Trap Configurations.

<u>Laser Shock from One Side Only</u>. Another experiment to evaluate laser shocking of configurations with limited laser access to the back side was to laser shock a specimen from one side with the opposite surface a free surface (no supporting material or momentum trap was present). The results are presented in Figure 7 and Table 5. The residual stresses were measured at the locations G, H, and I (Figure 2c) on the laser shocked side, but only at locations G and H on the back side.

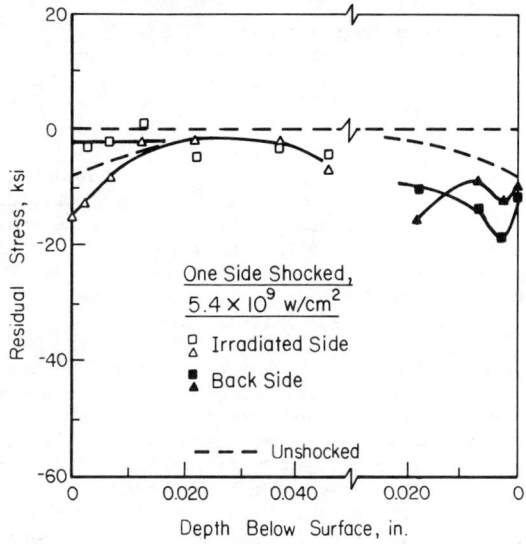

Figure 7. Residual Stress Profiles for Shocking one side only.

Table 5. Residual Stress Results for Specimens Without a Drilled Hole, Fused Quartz Overlay

Specimen Number	Power Density, W/cm^2	Depth Below Surface, in.	Residual Stress, ksi Location G	H	I
R7	Split Beam 6.0, 6.0 X 10^9	0.000	-43.9	-55.0	42.1
		0.002	-40.9	-50.5	7.8
		0.008	-32.6	-33.7	-4.2
		0.013	-31.2	-30.8	-11.3
		0.022	-24.0	-22.5	-9.6
		0.033	-16.2	-16.5	-12.4
		0.042	-10.7	-12.9	-6.1
		Laser Shock Side			
R10	One Side 5.4 X 10^9	0.000	0.3	-15.0	0.8
		0.002	-3.1	-12.8	4.2
		0.007	-2.2	-8.8	-22.0
		0.013	1.0	-2.4	-19.6
		0.022	-4.9	-2.2	-9.8
		0.037	-3.3	-2.6	-9.0
		0.046	-4.5	-7.5	-8.5
		Unshocked Side			
		0.000	-11.8	-9.9	
		0.002	-18.7	-12.8	
		0.007	-13.3	-8.7	
		0.018	-10.3	-15.9	

The residual stresses after laser shocking one side only (Figure 7) compared to the momentum trap (Figure 5) which was similarly shocked from one side at a time, are very different. The surface stresses on the irradiated surface of the one side only specimen are compressive but very low, about that of the unshocked material (Figure 5). The lower residual stress at the center of the laser shocked zone compared to those away from the center is consistent with the results for split beam shocking (Figure 3). There appears to be some increase in the compressive stress beyond 0.040 in. below the surface but the increase is no higher than the projected levels of the stresses at that depth for the split beam and momentum trap configurations.

The back side results are similarly surprising. The residual stresses are compressive, at levels higher than the unshocked material, and rise to a maximum at 0.002 in. below the surface before decreasing at greater depths.

From the momentum trap results, it is clear that the shocked side on the one-side shocked specimen had a significant residual compressive stress profile similar to that shown in Figure 5 after the initial shock wave had passed through the specimen. However, instead of passing out of the specimen and being carried away as in the momentum trap experiment, the shock wave reflected from the unsupported back surface as a tensile stress wave, producing the residual stress profile visible at the rear surface. However, it is not understood why the residual stresses at the back surface are compressive and not tensile as might be expected if conditions similar to those contributing to spalling were developed at this free surface.

When the reflected tensile stress wave returns to the original shocked surface, it had to have had enough amplitude to reduce the original compressive residual stresses to the levels observed in Figure 7. Previous measurements of the change in the amplitude of stress waves as they travel through aluminum showed considerable decrease in peak pressure after passage through 3 mm of aluminum (5). In the present case, the distance of travel from the irradiated surface to the back surface and return to the front surface would be 12 mm, a much longer distance, and in addition there would be a loss of amplitude through plastic interaction with the back surface as indicated by the residual stress pattern there. It might be expected that after this distance the peak stress would be at or below the dynamic yield strength of the unshocked material, i.e., the wave would be an elastic wave. The observation that it was able to modify the front surface residual compressive stresses, possible only by plastic deformation, suggests that a Bauschinger effect is operative. The Bauschinger effect is the decrease in flow stress shown by materials when the direction of plastic deformation is reversed, e.g., first in compression, then in tension. This phenomenon would explain the elimination of the shocked side residual compressive stresses by the reflected tensile stress wave.

If the reflected tensile wave is the cause for the low compressive residual stresses at the shocked surface, then there are many applications where surface residual compressive stresses can be created and retained by shocking one side only. One of these is in thicker specimens where the amplitude of the ringing elastic wave has decayed below the Bauschinger yield stress of the compressed material by the time it returns to the shocked surface. This effect is supported by recent results in thick steel specimens (8). Another is where the rear surface is not a flat reflecting surface. For example, in applications such as keyways and fillets in shafts and machine or structural parts, the rear surface may be cylindrical or highly irregular. These non-flat surface configurations will cause the reflected tensile wave to be diffuse making it unable to modify the front surface residual stresses.

Influence of Annular Shaped Shocked Region. Another approach was explored for the purpose of modifying crack initiation and propagation to render the cracks visible on the sheet surface soon after initiation (avoid crack tunneling) (7) but still inhibit crack propagation. This was an annular shaped shocked region which allowed the crack to initiate and grow from the starter notch similar to an unshocked specimen, but to later encounter a region of residual stress which would slow the propagation rate. This annular beam was configured to the hole as shown in Figure 1 and the split beam arrangement was used. The results for the residual stresses are presented in Table 6 and Figures 8 and 9.

Table 6. Residual Stresses in a Specimen Shocked With an Annular Beam. Specimen R11[a].

Depth Below Surface, in.	Residual Stress,[b] ksi					
	A	B	C	D	E	F
0.000	-23.3	-34.0	-64.5	-61.1	-56.2	41.8
0.003	-22.8	-34.0	-53.3	-52.3	-47.8	1.9
0.008				-46.1		
0.012	-16.8	-17.4	-34.2	-42.1	-32.1	-20.5
0.022				-25.6		
0.033				-24.1		

(a) Power density = 4.4 X 10^9 w/cm^2.
(b) The lettered locations correspond to those shown in Figure 8.

The results show that even in the unshocked region around the hole, there is a significant increase in the compressive residual stresses. The magnitude of the residual stress rises to a maximum across the laser shocked region itself (Figure 9). The surface stresses are even higher than those observed with the solid beam. However, there was some non-uniformity of the beam intensity around the annular region and the residual stresses were measured in a region that showed a significant depression of the laser shocked region; an indication of the highest power density in this region.

Immediately outside the laser shocked zone, a residual tensile stress should be present to compensate the compressive stress. In the case of Specimen R11, the tensile stress extends only a few thousandths of an inch below the surface before it becomes compressive. A similar result was found in Specimen R7 (Location I, Table 5, but not in Specimens R1 or R4 (Location C, Table 3). The measured stresses in this region may be quite sensitive to the location of the X-ray beam just inside or outside the edge of the laser-spot.

The residual stresses extending beyond the laser shocked zone raise interesting questions. How far outside the region do these stresses extend? Is there a thickness dependence? How will these stresses affect crack initiation or propagation from nearby edges or surfaces outside the laser shocked zone? These are questions which must be addressed at some point.

Acrylic Polymer Transparent Overlay. The identification of a variety of transparent overlays is vital to the continued development of laser shock processing. The fused quartz overlay is suitable only for flat surfaces. Water is better for conforming to non-uniform surface configurations and will often be simple and quick to apply in a production situation. Another approach is to use polymeric overlays which are flexible, cheap and may even ultimately have adhesive included as an integral part of the overlay system.

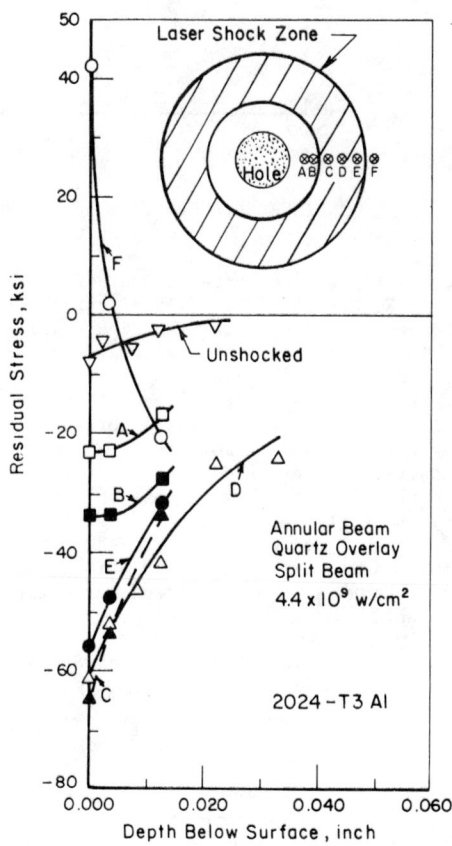

Figure 8. In-Depth Residual Stress Profiles for the Annular Shaped Beam. The Letters Designate the Positions of the Stress Measurements. Specimen R11.

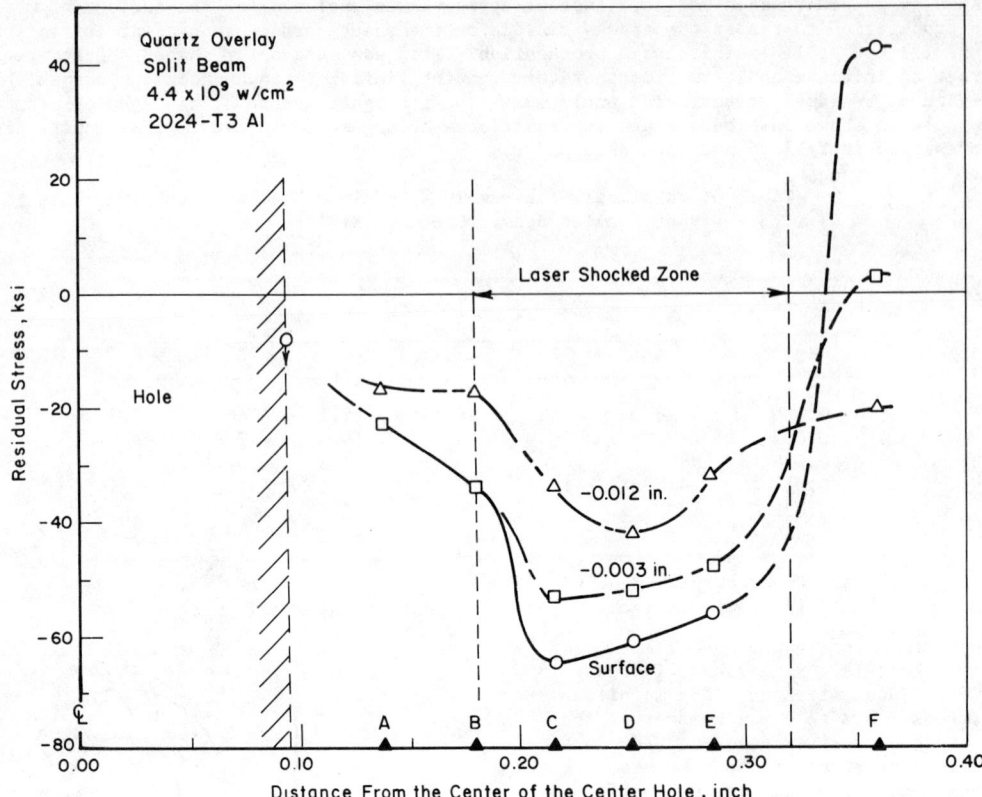

FIGURE 9. Profiles of the Residual Stresses Along the Laser Spot Radius at and Below the Surface for the Annular Shaped Beam. Specimen R11.

Two experiments were conducted using acrylic as a transparent overlay material. The results are presented in Table 7 and Figure 10. The residual stress profile midway between the edge of the hole and the edge of the laser shocked zone (Location B in Figure 2a) are shown in Figure 10. Two aspects of these profiles are noteworthy. The magnitudes of the surface stresses are comparable to those obtained with the fused quartz overlays. However, a comparison included in Figure 4b shows that there may be more scatter in these results or else that the surface stress may have a much different dependence on power density than was obtained for fused quartz. In addition, the residual stresses for the acrylic overlay may fall off more rapidly below the surface than for most of the split beam or momentum trap results using the fused quartz overlay. The stress gradient for both power densities is very similar to that of the 12 X 10^9 W/cm^2 shot for the fused quartz overlay (Figure 4a), showing extrapolations to zero residual stress at 0.030 to 0.050 in. below the surface. This may be caused either by the stress waves being of shorter duration at very high power densities or by a characteristic of the acrylic overlay itself. The shorter duration stress waves could be caused in either case by the formation of a reflecting plasma during the irradiation which uncouples the beam from the blow-off material. The peak pressure of shorter duration stress waves may be lower and attenuate more rapidly than that of longer duration stress waves.

Table 7. Residual Stress Results for Standard Split Beam Laser Shock Processing Around a Drilled Hole with an Acrylic Overlay.

Specimen Number	Power(a) Density, W/cm^2	Depth Below Surface, in.	Residual Stress, ksi Location of Measurement		
			A	B	C
R8	12, 12 X 10^9	0.000	-48.5	-59.7	4.4
		0.002	-41.7	-43.7	-11.8
		0.006	-33.3	-31.3	-6.8
		0.012	-26.3	-28.5	-11.1
		0.022	-8.0	-9.9	-3.2
R9	10.0, 9.7 X 10^9	0.000		-45.6	
		0.002		-37.6	
		0.007		-30.5	
		0.012		-22.1	
		0.022		-17.1	
		0.034		-7.2	
		0.044		-0.1	

(a) The power density on each side of the specimen is shown: Side 1, Side 2.

<u>In-Hole Residual Stresses</u>. It is possible that laser shocking around fastener holes might actually decrease the cycles to crack initiation at the side of the hole at mid-thickness (7), if residual tensile stresses were created at midthickness to compensate for the surface compressive stresses. Since crack initiation as well as crack propagation can influence fatigue life, it was necessary to establish what stress conditions were developed at the crack initiation sites on the hole surface by LSP.

For this purpose, specimens containing drilled holes were laser shocked with the standard solid, split beam using fused quartz overlays. Both laser shocked and unshocked specimens were then cut in half along the diameter of the holes. The residual stresses were then measured at the mid thickness region of the sheet on the surface of the hole and at two in-depth levels below the original surface.

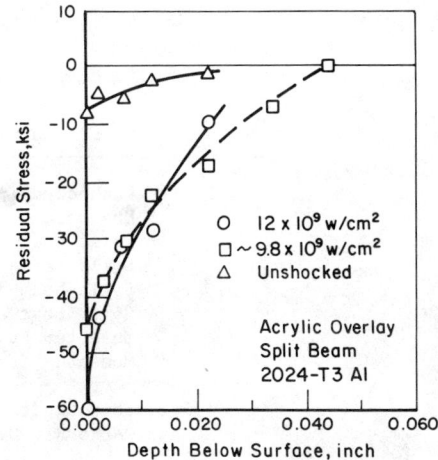

Figure 10. Residual Stress Profiles for the Acrylic Transparent Overlay at two Power Densities.

The results for the stress component normal to the sheet thickness direction (the stress tangential to the hole surface) are shown in Table 8 for Specimens R3 and R6. Specimen R3 was irradiated with a smaller spot size and a higher power density compared to Specimen R6 (Table 1). The as-drilled hole surface shows a near zero surface residual stress which drops to a small compressive stress extending to as far as 0.012 in. from the hole surface. Laser shocking definitely changes the residual surface stress to between 10 and 20 ksi compression to depths of at least 0.012 in. away from the hole surface. When these results are compared to the in-depth stresses below the laser irradiated surface there is a consistency between the two. The component of stress measured in the hole surface is the same stress as the tangential component measured in the laser shocked surfaces. For equivalent power densities some of the in-depth compressive stress profiles measured below the laser shocked surfaces (Figure 3) appear to extend well below the shocked surface, although at a low stress level. In these cases, the low level compressive residual stresses may extend completely through the thickness as suggested by the results from the hole surfaces.

Table 8. Residual Stresses on the Surface of the Drilled Holes at the Mid-Thickness of the Plate With and Without Laser Shocking.

Specimen R3		Specimen R6			
Drilled Plus Shocked		As Drilled		Drilled Plus Shocked	
Depth Below Surface, in.	Residual Stress, ksi	Depth Below Surface, in.	Residual Stress, ksi	Depth Below Surface, in.	Residual Stress, ksi
0.000	-11.2	0.000	0.2	0.000	-13.2
0.002	-14.2	0.002	-7.0	0.002	-19.6
0.006	-9.6	0.007	-3.4	0.007	-21.0
0.011	-18.3	0.012	-2.4	0.013	-24.3

The somewhat higher compressive stresses on the hole surface after laser shocking suggest that the cycles to crack initiation should be greater in the laser shocked compared to the unshocked specimens.

Fatigue Results

Fatigue crack initiation/propagation specimens were laser shocked in several conditions as described in Table 2. In addition, two unshocked specimens were tested for baseline comparisons. The fatigue life results are shown in Table 9 and the crack propagation results are presented in Figure 11. The critical crack length for this specimen configuration is about 2.2 in. Thus, the cycles at the sharp upturn in crack growth rate represents the nominal fatigue life for the specimens tested. Several reference crack lengths are indicated in Figure 11. These are a_s, the starting crack length equal to the notch root to notch root distance of the starting hole; a_{min}, the minimum crack length required to break through to the outside of the laser shocked zone, assuming the crack propagates only from one side of the hole through to the outside of the shocked zone on that side; a_{max}, the crack length required to break through the laser shocked zone if the crack propagates evenly from both sides of the hole (equal to laser shocked zone diameter); a_{ia}, the crack length required to reach the inside edge of the annular shocked zone assuming that cracks initiate and propagate evenly from both sides of the hole (equal to the inside diameter of the laser shocked region). Representative views of the fracture surfaces are shown in Figure 12 and maps of the propagating crack front contours are shown in Figure 13.

Table 9. Fatigue Crack Growth Data[a]

Specimen Number	Laser Shock Condition	Cycles to Failure	Improvement Factors
US 1	Unshocked	50,500	1
US 2	Unshocked	49,800	1
F1	Split beam-solid spot	2,020,000	40.3
F2	Split beam-solid spot	1,340,000	26.7
F3	Split beam-annular spot	94,040	1.9
F4	Split beam-annular spot	175,830	3.5
F5	Momentum trap	720,190	14.4
F6	Momentum trap	731,150	14.6

(a) All tests conducted at 15 ksi cross section stress. Stress Ratio R = 0.1.

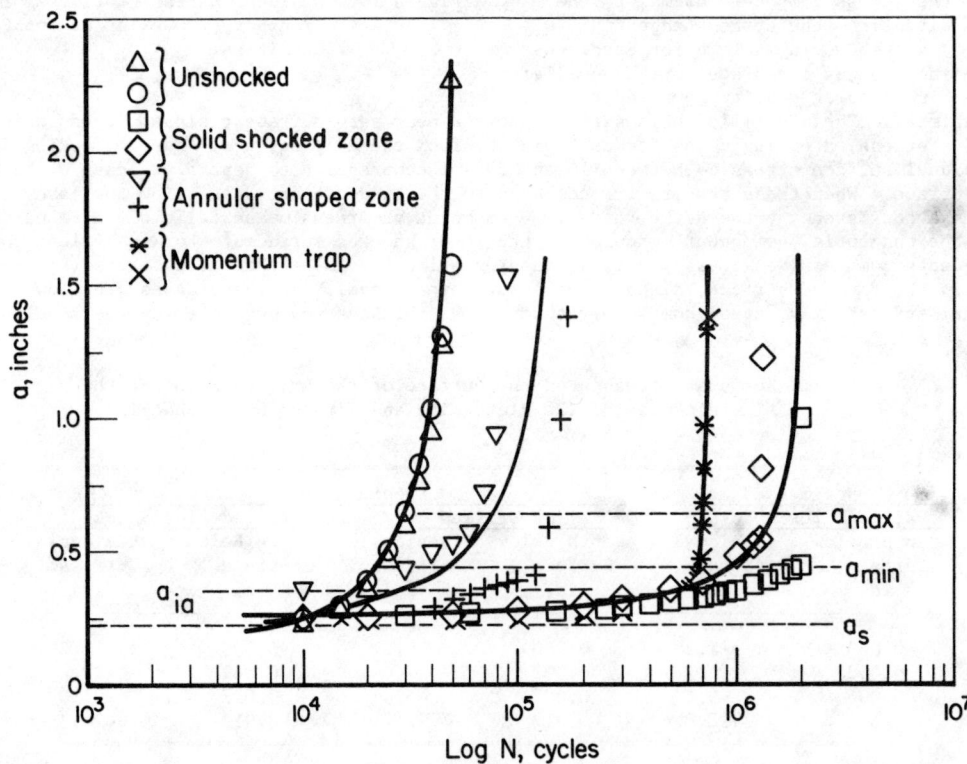

Figure 11. Crack Propagation Results for Unshocked and Laser Shocked Specimens for Different Laser Shocking Conditions Where the Plotted Crack Length is the Tip to Tip Crack Length From Both Sides of the Hole. The various meanings of a are described in the text.

The unshocked specimens have the shortest fatigue life and also the fastest crack propagation rates for all crack lengths less than the minimum length necessary to break through the laser shocked zone, a_{min} (Figure 11). For the laser shocked conditions, the split-beam annular shaped shocked zone showed a fatigue life of 2 to 3.5 times greater than the unshocked conditions while the momentum trapped specimens showed increases of 14.5 and the split-beam solid shocked zone showed an increase of 27 to 40 times (Table 9).

The life improvement for the solid split beam is nearly an order of magnitude greater than the improvement obtained in an earlier program (7). This is probably attributable to several differences in the laser shocking conditions used in this program compared to the previous one, i.e., different transparent overlays (fused quartz overlay vs. water overlay), slightly lower power densities (5 X 10^9 W/cm^2 vs. 6 X 10^7 W/cm^2), a different stress concentration/crack initiation configuration (EDM notch vs. a 0.081 in. radius) and a longer crack path through the laser shocked zone (0.207 vs. 0.132 in.). Each of these effects, except for the sharper notch, would significantly improve the fatigue life. The effect of the slightly lower power densities in this investigation is probably small (Figure 4b). However, the crack initiation behavior is significantly different between this investigation and the earlier one (7).

In the earlier investigation, crack initiation in the unshocked condition was reported to occur at nominally 10^5 cycles (Table A-10, Specimens 29 and 30, reference 7). This value was somewhat conjectural because of crack tunneling and the small, nearly invisible crack openings in the laser shocked zone. For both the shocked and unshocked conditions, the cracks tended to initiate at a single site on the edge of the hole in the central region of the plate thickness (7).

In this investigation the improved electrical crack monitoring method permits more confidence in the crack initiation data. Crack initiation appears to begin at 10^4 cycles or earlier (Figure 11); compare extrapolation of the a vs. N curves to the dashed line a_s. This is up to an order of magnitude earlier than the estimated crack initiation period of the previous tests (7) and is caused by earlier crack initiation at the sharper starting notch. Also, multiple crack initiation sites are visible along the length of the notch on the fracture surfaces shown in Figure 12. The edge of the notch is along the top of each micrograph. Because of early crack initiation, most of the fatigue life in this investigation was in the crack propagation mode. For the unshocked specimens crack initiation appears to be 20 percent or less of the observed life compared to possibly greater than 70 percent in the earlier program (7). In the laser shocked specimens it is an even lower percentage, less than 1 percent in some cases.

The crack growth rate of all the laser shocked specimens is much lower than that of the unshocked specimens for crack lengths less than a_{min}. Most of the laser shocked specimens had a low growth rate up to a_{min}, and then either an almost discontinuous increase in rate from a_{min} to failure or a rate similar to that of the unshocked specimens. The sharp increase in growth rate suggests that in these cases the cracks grew mostly from one side of the hole and breached the laser shocked zone at length a_{min}. Thereafter, they propagated rapidly and forced the crack from the other side of the hole to penetrate the laser shocked zone very quickly under the much higher stress concentration created by the longer crack length. This is consistent with the data showing that the specimens showing the abrupt crack acceleration also had lower apparent crack lengths at a given number of cycles compared to the specimens showing more uniformly increasing crack propagation rates. However, a comparison of the crack lengths on both sides of the hole with increasing cycles does not support this idea. In both Specimens F5 (momentum trap) and F1 (standard split beam) the cracks grew at about the same rate from both sides of the crack and the acceleration in rate would then have occurred when both cracks were 30 to 50 percent through the laser shocked zone.

It is useful to compare the crack propagation behavior derived from the fracture surface examination for the different LSP conditions (Figures 12 and 13) and the crack growth curves (Figure 11). In the unshocked specimens, the crack front was nominally straight across the thickness of the specimen at a very early stage, and continued to propagate in this fashion as the crack grew away from the hole (Figure 13a). The split, solid beam condition shows that the surface residual compressive stresses cause the crack growth rate at the surface to lag the rate in the mid-thickness region. This slows the crack propagation rate significantly thereby increasing the fatigue life (Figure 11).

In the momentum trapped specimens, the crack front shape (Figure 13c) showed the cracks propagated differently than they did in the split beam condition. The crack front along the first side shocked, propagated ahead of the front along the second side shocked, where it propagated similar to the split beam condition. The residual surface stresses on the second side shocked were more effective in slowing the crack than was the residual stress in the first surface shocked. This difference indicates that the stress wave from the second shot modified the stress field behind the first surface shocked enough to significantly influence the crack propagation rate even though the appearance of the residual stress profiles for each surface (Figure 5) are not much different. Interestingly, the crack length measurement was made on the first side shocked in this specimen (X in Figure 11) and it showed similar crack propagation rates to the split beam shocked specimens below a_{min}. The sudden crack acceleration leading to shorter life compared to the split beam condition probably occurred when the crack on the first side shocked (left side in Figure 13c) had broken through the laser shocked zone. This is supported by the contours of the crack front in Figure 13c being drawn out almost parallel to the second side shocked by the rapid advancement along the first side shocked. If this modification of the first side stress field had not occurred, the momentum trapped specimen would be expected to have much longer fatigue lives.

Another possible contribution to longer lives after split beam shocking compared to momentum trapping may be the deformation created in the center thickness region of the plate by the superposition of the shock wave (5). This effect contributes additional cold work to the specimen mid-thickness region and possibly modifies the residual stress pattern.

The effect of the annular shocked zone was to produce some improvement in fatigue life compared to the unshocked condition. One of the specimens appeared to show much earlier crack initiation than the other specimens (inverted triangles, Figure 11). Both specimens, but one in particular (crosses), showed the anticipated slow

Edge of Notch

Edge of Notch

a. Unshocked condition

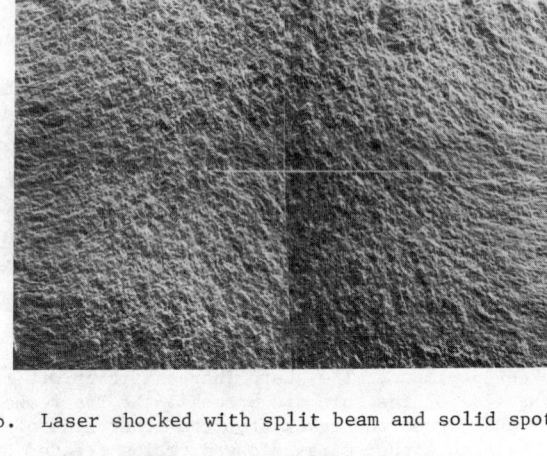

b. Laser shocked with split beam and solid spot

Edge of Notch

Edge of Notch

c. Laser shocked on both sides successively with a momentum trap

d. Laser shocked with split beam

Figure 12. Fracture Surfaces of Specimens for each Condition Fatigue Tested. The edge of the notch is located at the top of each micrograph with the crack propagating downward across the width of the specimen. Magnification: the specimen width is 0.25 in.

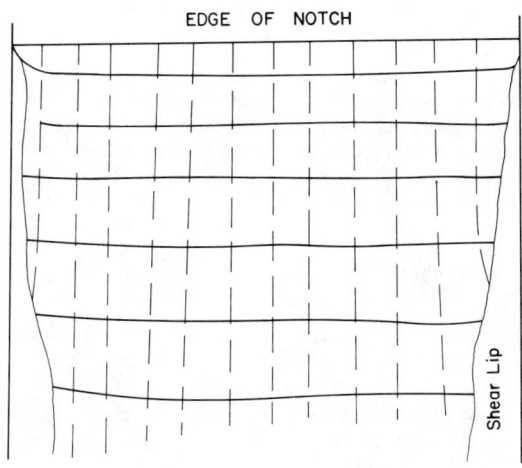

a. Unshocked condition

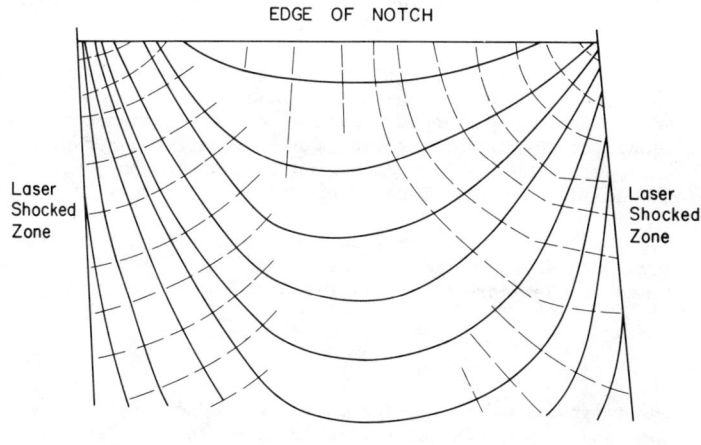

b. Laser shocked with split beam and solid spot

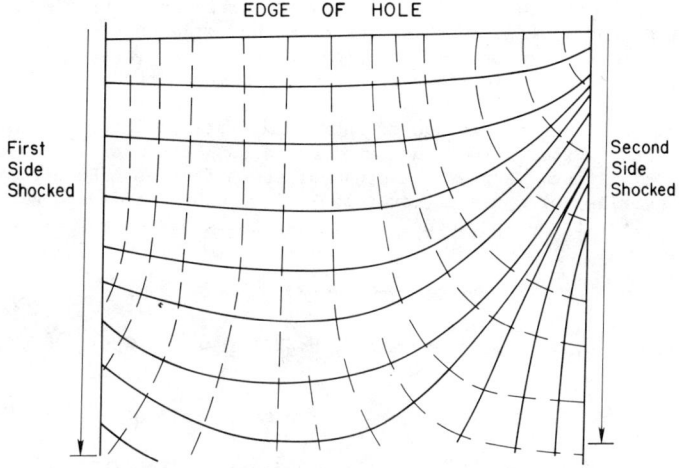

c. Laser shocked on both sides successively with a momentum trap

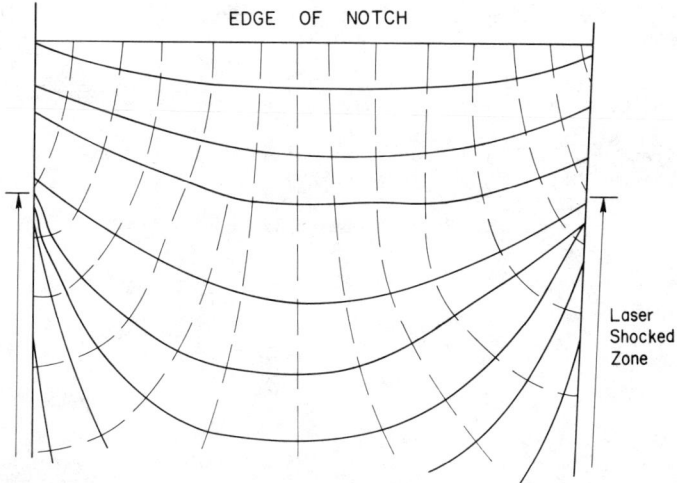

d. Laser shocked with split beam and an annular spot

Figure 13. Schematics of the Fracture Surfaces Shown in Figure 12 Showing the River Line Pattern Contours as Dashed Lines and successive Crack Front Contours as Solid Lines. These drawings are oriented identically to the micrographs in Figure 12.

21

down of crack propagation when the crack reached the length a_{ia}, at the edge of the laser shocked zone (compare the slope indicated by the data points between a_s and a_{ia} to that between a_{ia} and a_{min}).

The fracture surfaces (Figures 12d and 13d) clearly show the effects of the residual stresses (Figures 8 and 9). The cracks initiate all along the edge of the notch and start to propagate much as in the unshocked specimen. As the crack grows and approaches the inside diameter of the laser shocked zone, the portions of the crack front intersecting the surfaces slow down through the interaction with the residual compressive stresses. At this point the crack front begins to tunnel through the mid-thickness of the specimens in the same way as in the solid split beam shocked specimens.

The annular shaped shock beam zone is effective in decreasing the crack propagation rate. With a more uniform beam intensity and improved beam shaping methods it is expected that the observed life increase of two to three times can be increased still further.

CONCLUDING REMARKS

The improvements in the fatigue life produced by laser shock processing are clearly a result of the residual compressive stresses developed in the irradiated surfaces slowing the crack propagation rate. The change in the shape of the crack front and the slowing of the crack growth rate when the crack front encounters the laser shocked zone in the annular beam LSP condition support this contention.

The compressive residual stresses initially produced below the irradiated surface can be modified or eliminated by the passage of a tensile stress wave. For this reason, if one side only shocking is used, it is necessary to either process thicker specimens to decrease the intensity of the reflected wave, use specimens with an irregular back surface to diffuse the reflected wave, or use a momentum trap.

Laser shock processing is a very effective method of increasing fatigue life in aluminum by treating fatigue critical regions. It is not known what effect the tensile residual stresses surrounding the laser shocked region would have on crack initiation. A previous study of the effect of LSP on fatigue of welded aluminum specimens used a series of overlapping spots to cover the weld and heat affected zones. Significant increases in fatigue life were observed for these specimens, indicating that this effect may not be a serious problem (6).

REFERENCES

1. B. P. Fairand, B. A. Wilcox, W. J. Gallagher, and D. N. Williams, J. Appl. Phys., Vol. 43, p. 3893-3895 (1972).

2. A. H. Clauer, B. P. Fairand, and B. A. Wilcox, Met. Trans. A., Vol. 8A, p. 119-125 (1977).

3. A. H. Clauer, B. P. Fairand, and B. A. Wilcox, Met. Trans. A., Vol. 8A, p. 1871-1876 (1977).

4. B. P. Fairand and A. H. Clauer, J. Appl. Phys., Vol. 50, p. 1497-1502 (1979).

5. A. H. Clauer and B. P. Fairand, Applications of Lasers in Materials Processing, ed. by E. Metzbower, American Society for Metals, Metals Park, Ohio (1979).

6. A. H. Clauer, J. H. Holbrook, and B. P. Fairand, Shock Waves and High Strain Rate Phenomena in Metals, ed. by M. A. Meyers and L. E. Murr, Plenum Press, New York (1981), p. 675-702.

7. S. C. Ford, B. P. Fairand, A. H. Clauer, and R. D. Galliher, Investigation of Laser Shock Processing, Final Report, AFWAL-TR-80-3001, Vol. II (August, 1980).

8. A. H. Clauer, C. T. Walters, and S. C. Ford, to be published.

NONLINEAR DYNAMIC TEMPERATURE CHARACTERIZATION OF PULSED LASER ANNEALING OF SEMICONDUCTORS

Dim-Lee Kwong, R. Kwor, C. Paz de Araujo
Dept. of Electrical Engineering
Univ. of Notre Dame, Notre Dame, IN 46556

ABSTRACT

The dynamic characteristics of transient laser annealing of semiconductors is examined extensively and the rise in lattice temperature specifically attained in silicon under the influence of a high power laser beam irradiation is characterized analytically over a wide range of laser wavelength, intensities, and durations. For the case of silicon, with increasing lattice temperature the optical absorption coefficient is drastically enhanced, while the thermal conductivity is considerably reduced. These two temperature dependent material parameters are, therefore, strongly coupled during the laser beam irradiation and the effects of this coupling on the thermal evolution of the sample are examined comprehensively. Specifically, the kinetics of the depth profile of lattice temperature, the threshold pulse energy for the onset of surface melting are characterized explicitly in terms of both material parameters and operating laser beam characteristics.

I. INTRODUCTION

Laser beam processing techniques have a great potential for advancing state of the art in fabrication of semiconductor devices and integrated circuits. The unique characteristics of this novel technology, due to its ability to heat a spatially localized region to a high temperature for a short time duration, may open up various possibilities of microelectronic device fabrication that are novel, and could prove to be essential for the realization of VLSI and/or VHSI. Numerous reports [1-5] have demonstrated a high power laser can (1) anneal completely the ion implantation damage without introducing deleterious effects in the substrates, (2) incorporate substitutionally dopants into an epitaxially regrown perfect silicon lattice at concentrations far exceeding the solid equilibrium solubility without any precipitates being generated, (3) recrystallize polycrystalline silicon films deposited on amorphous insulating substrates for three-dimensional silicon-on-insulator (SOI) device structures, (4) induce the formation of metal silicides for making ohmic and Schottky barrier contacts in semiconductor devices fabrication and gate electrodes and interconnections in metal-oxide-semiconductor (MOS) integrated circuit fabrication, (5) improve the dielectric properties of thin oxides thermally grown on polysilicon, and (6) improve the resistivity control of polysilicon resistor load elements. In each of these applications, laser pulses with durations of 20-150 nsec and energy densities of several Joules/cm^2 were used to melt the samples to a desired depth to achieve material modifications needed for device fabrication. Full utilization of the inherent advantages offered by the use of directed energy radiation is thus contingent upon a quantitative understanding of the transformation of ion-implanted amorphous and polycrystalline silicon under the influence of laser irradiation. The various processes involved in the coupling of laser energy to the lattice and the subsequent transport of the deposited energy in the semiconductor layer can be examined based on the following parameters:

(i) the lattice temperature profile attained in Si for a given rate of energy input,

(ii) the time duration of the elevated temperature, and

(iii) the rate of cooling after the incident radiation has been turned off.

These key parameters are of crucial importance to understanding the grain growth mechanism, the redistribution of impurity dopants and defects, and, therefore, ultimately determining the electrical characteristics of laser-processed semiconductor devices. Several numerical treatments of lattice heating during laser annealing have been carried out [6-12], and the time evolution of the depth profile of the lattice temperature, melt front generation and propagation, and dopant redistributions were described numerically. However, to gain physical insights and information concerning the initial choice of variables required for controlling the final parameters of laser-annealed devices, it is important to have an analytical model of lattice temperature, instead of having to rely on extensive computer simulations. Previous work relating to this problem includes (1) an analytical description of CW laser annealing in the steady state, including the temperature dependence of thermal diffusivity [13] and (2) an approximate solution of lattice temperature rise near the surface of material for the case of short heating time, where the heat diffusion into the bulk material can be neglected [14-15]. The transient depth temperature (T) profile generated in Si annealed by a high-power laser beam is mainly determined by two parameters (i) the energy deposition depth, $\alpha^{-1}(T)$ and (ii) the heat diffusion length, $(D\tau)^{1/2}$, where $\alpha(T)$ is the optical absorption coefficient, $D(T)$ is the thermal diffusivity, and τ is the laser heating time. For the case of Si, these two parameters drastically decrease with increasing T [16-17]. Hence the deposition depth of the laser energy is reduced rapidly during the pulse, and the heat conduction into the bulk of the sample is concomitantly slowed down. This nonlinear dynamic feedback effect and the strong coupling of $\alpha(T)$ and $D(T)$, therefore, drive the surface temperature to the melting point in a very short time. Hence, to obtain the thermal evolution in the material for a given pulse energy and duration, one has to incorporate the sensitively temperature-dependent α and D in analyzing this non-linear, rapid heating and transport process during pulsed laser annealing.

In this paper, the dynamics of pulsed laser annealing of Si are _analytically_ examined over a wide range of pulse durations, densities, and laser wavelengths. The strong temperature dependencies of both thermal conductivity and optical absorption coefficient are incorporated explicitly. The time-dependent, nonlinear, inhomogeneous heat diffusion equation is solved by a new technique, which is based on a parametrized perturbation scheme in a Green's function formulation. The technique can describe analytically, to a desired degree of accuracy, the dynamics of threshold pulse energy for surface melting, melt front generation and propagation, and the recrystallization process due to a high power laser irradiation. Because of the analytical nature of our approach we can bring out some important physical quantities pertinent to laser processing experiments and discuss the effects of various parameters of the laser beam on these quantities. In Section II we summarize the dynamic theory of laser-induced lattice heating and discuss the disagreements between the thermal melting model and the plasma annealing model using the experimental results published so far. In Section III we present the parametrized perturbation scheme which has been developed for analyzing the nonlinear diffusion processes. The technique is used in Section IV for specifically analyzing the dynamics of temperature rise in single crystals of silicon as well as in an implant amorphorized silicon layer irradiated by a pulsed laser beam. In Section V a few concluding remarks are made.

II. DYNAMICS OF LASER-INDUCED LATTICE HEATING

The incident laser energy initially generates electron-hole pairs and increases the kinetic energy of these excess charge carriers. These hot, photo-excited carriers thermalize subsequently with each other and eventually with the lattice. A basic question concerning the ultimate lattice temperature profile achieved in laser-annealed devices is to what extent and on what time scale the absorbed energy from laser beam transferred to the lattice through phonon carrier collisions. In the conventional thermal melting model, it is assumed that the carrier diffusion length before carrier-phonon collisions occur is negligible compared with the absorption depth of the light, and that the latter parameter entirely determines the depth profile of the heat source. For typical pulsed laser annealing, the near-surface region of a sample can melt and stay molten for times of the order of 100 nsec. During this time, dopant diffusion in the liquid state and thermodynamically nonequilibrium segregation associated with ultrarapid recrystallization can occur. Van Vechten et al have suggested an alternative concept, namely that the laser annealing of displacement damage is achieved in the presence of a dense electron-hole plasma while the lattice remains relatively cool [18]. According to this plasma-annealing model, the energy transfer from photo-generated hot carriers to the lattice should be delayed (~ 200 nsec) through the screening of the phonon rate. These hot carriers may cause annealing via enhanced diffusion of point defects and glide and climb of dislocations. As the plasma density exceeds $8*10^{21}$ cm^{-3}, electrons excited into antibonding states lead to a second-order phase transition resulting in softening of the lattice. The semiconductor would then be in a fluid-like state with energy stored primarily in the electronic system. This energy would then be distributed over a much greater depth as carrier diffusion becomes important. As the plasma density decreases, the material passes back through the phase transition, leading to recrystallization. In this section, we investigate the response of a semiconductor under high power optical excitation and consequently under a high electron-hole pair generation rates, and discuss the transport phenomena of the deposited energy from the stand point of the lattice temperature rise. The incident laser energy is absorbed either by electron-hole pair creation or by free-carrier excitation. Initially, with few carriers present, the former process dominates, so that near the surface of the semiconductor the photon absorption rate g is given by

$$g = I(1 - R)\alpha/\hbar\omega_{\ell} \qquad (1)$$

where I is the incident laser power per unit area, R, the surface reflectivity of the sample, $\hbar\omega_{\ell}$, the photon energy, and α the optical absorption coefficient. The rise in time of the carrier density leads, in turn, to increased free carrier absorption of the incident laser beam energy. The net result is thus the production of hot, dense electrons and holes which thermalize with the rest of the carriers and eventually transfer their energy to the lattice via collision with phonons. For the case, where the optical absorption coefficient is ~ 10^5 cm^{-1}, pulse intensity ~ 100 MW cm^{-2}, and duration ~ 10 nsec, carrier densities, N exceeding 10^{19} cm^{-3} are attained in times much less than the laser pulse duration. Under these conditions, Yoffa pointed out that [19]:

(1) the photo-generated carriers instantaneously establish a thermodynamic equilibrium among themselves, with a Maxwell-Boltzman distribution characterized by a temperature, T_e which is much larger than the lattice temperature, T;

(2) the release of energy from the carrier system to the lattice via phonon emission is characterized by a relaxation time τ_e, which is estimated to be about 10^{-13} sec;

(3) Auger recombination is the dominant process, in which an electron and hole recombine and give their recombination energy to a third carrier. This process does not remove energy from the plasma system, but merely converts the band gap energy into excess kinetic energy of free carriers. During the subsequent relaxation, the energy is transferred to the lattice; and

(4) when a critical carrier concentration N*(10^{21} cm^{-3}) is exceeded, the rate of energy loss from carriers to the lattice is reduced due to the screening of the phonon emission rate. Under this circumstance, the carriers can diffuse away from the site of excitation before losing energy to the lattice. Thus the region in which the laser energy is transferred to the lattice extends further into the bulk material than the light absorption depth. This increase in the effective heating volume of the material leads to the requirement of increased pulse energy for annealing.

According to Yoffa's analysis, the change in time of the energy in plasma per unit volume is given by [19].

$$\frac{\partial E}{\partial t} = I(1 - R)\alpha e^{-\alpha x} - \frac{N\hbar\omega}{\tau_e} + D_a\left(\frac{E}{N}\right)\frac{\partial^2 N}{\partial x^2}, \qquad (2)$$

where the first term represents the input rate of laser energy via the single photon absorption.

Two-photon absorption may become important for pulse intensities greater than 500 MW cm^{-2} and is not included here. The second term accounts for energy loss due to phonon emission (energy $\hbar\omega$) by electrons with emission frequency $\tau_e \sim 10^{-13}$ sec. The last term represents the diffusion of energy associated with the ambipolar diffusion of excess charge carriers; N is the total carrier density and the ambipolar diffusivity

$$D_a = 2kT_e[\tau_e\tau_h/(m_e^*\tau_h + m_h^*\tau_e)] \quad (3)$$

is given in terms of electron (hole) effective mass $m_e^*(m_h^*)$.

In terms of the nanosecond time scale, N reaches steady state almost instantaneously at a value

$$N_{ss}(x) = I(1-R)\alpha(\tau_e/\hbar\omega)\frac{\gamma/\alpha}{1-\gamma^2/\alpha^2} \cdot [\exp(-\gamma x) - (\gamma/\alpha)\exp(-\alpha x)], \quad (4)$$

where the effective diffusion length of free charge carriers, γ^{-1} is given by

$$\gamma^{-1} = [D_a\tau_e(E/N\hbar\omega)]^{1/2} \quad (5)$$

In obtaining Eq. (4), Yoffa used the energy conservation relation:

$$\int_0^\infty g\hbar\omega_\ell e^{-\alpha x}\,dx = \int_0^\infty \frac{N_{ss}(x)}{\tau_e}\hbar\omega\,dx \quad (6)$$

and assumed that the energy per carrier, E/N, and τ_e are weakly dependent on N. The local lattice heating rate is, therefore, given by

$$S \equiv (\hbar\omega/\tau_e)N_{ss}(x)$$
$$= I(1-R)\alpha\frac{\gamma/\alpha}{1-\gamma^2/\alpha^2} \quad (7)$$
$$\cdot [\exp(-\gamma x)-(\gamma/\alpha)\exp(-\alpha x)].$$

Owing to the extreme nonlinearity of the hot-carrier effects, the uncertainty in the parameters chosen for Yoffa's example along with the simplifying assumptions made in her calculations prevent us from making an accurate calculation of the temperature to which the lattice is heated or from specifying precisely the threshold laser energy for the onset of surface melting. However, the effect of carrier diffusion on the coupling of laser beam energy to the lattice can be seen from Eq. (7). It is worth observing that, with increasing diffusion length, γ^{-1}, the heating rate near the surface is significantly reduced, while the effective heated volume of the material increases well beyond the absorption depth, α^{-1}. Furthermore, with increasing laser power, the laser-induced plasma density as given by Eq. (4) becomes increasingly dense. At such high carrier density, screening by the plasma becomes more important and the carrier-phonon collision rate, which is inversely proportional to N^2, falls rapidly with N. Thus, at extremely high excitation rates (> 10^{23} cm^{-3}) the laser energy is given to the lattice within a characteristic depth determined primarily by carrier diffusion length, and carrier diffusion could play a dominant role in determining the ultimate lattice temperature rise.

Although Eq. (7) incorporated formally the effects of ambipolar carrier diffusion during high power laser irradiation, the diffusion length γ^{-1} has to be examined in more detail. Specifically, γ^{-1} is the average distance a hot carrier traverses before it gives up its energy to the lattice by phonon emission. If γ^{-1} is comparable to or even exceeds the absorption length, α^{-1}, the diffusion of hot carriers plays an important role. On the other hand, if the plasma density does not exceed the critical carrier concentration for the screening effect to be appreciable so that the rate of energy release from carriers to the lattice is not delayed, then the carrier temperature does not become very high nor does the lattice remain cool. Under this situation the ambipolar diffusivity of carriers is mainly determined by the lattice temperature and the value D_a remains rather small.

The change in time of the surface reflectivity of silicon (and other semiconductors) has been used as a means of inferring the dynamic characteristics of pulsed laser annealing processes. Liu et al, using conventional "pump and probe" technique, measured the changes in reflectance and transmittance of silicon wafers and silicon films on sapphire after irradiation with a pump pulse of 20 picoseconds duration at λ = 532 nm [20]. They found that a first-order phase transition of melting occurs at 0.2 Joul/cm^2, and the electron-hole plasma density never exceeds 10^{21} cm^{-3} before melting.

Lietoila and Gibbons [21] numerically solved a set of simultaneous coupled equations for lattice temperature, carrier concentration and temperature for typical nanosecond and picosecond laser pulses used for annealing experiments. They found that the maximum carrier concentration achieved during a nanosecond laser pulse was 2-3*10^{20}/cm^3. At this level, the surface reflectivity is not significantly influenced by the presence of plasma and the screening effect on the phonon emission rate is negligible. However, in the picosecond regime, almost all of the laser energy is at first stored in the carrier system, and a carrier temperature exceeding 30,000°K achieved for a 3 Joul/cm^2, 30 picosecond pulses at 1.06 um. In this case, the surface carrier concentration increases enough to cause an enhanced reflectivity from the plasma lasting for at least 0.1 nsec. The lattice temperature reaches 1410°C as carriers distribute their energy to the lattice after the laser pulse, and the energy stored in the plasma system would be enough to supply the heat of fusion at the silicon surface.

Recently Lo and Compaan [22] have measured the lattice temperature in silicon during the irradiation by pulsed dye laser beams (λ = 485 nm, pulse duration = 9 nsec) with densities in the range 0.7 -1.1 J cm^{-2}, using nanosecond time-resolved Raman scattering technique. They reported a temperature rise of ~ 300°C for a 1.0 Joul cm^{-2} dye laser pulse, thus lending support to the plasma annealing model. Narayan et al, using the same dye laser pulses as Lo and Compaan in their Raman temperature measurements, studied the annealing mechanism of displacement damage, dissolution of boron precipitation, dopant profile broadening, and the formation of constitutional supercooling cells [23]. The results of their investigation were interpreted on the basis of a thermal melting model. The discrepancy in the two sets of data for the same laser pulse is attributed by Narayam et al. to pulse-to-pulse variations in energy density for at least about 50% of the laser spots. The probe beam size in Lo and Compaan's experiments varied from 70 to 100 µm. Since the size of the melted region varied from 20 to 60 µm, the beam is probing a substantial amount of unmelted region. This may explain the low temperature rise measured in the Raman experiments.

In conclusion, for nanosecond laser annealing, it appears resonable to assume that the screening of phonon-carrier collision does not play a major role in the rate of energy transfer, and the absorbed energy of the laser beam is almost instantaneously coupled to the lattice. As a consequence, the absorption depth of the laser beam determines the heating profile and the power density at depth x is given by

$$S(\gamma^{-1} = 0) = I(1 - R)\alpha e^{-\alpha x} \qquad (8)$$

The thermal effects induced by the laser beam irradiation can then be described by the usual heat diffusion equation with a source term given by Eq. (8).

For typical nsec pulsed laser annealing, the diameter of the incident laser is ~ 200 μm, whereas the thermal diffusion length is less than 1 μm. Therefore, the lateral diffusion in the plane of the incident surface may be neglected and the heat diffusion equation in this case reduces to an one-dimensional equation. Furthermore, since the temperature of the sample can be raised in a few nanoseconds to a high value, the strong temperature dependence of thermal conductivity and optical absorption coefficient has to be incorporated in annalyzing this nonlinear dynamic heat transport process. For an indirect band-gap semiconductor, such as Si, the absorption coefficient is given by [24]

$$\alpha = A(\hbar\omega_\ell - E_g)^2 \qquad (9)$$

Here A is an increasing function of T, $\hbar\omega_\ell$ is the input photon energy, and E_g is the energy gap of the material. With increasing T, the change in the periodic potential experienced by the electrons due to thermal expansion and lattice vibration change results in a significant narrowing of the band gap. The forbidden band gap of silicon, E_g, is given by [25]

$$E_g = 1.16 - \frac{7.02 \times 10^{-4} T^2}{T + 1108} \qquad (10)$$

Therefore, with increasing T, E_g is significantly reduced and α is drastically increased. For example, at a wavelength of 693 nm α(T) can be approximated as [17]

$$\alpha(T) = \alpha_R e^{T/T_R} \qquad (11)$$

where

$$\begin{array}{lll}
\alpha_R = 1340 \text{ cm}^{-1} & T_R = 427°C & \text{at } 694 \text{ nm} \\
\alpha_R = 5020 \text{ cm}^{-1} & T_R = 430°C & \text{at } 532 \text{ nm} \qquad (12)\\
\alpha_R = 9310 \text{ cm}^{-1} & T_R = 434°C & \text{at } 435 \text{ nm}
\end{array}$$

On the other hand, the thermal conductivity, K(T) is given by [26]

$$K = cv^2 \tau/3 \qquad (13)$$

Here c is the specific heat of phonons, v, the phase velocity and τ, the phonon collision rate. At high temperature the phonon density can be approximated by

$$n = \frac{1}{e^{\hbar\omega/kT} - 1} \cong \frac{kT}{\hbar\omega} \qquad (14)$$

and the total number of phonons is proportional to T. Since a given phonon that contributes to the thermal flux is more likely to be scattered with increases in the number of thermally-excited phonons present, the relaxation time should decrease with increasing T. In addition, at high temperatures the phonon specific heat obeys the law of Dulong and Petit and is temperature-independent. Hence, the thermal conductivity decreases with rising lattice temperature. For instance, the measured thermal diffusivity of silicon, D(T), which is defined as K/cp, can be accurately fitted by [16]

$$D(T) = \frac{D_R}{1 + aT} \qquad (15)$$

with D_R = 0.94 cm^2/sec and a = 0.0072/°C. With α(T) and D(T) incorporated in the heat transport equation, one can realistically examine the dynamics of the depth profile of T under pulsed laser irradiation. Hence, we can write the equation for the lattice temperature rise as

$$\frac{\partial T}{\partial t} = \frac{\partial}{\partial x}(D(T)\frac{\partial T}{\partial x}) + \frac{I(1-R)}{c\rho}\alpha(T)e^{-\alpha(T)x} \qquad (16)$$

Here, c(T) is the heat capacity, ρ(T) is the mass density, α(T) is the optical absorption coefficient and D(T) is the thermal conductivity. The boundary conditions are

$$\frac{\partial T}{\partial x}\bigg|_{x=0} = 0, \quad T(x\to\infty,t) = 0 \qquad (17)$$

if the thickness, d of the semiconductor layer is large enough to be approximated by a semi-infinite medium.

III. PARAMETRIZED PERTURBATION THEORY

In this section we present the parametrized perturbation scheme we have developed to analytically examine the strongly nonlinear heat transport process occuring in high power laser annealed semiconductors [27,28]. Since the specific heat c(T) and the mass density ρ(T) of the material are in most cases weakly dependent on temperature, we regard these two quantities as constants. The rise in lattice temperature, T is investigated analytically by partitioning it into two parts:

$$T(x,t) = T_o(x,t) + \Delta T_o(x,t) \qquad (18)$$

Here, T_o is taken to grow in time due to the primary laser heating with a <u>constant</u> optical absorption coefficient, α_o and to <u>diffuse</u> in space with a <u>constant</u> thermal diffusivity, D_o; viz.

$$\frac{\partial T_o}{\partial t} = D_o \frac{\partial^2 T_o}{\partial x^2} + \frac{I(1-R)}{c\rho}\alpha_o e^{-\alpha_o x} \qquad (19)$$

The values of D_o and α_o are yet unknown and are regarded, therefore, as two basic parameters. The solution of T_o satisfying the boundary conditions given by Eq. (17) can be obtained by using the Green's function technique.

The nonlinearity of the problem arising from the temperature dependence of both thermal diffusivity, D(T) and optical absorption coefficient α(T) can be described in the form of the secondary source function, which is a function of T_o and drives the correction term ΔT_o. Inserting Eqs. (18) and (19) into Eq. (16) yields,

$$\frac{\partial}{\partial t} \Delta T_o = [D(T_o + \Delta T_o) - D_o] \frac{\partial^2}{\partial x^2} (T_o + \Delta T_o)$$

$$+ [\frac{\partial}{\partial x} D(T_o + \Delta T_o)][\frac{\partial}{\partial x}(T_o + \Delta T_o)] \quad (20)$$

$$+ \frac{I(1-R)}{c\rho}(\alpha(T_o)e^{-\alpha(T_o)x} - \alpha_o e^{-\alpha_o x})$$

We quasilinearize Eq. (20) by partitioning ΔT_o further as

$$\Delta T_o(x,t) = T_1(x,t) + \Delta T_1(x,t) \quad (21)$$

where T_1 obeys the equation

$$\frac{\partial T_1}{\partial t} = D_o \frac{\partial^2 T_1}{\partial x^2} + S_1(T_o) \quad (22)$$

with

$$S_1 = [D(T_o) - D_o]\frac{\partial^2 T_o}{\partial x^2} + \frac{\partial D(T_o)}{\partial x}\frac{\partial T_o}{\partial x}$$

$$+ \alpha(T_o)e^{-\alpha(T_o)x} - \alpha_o e^{-\alpha_o x} \quad (23)$$

It is important to point out that the secondary source term S_1 for the first-order temperature correction term T_1 results from

(1) the net local departure of influx of heat associated with the temperature dependence of thermal diffusivity from D_o, (i.e., $D(T_o) - D_o$) and

(2) the departure of local lattice heating rate $\alpha(T_o)\exp(-\alpha(T_o)x)$ from $\alpha_o \exp(-\alpha_o x)$, (namely $\alpha(T_o)\exp(-\alpha(T_o)x) - \alpha_o\exp(-\alpha_o x)$).

In as much as T_o can be obtained explicitly, S_1 is an explicit function of material depth x, heating time t, laser power density I, and the parameters, α_o, D_o. Eq. (22) is identical in structure to Eq. (19), so that T_1 can be treated exactly with the use of the Green's function technique for the same boundary conditions.

A few comments are due at this point concerning the optimal choice of thermal diffusivity, D_o and absorption coefficient, α_o. Recall that the zeroth-order temperature term T_o is specified in terms of α_o and D_o, and the secondary source function S_1 is, in turn, expressed explicitly in terms of T_o and thus D_o and α_o. Mathematically, α_o and D_o are chosen in such a way that the resulting T minimizes the contribution of S_1. Thus T_1 is much less than T_o. Physically, D_o and α_o can be respectively interpreted as the effective dynamic mean thermal diffusivity and absorption coefficient, which are determined in a self-consistent way, given pulse intensity and duration. An inhomogeneous heat generating source distributes the temperature in space. The exact nature of this temperature distribution inside the material is governed by the cumulative effect of local heating rates as well as the local heat diffusion rates. Because the optical absorption coefficient and thermal diffusivity are strongly dependent on the local temperature rise, the depth profile of temperature is significantly affected by the distributed diffusivities and absorption coefficients. The optimal values of D_o and α_o are the particular values chosen from this distribution so that the cumulative effect of the departures of local diffusivity and absorption coefficient from these values is minimal.

The only way to determine these optimal values of D_o and α_o is to formally obtain T_o as a function of D_o and α_o and to minimize T_1. One can likewise carry out the higher-order iterations to improve the accuracy of the solutions to the desired degree. One of the advantages of this perturbation scheme is the rapid convergence of the solution. Furthermore, the analytical expression of the depth profile of lattice temperature thus obtained is explicit with respect to space and time, as will be illustrated in the next section.

IV. LASER HEATING OF SILICON

We now consider the optical heating of ion-implanted silicon, using the parametrized perturbation technique detailed in the previous section. The solution for the zeroth-order dominant term, T_o, satisfying Eq. (19), can be obtained by using the Green's function

$$T_o = \int_0^t dt_o \int_0^\infty dx_o \frac{I(t_o)(1-R)}{c\rho} \alpha_o e^{-\alpha_o x_o} \quad (24)$$

$$\cdot [G(x,t|x_o,t_o;D_o)]$$

with

$$G(x,t|x_o,t_o;D_o) = [4\pi D_o(t-t_o)]^{-1/2}$$

$$\cdot \sum_\pm \exp - \frac{(x \pm x_o)^2}{4D_o(t-t_o)} \quad (25)$$

Here, T_o satisfies the boundary conditions specified by Eq. (17). Taking the laser pulse, $I(t)$, to be rectangular in time and carrying out the integrations yields

$$T_o(x,t) = [I(1-R)/c\rho\alpha_o D_o]F(\xi,\tau), \quad (26)$$

where the response function F

$$F(\xi,\tau) = -\xi \text{erfc}(\frac{\xi}{\sqrt{\tau}}) + (\frac{\tau}{\pi})^{1/2} e^{-\xi^2/\tau} - e^{-\xi}$$

$$+ \frac{1}{2}e^{\tau/4}[e^{-\xi}\text{erfc}(\frac{1}{2}\sqrt{\tau} - \frac{\xi}{\sqrt{\tau}}) \quad (27)$$

$$+ e^\xi \text{erfc}(\frac{1}{2}\sqrt{\tau} + \frac{\xi}{\sqrt{\tau}})]$$

is expressed in terms of a scaled material depth ξ and a scaled thermal diffusion time τ:

$$\xi = \alpha_o x$$
$$\tau = 4\alpha_o^2 D_o t \quad (28)$$

We take the values of R = 0.37, ρ = 2.31 g/cm³, and c = 0.84 Joul/g°C for Si. In Fig. 1 and 2, we have plotted $F(\xi,\tau)$ as a function of both ξ and τ. Although the response function F depends in a rather complicated way on ξ and τ via the complimentary error functions, it behaves smoothly and monotonically with respect to ξ and τ; it decreases with increasing ξ and increases with increasing τ. T_o can therefore be accurately reexpressed in terms of simple functions of τ and ξ as follows. From Fig. 1 one can determine for a fixed ξ the depth over which the temperature decreases by a factor 1/e. This depth, $w(\tau)$, obtained as a function of τ is again seen to be smooth and monotonically increasing with τ [see Fig. 3]; $w(\tau)$ can be fitted accurately by

$$\omega(\tau) = 0.55 + 0.32\tau^{1/4} + 0.42\tau^{1/2} \quad (29)$$

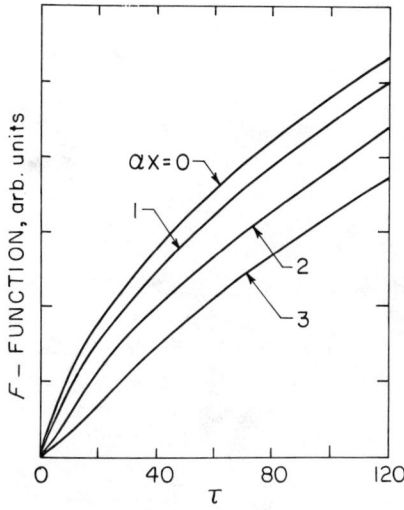

Fig. 2 The response function F vs. scaled time τ for different depths

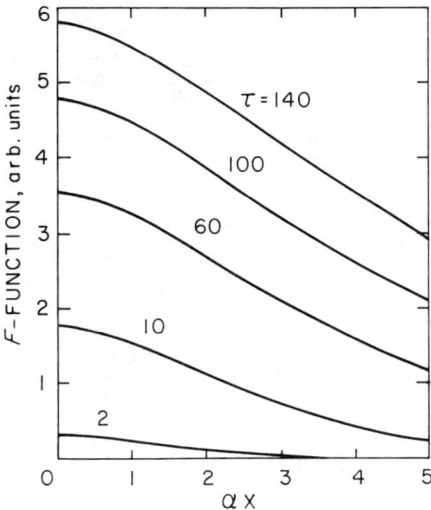

Fig. 1 The response function F vs. depth αx for different times

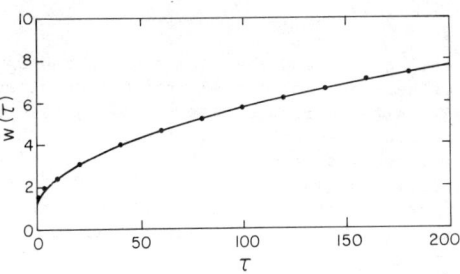

Fig. 3 Width is depth of F vs. τ: a filled circle represents the width w of F at a fixed τ and the solid line is fitted by Eq. (29)

In Fig. 4 we have replotted the depth profile of F for several τ values against $\xi/w(\tau)$. The similarity of these curves with the Gaussian profiles, $\exp(-[\xi/w(\tau)]^2)$ is obvious and the minor departure from the Gaussian shape can be corrected for by introducing a low-order polynomial in $\xi/w(\tau)$. For practical purposes, one can therefore approximate T_o by

$$T_o \simeq \frac{I(1-R)}{c\rho\alpha_o D_o} g(\tau) e^{-[\xi/\omega(\tau)]^2} \quad (30)$$
$$\cdot \{1 - 0.8[\xi/\omega(\tau)]^2 + 0.8[\xi/\omega(\tau)]^3\},$$

where $g(\tau)$ corresponds to the value of the F function at the surface ($\xi=0$):

$$g(\tau) = \left(\frac{\tau}{\pi}\right)^{1/2} - 1 + e^{\tau/4}\text{erfc}\left(\frac{1}{2}\sqrt{\tau}\right) \quad (31)$$

As mentioned earlier, the optimal values of thermal diffusivity (D_o) and optical absorption coefficient (α_o) should now be determined self-consistently. Note in Eq. (23) that S_1 is specified in terms of T_o, which, in turn, depends on D_o and α_o. Now D_o and α_o are chosen in such a way that the resulting T_o minimizes the contribution of S_1 to T_1, i.e. $T_1 \ll T_o$. The detailed algebra for this determination together with the explicit solution of T_1 is relegated to Appendix A and the final results are presented here. T_1 is minimized with the choice of D_o equal to the local diffusivity at a depth ξ_m and also with the choice of α_o equal to the local absorption coefficient at a depth ξ'_m, <u>for a given pulse intensity and duration</u>:

$$D_0 = \frac{D_R}{aT_0(\xi_m, \tau/2)} \ln[1 + aT_0(\xi_m, \tau/2)] \quad (32)$$

$$\alpha_0 = \alpha(T_0(\xi'_m, \tau/2)) = \alpha_R e^{T_0(\xi'_m, \tau/2)/T_R} \quad (33)$$

with ξ_m and ξ'_m given by

$$\xi_m = \frac{1}{\sqrt{\pi A}} \quad (34)$$

$$\xi'_m = -\frac{B}{2A} + \frac{\xi_m}{e^{\frac{B^2}{4A}} \operatorname{erfc} \frac{B}{2\sqrt{A}}} \quad (35)$$

and

$$A = \frac{1}{w^2(\tau/2)} + \frac{1}{\tau} \quad (36)$$

$$B = \frac{1.54}{\sqrt{\tau}} \quad (37)$$

When the values of D_0 and α_0 thus found is inserted back into Eq. (26), one obtains the complete, explicit expression for the dominant temperature term, T_0. The values of α_0 and D_0 as given by Eq. (32) and (33) in terms of both material parameters (a, c, R, ρ, α, D) and laser beam parameters (I, t) are significant results of our analysis.

Figures 5 and 6 show the values of α_0 and D_0 as functions of heating time for several different pulse intensities. With increasing heating time, the lattice temperature rises and the effective mean thermal diffusivity decreases with the effective optical absorption coefficient increasing rapidly. Furthermore, the increase in α_0 results in raising the temperature near the surface of the sample due to the enhanced heating rate therein. This, in turn, decreases D_0 and the heat diffusion into the bulk material is slowed down. This slowing down of the heat diffusion will further raise the surface temperature and therfore will significantly enhance the absorption coefficient. The dynamic feedback of these two processes reduces substantially the energy deposition depth during the irradiation and also confines the absorbed energy near the surface region of the material. Under these conditions, the surface heating efficiency is considerably enhanced and the threshold pulse energy for the surface melting is drastically reduced. The other important feature is that the rate with which D_0 decreases and α_0 increases in time depends sensitively on the operating pulse intensity. For example, at I = 40 MW cm^{-2} and pulse duration 1 nsec, D_0 = 0.88 cm^2/sec., and α_0 = 5*10^3 cm^{-1}, while at I = 100 MW cm^{-2} with the same pulse duration D_0 is reduced to 0.3 cm^2/sec. and α_0 is increased to 2*10^5 cm^{-1}. Because T_0 becomes large with a small D_0 and a large α_0 (Eq. (27)) one can observe clearly that the efficiency of lattice heating depends sensitively on the pulse intensity used.

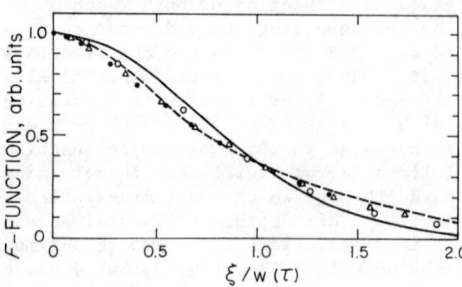

Fig. 4 F function vs. $\xi/w(\tau)$: open circle (o) represents the F function at $\tau=20$, cross (x) at $\tau=100$, and filled circle (·) at $\tau=180$; the solid line is a fit by $\exp[-\xi^2/w^2(\tau)]$ and the dashed line is fitted by the gaussian modulated polynomial, Eq. (30)

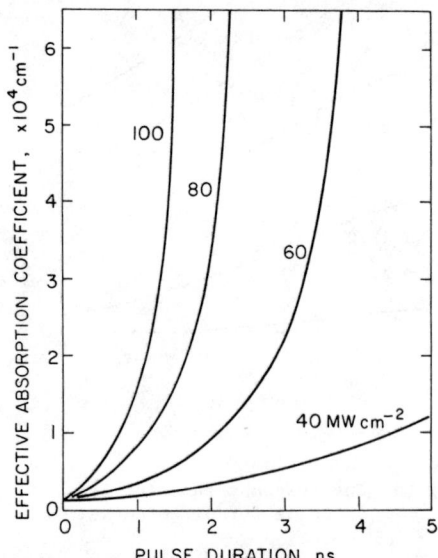

Fig. 5 Effective optical absorption coefficient, α_0 vs. heating time for different pulse intensities at 694 nm.

The rise in lattice temperature up to the first-order perturbation is given by

$$T(x,t) = T_0(x,t) + T_1(x,t) \quad (38)$$

In Fig. 7 we present the depth profile of $T_0 + T_1$, T_0 and T_1 for a fixed pulse intensity and energy. As can be observed, the first order correction term T_1 is much smaller than T_0. The maximum contribution of T_1 never exceeds 20% of T_0 for pulse intensities varing from 10 to 100 MW/cm^2. This rapid convergence is clearly due to the optimal determination of the effective absorption coefficient and thermal diffusivity in our perturbation scheme. For practical purposes one may, therefore, approximate the laser-heated lattice temperature rise by T_0 with α_0 and D_0 used as the effective optical absorption coefficient and thermal diffusivity, respectively.

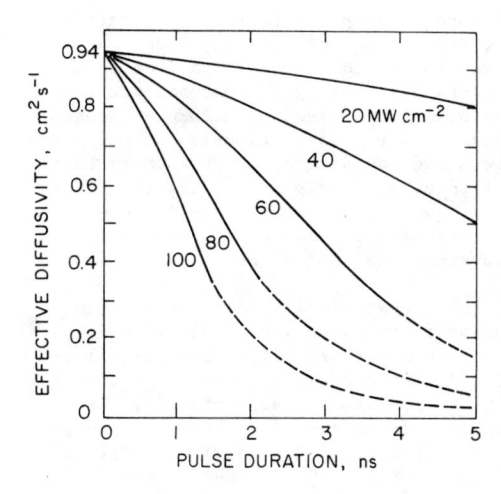

Fig. 6 Effective diffusivity, D_0 vs. heating time for different pulse intensities at 694 nm. The dashed line is the extrapolation of the single phase annealing

The dynamic feedback of the rapidly reduced local thermal diffusivity and energy deposition depth of laser beam affecting, in turn, the resulting lattice temperature profile is more clearly seen in Fig. 8, where we have plotted T as a function of depth for a fixed pulse energy with different intensities and durations. At 10 MW cm^{-2} for 10 nsec, the effective thermal diffusion time, $\tau = 4\alpha_0^2 D_0 t$ is long enough to increase the heated volume beyond the energy deposition depth, α^{-1}. Because of this rapid transfer of energy into the bulk material the rise in temperature

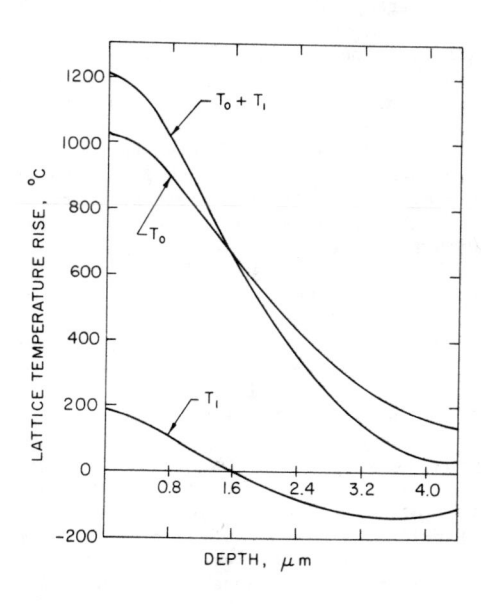

Fig. 7 Depth profile of temperature rise

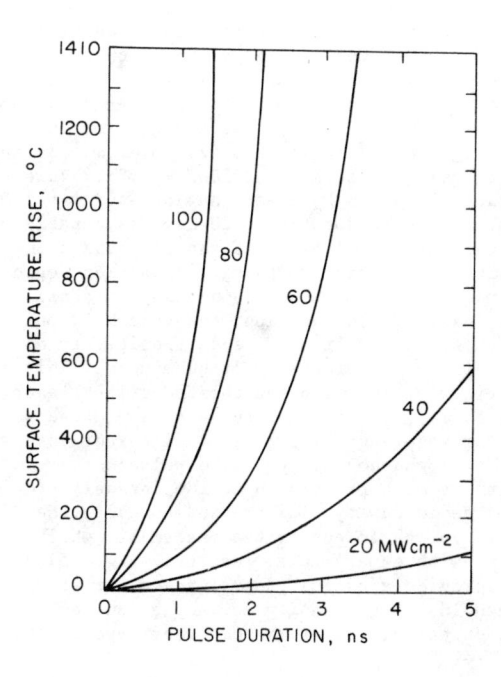

Fig. 8 Surface temperature rise vs. heating time for different pulse intensities at 694 nm.

near the surface is small. With increasing pulse intensity, a rapid reduction in D_0 ensues and the heat diffusion time slows down considerably. Furthermore, due to the shrinking of the energy deposition depth, the laser energy is mostly confined near the surface. These two effects, therefore, drastically enhance the growth rate of the surface temperature.
This nonlinear dynamic heating effect is again obvious in Fig. 8, in which we have plotted the surface temperature rise as a function of heating time for several different pulse intensities. As can be clearly seen, the rise in T is strongly dependent on the operating pulse intensity.

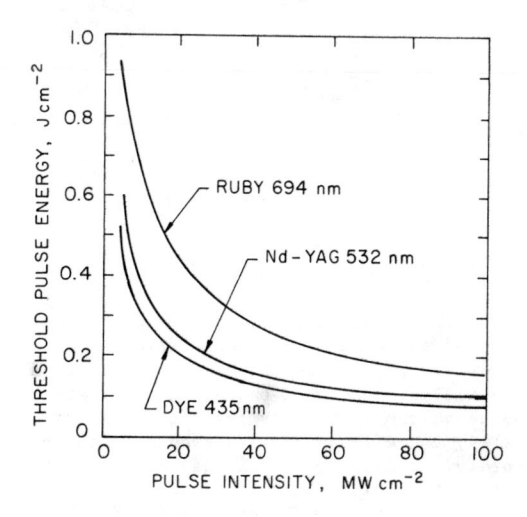

Fig. 9 Threshold pulse energy for surface melting vs. pulse intensity at several wavelengths

The characteristics of pulsed laser heating of Si lattice are summarized in Fig. 9, in which the threshold laser energy for the onset of surface melting is plotted as a function of pulse intensity and different laser wavelengths. At 694 nm wavelength for 20 MW cm^{-2}, the energy needed is about 0.42 Joule cm^{-2} is sufficient to melt the surface at 100 MW cm^{-2}. This suggests that a pulsed ruby laser, having energy $\sim$ 1.0 Joule cm^{-2} and time duration $\sim$ 20 nsec is capable of melting the surface of Si wafer and of propagating the melt front well beyond the energy deposition depth. The thermal melting model for annealing is consistent with our results. In addition, the threshold pulse energy for surface melting depends sensitively on the operating frequency of the laser beam used. This is seen in Fig. 10, in which the threshold laser energy required to melt the silicon surface is plotted as a function of wavelength for several different intensities. The dependency of threshold pulse energy on wavelength is most pronounced at low intensity region. As the pulse intensity is increased however, the difference in threshold energy becomes small, which is consistent with experimental results. Fig. (5) and Fig. (6) provide a useful general guide for predicting the threshold energy density for any given combination of pulse duration, intensity and laser wavelength.

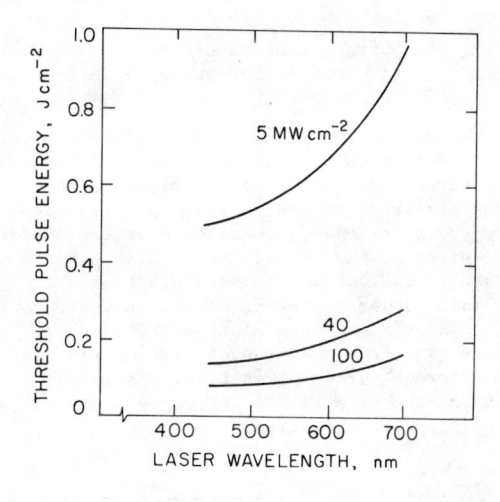

Fig. 10 Threshold pulse energy for surface melting vs. wavelength for different pulse intensities

V. DISCUSSION

We have presented in this paper an analytical description of the dynamic characteristics of lattice temperature rise in Si annealed by a high power laser beam. The analysis is based on the parametrized perturbation technique in Green's function formulation. Specifically, the effect of the strong coupling between rapidly reduced thermal diffusivity and optical absorption depth during the laser beam irradiation has been discussed and the kinetic of the depth profile of the lattice temperature, the threshold pulse energy for the onset of surface melting, melt front generation and propagation, and the recrystallization velocity are characterized in terms of both the material and laser beam parameters. This analysis provides considerable insight into the transient heating phenomena and localized material modifications that can be achieved by exposing a semiconductor to pulsed laser irradiation, and gives an accurate temperature distribution and a precise estimation of the pulse duration required for surface of the sample to reach melt condition at a given pulse intensity. These quantities are of crucial importance for the fabrication of microelectronic devices and circuits. The results are summarized as follows:

(1) For the case of silicon, in which the thermal conductivity and energy deposition depth of the sample decrease drastically with rising lattice temperature, T, the surface temperature can be raised efficiently by increasing laser intensity for a fixed pulse energy. This is due to the dynamic feedback effect of the rapid slowing down of heat diffusion into the bulk material and the concomitant shrinking of laser energy deposition depth near the surface. This can be qualitatively seen as follows. Initially α is small and $\tau \equiv 4\alpha^2 Dt \ll 1$, hence from Eq. (26), $T \sim I(1-R)(\alpha t/c\rho)\exp{-\alpha x}$, indicating that T grows linearly in time according to the local input energy profile. The ensuing increase in α as a result of the rise in T shifts the depth profile of T toward the surface. Simultaneously, τ value is increased and in the limit $\tau \gg 1$, again from Eq. (26), $T \sim I(1-R)t^{1/2}$. Expressed in terms of the deposited energy during t, $E = I(1-R)t$, $T \sim E/t^{1/2}$. Thus, it is clear that for a fixed amount of deposited energy T increases more efficiently if E is compressed in a shorter time, i.e. if the operating laser intensity is large.

(2) Our calculation did not incorporate the sudden jump in surface reflectivity value, which occur during the laser pulse. This may raise the values of our threshold pulse energy for the onset of surface melting approximately by a factor of 2.

Our analytical method can be applied to the real-life process implementation for integrated circuit fabrication with the following generalizations:

(1) Two-phase annealing: We have thus far analytically followed in time the depth profile of T for the case of single phase laser annealing. Our technique can be generalized to describe the two phase annealing. With the onset of surface melting of the silicon slice, for example, the reflectivity increases to a value 0.72, while the absorption depth shrinks down to about 100 Å. The molten layer thickness will grow in time and store a large fraction of the laser energy deposited as the fusion heat for each increment of melt depth. This process is self-limiting in that the heat of fusion consumes a large amount of energy, while the local temperature remains at the melting point. Because of the small temperature gradient near the vicinity of the heat source, the thermal diffusion is minimized. Consequently, near the end of the laser pulse the depth profile of T in the bulk will remain approximately the same as that at the onset of surface melting. The dynamics of the melt front penetration can be described, using our technique, coupled with the heat-balance integral, the heat diffusion equation in the liquid, the equation of motion of the melt front and the energy conservation principle at the fusion front. The results for the two-phase annealing will be reported elsewhere.

(2) Cooling of lattice temperature: An explicit investigation of cooling of lattice after the laser beam irradiation is important. The local cooling rate of lattice temperature, together with its heating rate determines the time period of elevated lattice temperatures, in which the annealing process is believed to occur. For the case of two-phase radiation annealing, the velocity of melt front propagating back to the surface is strictly governed by the local cooling rate. An accurate, analytic description of this velocity will be extremely valuable for the discussion of epitaxial regrowth dynamics of polysilicon. The lattice temperature is governed in this case by the heat diffusion equation without source term, where the temperature dependent diffusivity now increases rapidly with increasing cooling time. The nonlinear equation can be readily treated by our technique.

(3) CW laser annealing and the lateral thermal diffusion: CW laser annealing techniques have thus far generated an impressive list of interesting results, such as the complete electrical activation of dopant atoms, generation of large single crystal islands or crystal lographically oriented single crystal films in polysilicon. It is, therefore, important to characterize analytically the cw laser annealing on a μs or ms time scale. Our technique, when generalized to include the two-phase heat diffusion processes in a manner discussed above, is well suited to treat this problem. The additional generalization required in this case is to include the effect of lateral heat diffusion during the irradiation time. This amounts to considering three-dimensional, parametrically linearized heat diffusion equation, whose solutions can be obtained by means of a Green's function technique.

(4) Finally pulsed laser processing technique has shown to improve the electrical characteristics of oxides grown on top of laser-annealed polysilicon films, and this result obviously enhances the possibility for the fabrication of 3-dimensional IC structures. However, there are several major considerations to be taken into account for a full utilization of this interesting result: (a) laser heating of the polysilicon films should be limited to the top surface regions, and (b) the input laser pulse energy should be controlled to match precisely the value required to melt only the top surface region. The pulse energy exceeding this value will cause deleterious effects o heating the underlying SiO_2/Si interface. For a fixed pulse intensity, the duration should be such that $t_p \leq d/4D$, where t_p is the laser pulse duration and D is the thermal diffusivity of polysilicon. For this, it is important to evaluate the extent of melting of the polysilicon films and the transient heat distribution in the multilayer structure. Our technique, when generalized to include the two-phase heating processes in a manner discussed in (1), is well suited to treat this problem. The additional generalization required in this case is to include the sample depth dependence of optical absorption coefficient and thermal diffusivity. The results for this multilayer annealing will be reported elsewhere.

APPENDIX

In this appendix we determine α_0, D_0 self-consistently. The solution of the first order correction term T_1 obeying the linear diffusion equation (21) can be expressed in terms of the Green's function as

$$T_1 = \int_0^t dt_0 \int_0^\infty dx_0\, S_1(x_0,t_0) G(x,t|x_0,t_0) \quad (A-1)$$

where

$$G(xt|x_0 t_0) = [4\pi D_0(t-t_0)]^{-1/2} \sum_\pm \exp - \frac{(x_0 \pm x)^2}{4D_0(t-t_0)} \quad (A-2)$$

and t denotes the irradiation time. The source term, S_1 which is given explicitly in terms of x_0, t_0 via $T_0(x_0,t_0)$ varies slowly and monotonically in time. Hence we may put

$$S_1(x_0, t_0) = S_1(x_0, t/2) \quad (A-3)$$

and bring it out of the t_0-integration (mean value theorem). The remaining time-varying part is readily integrated to give

$$T_1 = \frac{1}{2D_0\sqrt{\pi}} \sum_\pm \int_0^\infty d\xi_0\, f(\xi_0,\xi)\, S_1(\xi_0, 1/2\tau) \quad (A-4)$$

where

$$f(\xi_0,\xi) = \sqrt{\tau}\, e^{-\frac{(\xi+\xi_0)^2}{\tau}} - \sqrt{\tau}\,|\xi_0 \pm \xi|\, \mathrm{erfc}\, \frac{|\xi_0 + \xi|}{\sqrt{\tau}} \quad (A-5)$$

and ξ, τ are defined in the text by Eq. (16). Inserting Eq. (13) for S_1 into Eq. (A-3) we have

$$T_1 = \frac{1}{2D_0\sqrt{\pi}} \sum_\pm \int_0^\infty \int_0^\infty d\xi_0 f(\xi_0,\xi) \left[(D(T_0) - D_0) \frac{\partial^2 T_0}{\partial \xi_0^2} \right.$$
$$\left. \cdot \frac{\partial D(T_0)}{\partial \xi_0} \frac{\partial T_0}{\partial \xi_0} \right] + \frac{I(1-R)}{\alpha_0 c\rho} \int_0^\infty dx_0\, f(\xi_0,\xi)$$
$$\cdot [\alpha(T_0)e^{-\alpha(T_0)x_0} - \alpha_0 e^{-\alpha_0 x_0}] \quad (A-6)$$

Throughout the analysis in this appendix we make the following convention: T_0 as a function of time and depth (Eq. (26)) will be taken as $T_0(\xi_0, \tau/2)$ inside an integral and as $T_0(\xi, \tau/2)$ otherwise. We next perform a few integrations by parts with respect to ξ_0 for the first term of Eq. (A-5). With the use of the boundary conditions for T_0 and identification

$$\frac{1}{2D_0\sqrt{\pi}} \sum_\pm \int_0^\infty d\xi_0\, f(\xi_0,\xi)\, \left[(D(T_0)-D_0)\frac{\partial^2 T_0}{\partial \xi_0^2} + \frac{\partial D(T_0)}{\partial \xi_0}\frac{\partial T_0}{\partial \xi_0} \right] = \frac{-1}{2D_0\sqrt{\pi}} \sum_\pm \int_0^\infty \frac{\partial T_0}{\partial \xi_0} [D(T_0)-D_0]$$

$$\frac{\partial}{\partial \xi_0} f(\xi,\xi_0) = \frac{H(\xi)T_0}{D_0} - \frac{1}{D_0\sqrt{\pi\tau}}$$

$$\int_0^\infty T_0 H(\xi_0) e^{-\frac{(\xi_0+\xi)^2}{\tau}} d\xi_0 \qquad (A\text{-}7)$$

where the H-function of argument ξ_0, for example, is defined as

$$H(\xi_0) = D_0 - \frac{D_R}{aT_0(\xi_0, \tau/2)} \ln[1 + aT_0(\xi_0, \tau/2)] \qquad (A\text{-}8)$$

Thus T_1 becomes

$$T_1 = \frac{H(\xi)T_0}{D_0} - \frac{1}{D_0\sqrt{\pi\tau}} \sum_\pm \int_0^\infty T_0 H(\xi_0) e^{-\frac{(\xi_0+\xi)^2}{\tau}} d\xi_0$$

$$+ \frac{1}{2D_0} \frac{I(1-R)}{\alpha_0 c\rho} \int_0^\infty dx_0 T_0 \, f(\xi_0,\xi) G(\xi_0) \qquad (A\text{-}9)$$

with

$$G(\xi_0) = \frac{\alpha(T_0) e^{-\alpha(T_0)x_0} - \alpha_0 e^{-\alpha_0 x_0}}{T_0} \qquad (A\text{-}10)$$

Since H and G vary smoothly w.r.t. ξ_0, one may Taylor expand H and G centered at ξ_m and ξ'_m, respectively:

$$H(\xi_0) = H(\xi_m) + (\xi_0 - \xi_m) H'(\xi_m) + \ldots \qquad (A\text{-}11)$$

$$G(\xi_0) = G(\xi'_m) + (\xi_0 - \xi'_m) G'(\xi'_m) + \ldots \qquad (A\text{-}12)$$

$$T_1 = \frac{T_0 H(\xi)}{D_0} - \sum_\pm T_0(\xi=0,\tau/2) e^{-\frac{\xi^2}{2\tau}} [\beta_0 G(A, \frac{\xi}{\tau}, \pm \xi_m, 0) + \frac{1}{A}(\beta_0 - \beta_1 e^{\frac{\xi^2}{\tau} + B\xi})]$$

$$+ \sum_\pm \beta_1 T_0(\xi=0,\tau/2) \cdot [2G(A, \frac{2\xi}{w^2} + B, \xi_0, \xi) \pm G(A, \frac{+\xi}{w^2} + B, \pm(\xi'_m - \xi), 0)] \qquad (A\text{-}23)$$

Now we chose D_0 and α_0 in such a way that the dominant terms in T_1 corresponding to the first terms in the above Taylor expansions vanish; that is

$$H(\xi_m) = 0 \qquad (A\text{-}13)$$

$$G(\xi'_m) = 0 \qquad (A\text{-}14)$$

Upon combining Eqs. (A-8), (A-13) and Eqs. (A-10), (A-14), (3), there results

$$D_0 = \frac{D_R}{aT_0(\xi_m, \tau/2)} \ln[1 + aT_0(\xi_m, \tau/2)] \qquad (A\text{-}15)$$

$$\alpha_0 = \alpha(T_0(\xi'_m, \tau/2)) = \alpha_R \exp(T_0(\xi'_m, \tau/2)/T_R) \qquad (A\text{-}16)$$

These two equations are used in the text as Eqs. (31) and (32). Note that the values of ξ_m, ξ'_m are not

specified thus far. We may choose these values to reduce further the magnitude of T_1 by requiring that the second dominant contributions in the Taylor expansion equal to zero near the surface ($\xi = 0$):

$$\int_0^\infty T_0(\xi_0 - \xi_m) e^{\frac{-\xi_0^2}{\tau}} d\xi_0 = 0 \qquad (A\text{-}17)$$

$$\int_0^\infty T_0 \, f(\xi_0,\xi=0)(\xi_0 - \xi'_m) d\xi_0 = 0 \qquad (A\text{-}18)$$

The values of ξ_m, ξ'_m as determined from Eqs. (A-17), (A-18) are given by

$$\xi_m = \frac{1}{\sqrt{\pi A}} \qquad (A\text{-}19)$$

$$\xi'_m = -\frac{B}{2A} + \frac{\xi_m}{e^{\frac{B^2}{4A}} \text{erfc} \frac{B}{2\sqrt{A}}} \qquad (A\text{-}20)$$

with

$$A = \frac{1}{w^2(1/2\tau)} + \frac{1}{\tau} \qquad (A\text{-}21)$$

$$B = \frac{1.54}{\sqrt{\tau}} \qquad (A\text{-}22)$$

Upon inserting Eqs. (A-11), (A-12), (A-19), (A-20), in Eq. (A-9) and performing the ξ_0-integration, one finds

where

$$G(a,b,d,y) = (b/2a - d)(\sqrt{\pi}/2a) e^{\frac{b^2}{4a}} \text{erfc}(\sqrt{a}\, y - b/2\sqrt{a})$$

$$\beta_0 = \frac{\sqrt{2}}{\sqrt{\pi\tau}} \frac{T'_0(\xi_m)}{T_0(\xi_m)} (D_0 - \frac{D_R}{1 + aT_0}) \qquad (A\text{-}24)$$

$$\beta_1 = \frac{I(1-R)}{c_\rho \alpha_0 D_0} - \frac{\sqrt{\tau}\, T'_0(\xi'_m)(1 - \xi'_m)}{\sqrt{2\pi}\, T_0(\xi'_m) T_R}$$

34

REFERENCES

1. *Laser-Solid Interactions and Laser Processing - 1978* (Material Research Society, Boston), edited by S.D. Ferries, H.J. Leamy, and J.M. Poate (AIP Conference Proceedings, New York, 1979), vol. 50.

2. *Laser and Electron Beam Processing of Materials*, edited by C.W. White and P.S. Peersey (Academic, New York, 1980).

3. *Laser and Electron-Beam Solid Interactions and Material Processing*, edited by J.F. Gibbons, L. Hess, and T. Sigmon (North Holland, New York, 1981).

4. *Laser and Electron-Beam Interactions with Solids*, edited by B.R. Appleton and G.K. Celler (North Holland, New York, 1982).

5. *Laser-Solid Interactions and Transient Thermal Processing of Materials*, edited by J. Narayan, W.L. Brown, and R.A. Lemons (North Holland, New York, 1983).

6. J.C. Wang, R.F. Wood, and P.P. Pronko, Appl. Phys. Lett. 33, 455 (1978).

7. P. Baeri, S.U. Campisano, G. Floti, and E. Rimini, J. Appl. Phys. 50, 788 (1979).

8. A.E. Bell, RCA Review, 40, 295 (1979).

9. R.O. Bell, M. Toulemoude, and P. Siffert, Appl. Phys. 19, 313 (1979).

10. C.M. Surko, A.L. Simons, D.H. Auston, J.A. Golovchenko, and R.E. Slusher, Appl. Phys. Lett. 34, 635 (1979).

11. Y.S. Liu and K.L. Wang, Appl. Phys. Lett. 34, 363 (1979).

12. J.C. Schultz and R.J. Collins, Appl. Phys. Lett. 34, 84 (1979).

13. M. Lax, Appl. Phys. Lett. 33, 786 (1978).

14. J.R. Meyer, F.J. Bartoli, and N.R. Kruer, Phys. Rev. B21, 1559 (1980).

15. J.R. Meyer, N.R. Kruer, and F.J. Bartoli, J. Appl. Phys. 51, 5513 (1980).

16. *JNAF Thermodynamic Tables*, 2nd ed. (US GPD, Washington, D.C., (1971).

17. G.E. Jellison, Jr. and F.A. Modine, ORNL/TM 8002 (1982).

18. J.A. Van Vechten, R. Tsu, and F.W. Saris, Phys. Lett. 74A, 422 (1979).

19. E.J. Yoffa, Phys. Rev. B21, 2445 (1980).

20. J.M. Liu, H. Kurz and N. Bloembergen in Ref. 5.

21. A. Lietola and J.F. Gibbons, Appl. Phys. Lett. 34, 332 (1979).

22. H.W. Lo and A. Compaan, Phys. Rev. Lett. 44, 1604 (1980).

23. J. Narayan, J. Fletcher, C.W. White, and W.H. Christie, J. Appl. Phys. 52, 7121 (1981).

24. F.J. Blatt, *Physics of Electronic Conduction in Solids*, McGraw, New York (1968).

25. S.M. Sze, *Physics of Semiconductor Devices*, Wiley, New York, 1969, Ch. 1, 4, 8.

26. N.W. Ashcroft and N.D. Mermin, *Solid State Physics*, Holt, Rinehart and Winston, New York (1976).

27. D.M. Kim, D.L. Kwong, R.R. Shah, and D.L. Crosthwait, Jr., J. Appl. Phys. 52, 4995 (1981).

28. D.M. Kim and D.L. Kwong, IEEE J. Quantum Electron, QE-18, 224 (1982).

HEAT FLOW-ACOUSTIC EMISSION-MICROSTRUCTURE CORRELATIONS IN RAPID SURFACE SOLIDIFICATION

R. B. Clough, H. N. G. Wadley, and R. Mehrabian
National Bureau of Standards
Washington, DC 20234

ABSTRACT

Heat flow models are now available for the one- and two-dimensional rapid melting and resolidification of metallic substrates subjected to high energy laser and electron-beam sources. The models can account for both stationary and moving heat sources. In the past, experimental observations have been limited to post-solidification examinations of the microstructures and have resulted in establishment of correlations between fineness of structure, interface stability and extent of altered microstructure (e.g. melt depth) with variables such as the heat flux distribution in space and time. Techniques are accordingly needed for in situ measurement of the dynamics of laser and electron beam material interactions, possibly leading to in-process control applications and the detection of defective conditions.

Acoustic emission methods show promise for this. Acoustic emission accompanying absorption of 100 ms stationary electron beams of variable flux density have been measured from 1100 and 2219 aluminum alloys. The acoustic emission appears, in these alloys, to be generated by sudden stress relaxations accompanying plastic deformations and hot tearing, both induced by solidification stresses. The more gradually evolving solidification and thermal contraction stresses themselves appear incapable of generating detectable elastic radiation, at least for the longer duration pulses used here.

INTRODUCTION

Directed high energy sources (laser and electron beams) are currently being evaluated for rapid surface melting and resolidification of a range of engineering alloys [1]. In these processes, the bulk substrate, in intimate contact with the molten surface layer, acts as the quenching medium resulting in high solid/liquid interface velocities and short solidification times during cooling. These rapidly solidified alloys exhibit enhanced wear and corrosion resistance, arising, it is believed, from improved chemical homogeneity and the fine scale of microstructure. Such materials offer promise for applications such as bearing surfaces, cutting tool faces, and corrosion resistant parts.

At present, critical process variables such as solid/liquid interface velocity or the cooling rate can only be deduced using one- and two-dimensional heat flow models for melting and resolidification. Experimental methods to measure, in situ, process variables are required so that confirmation of microstructure-process variable relationships may be made. Furthermore, surface modification of some alloys is severely limited by the occurrence of cracking during solidification.

In this work, we report the use of acoustic emission techniques to monitor rapid solidification of electron beam surface melted 1100 commercially pure aluminum and a 2219 aluminum alloy in several different heat treated conditions. Heat flow theory has been used to set up controlled experiments in which shallow melts were formed at the surface of the aluminum plates. After attainment of steady state conditions, the energy source was terminated, resulting in rapid solidification for which the acoustic emission response was measured. Effects of varying both the substrate composition and electron beam parameters on the acoustic emission response have been investigated and linked with post-mortem metallographic observations of microstructure to deduce the origin of the elastic waves.

HEAT FLOW CALCULATIONS

The problem considered was the rapid melting and resolidification of an alloy substrate subjected to a high-intensity stationary heat source over a circular region on its bounding surface. The numerical solutions were developed for an Al-4.5 wt.% Cu alloy which is close to the composition of the 2219 aluminum alloy (whose principle solutes were 6.6 wt.% Cu, 0.34 wt.% Mn) used in the experiments reported in this paper. It was assumed that the substrate had previously been subjected to conventional solidification--it segregated and precipitated a weight fraction of eutectic calculated from the Scheil equation [2].

The generalized two-dimensional heat flow model and solution technique developed in an oblate coordinate system [3] can readily take both space and time variations of the heat flux into consideration. For the purpose of this paper, it is reasonable to assume that a step function heat flux is applied at time zero with a uniform circular distribution. The shape and location of the "mushy" zone, bounded in space by the liquidus and solidus isotherms, are then calculated for various combinations of process variables. Special emphasis is placed on conditions that result in steady state heat flow. That is, conditions under which the total amount of heat absorbed at the surface circular region is exactly offset by the conduction of heat into the substrate interior. This results in a steady-state thermal field with a maximum temperature at the center of the circular surface region. This enables a quiescent condition (i.e., with static stresses) to be established prior to solidification. Two dimensionless variables are used to describe heat flow:

$$k(T(0,0) - T_o)/qa \quad \text{and} \quad a/\sqrt{4\alpha_s t} \quad (1)$$

where k is thermal conductivity, $T(0,0)$ is the temperature in the center of the circular region; T_o is initial substrate temperature, q and a are absorbed heat flux and radius of the circular region, respectively. $a/\sqrt{4\alpha_s t}$ denotes dimensionless inverse time, α_s is thermal diffusivity of the solid, and t is time. Values of the variables are given in table I.

example, for a heat flux of $q = 2.7 \times 10^8$ W/m² absorbed over a circular region of radius $a = 10^{-3}$m ($qa = 2.7 \times 10^5$ W/m) temperatures of $T(0,0) = 1384$ K and $T(0,0) = 2054$ K are reached in $t \sim 9.2$ ms ($a/\sqrt{4\alpha_s t} \sim 0.6$) and $t = 52.7$ ms ($a/\sqrt{4\alpha_s t} \sim 0.25$), respectively. Note that for $t > 52.7$ ms ($a/\sqrt{4\alpha_s t} < 0.25$), the plot in Figure 1 approaches unity on the vertical axis, and steady state conditions are obtained, i.e. a maximum temperature $T(0,0) = 2054$ K is reached. Therefore, for products 1.56 x

Table I

Properties of Al-4.5% Cu Alloy Used in Calculations [4]

Notation	Name	Value
k_s	Thermal conductivity of solid	180.6 Wm⁻¹k⁻¹
k_ℓ	Thermal conductivity of liquid	100.8 Wm⁻¹k⁻¹
α_s	Thermal diffusivity of solid	7.6×10^{-5} m²s⁻¹
T_E	Solidus temperature	821 K
T_L	Liquidus temperature	923 K
T_v	Vaporization temperature	2744 K

In order to account for the release of heat of fusion, the first term of expression (1) was enlarged to:

$$(k_\ell(T(0,0) - T_L) + \Delta\zeta + k_s(T_E - T_o))/qa \quad (2)$$

where the subscripts ℓ and s denote liquid and solid, respectively and T_L and T_E are the liquidus and solidus temperatures, respectively. $\Delta\zeta$ is an equivalent term accounting for heat conduction in the "mushy" region and is calculated to be $\sim 6.2 \times 10^4$ W/m for the Al-4.5 wt.% Cu alloy [4].

It has been verified that for a given value of expression (2) all isotherms in the alloy substrate are located at the same dimensionless distance for identical dimensionless times. Figure 1 shows a plot of results obtained from computer calculations relating the two dimensionless variables noted above. A number of important facts are deduced from this figure. First, there is a minimum value of the product qa required if the center of the circular region is to reach a given temperature $T(0,0)$, e.g. the liquidus T_L or the vaporization T_v temperatures, of the alloy. That is, when the power is maintained for long times (for small values of $a/\sqrt{4\alpha_s t}$) the vertical axis approaches unity--steady state conditions then prevail if $T(0,0) < T_v$. The sum of the second and third terms in the numerator of the ordinate $\Delta\zeta + k_s(T_E - T_o) \cong 1.56 \times 10^5$ W/m, where T_o is ambient temperature. Therefore, for the Al-4.5 wt.% Cu alloy, a minimum value of the product qa $\sim 1.56 \times 10^5$ W/m is required if the center of the circular region is to reach the liquidus temperature of the alloy, $T(0,0) = T_L$. Similarly, $k_\ell(T_v - T_L) \sim 1.84 \times 10^5$ W/m, a minimum value qa $\sim 3.4 \times 10^5$ W/m is deduced for $T(0,0) = T_v$. The temperature at the center of the circular region for any dimensionless time can also be readily deduced from Figure 1. For $10^5 \leq qa \leq 3.4 \times 10^5$ W/m, $a = 10^{-3}$ m, and times $t \gtrsim 52.7$ ms temperatures are likely to be in the range $T_L \leq T(0,0) \leq T_v$ and steady state conditions are expected to prevail for the Al-4.5 wt.% Cu alloy substrate.

Figure 2 shows the location of the "mushy" zone (the region bounded by liquidus and solidus isotherms) during the approach to steady-state conditions, followed by rapid solidification. The isotherms here represent the situation in which heat was applied until almost a steady-state condition was achieved, the solidus isotherm reached a plateau, at which point beam power was turned off resulting in rapid solidification. The vertical distance between the two curves denotes the size of the "mushy" zone. Upon attainment of a steady-state condition the "mushy" zone extended $\sim 1/4$ of the maximum melt depth $z/a = 0.4$. Loss of superheat and shallower temperature gradients during solidification result in significantly larger "mushy" zone sizes. For example, when the substrate surface reaches the liquidus temperature, the solidus isotherm is at approximately $z/a \sim 0.3$--the size of the "mushy" zone is three times larger than when solidification started.

Average solidus and liquidus interface velocities and solidification times can be readily calculated from the data shown in Figure 2. Assume that $a = 10^{-3}$ m, $q = 2 \times 10^8$ W/m². Δt for the liquidus and solidus to move from their maximum depths, $z_L = 3 \times 10^{-4}$ m and $z_s = 4 \times 10^{-4}$ m, to the surface are ~ 3.7 ms and ~ 4.75 ms, respectively. This translates into average liquidus and solidus velocities of 8.1×10^{-2} m/s and 8.5×10^{-2} m/s, respectively. Note that while the two average interface velocities are almost equivalent, initially the solidus moves much slower than the liquidus resulting in the large "mushy" zone size noted above.

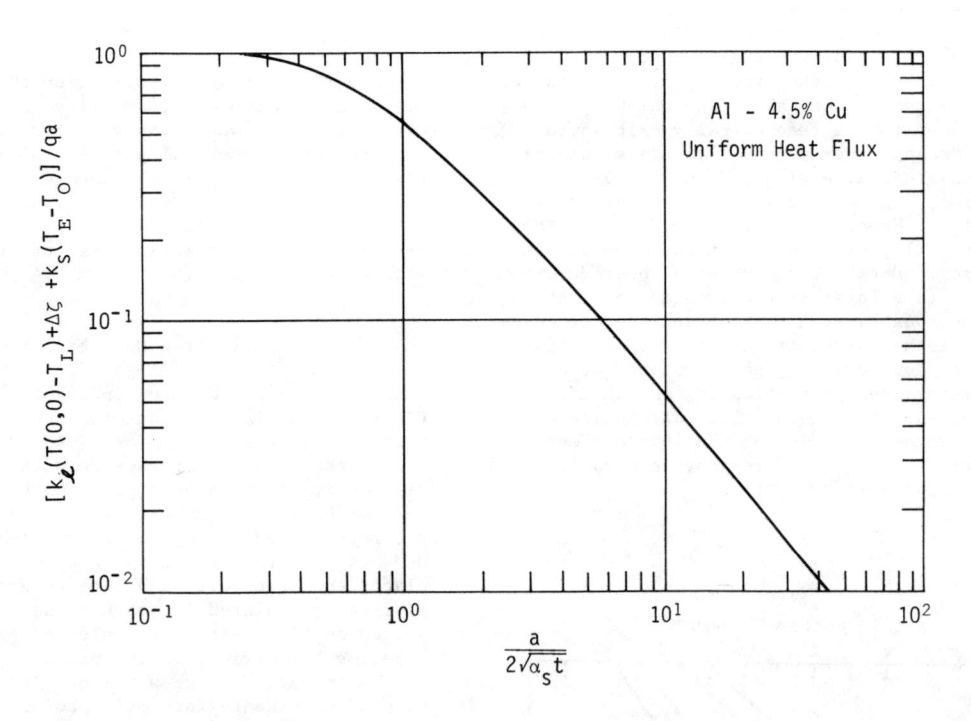

Figure 1. Temperature at the center of the circular region during melting of a semi-infinite Al-4.5 wt.% Cu alloy substrate as a function of uniform absorbed heat flux, radius of the circular region and time [4].

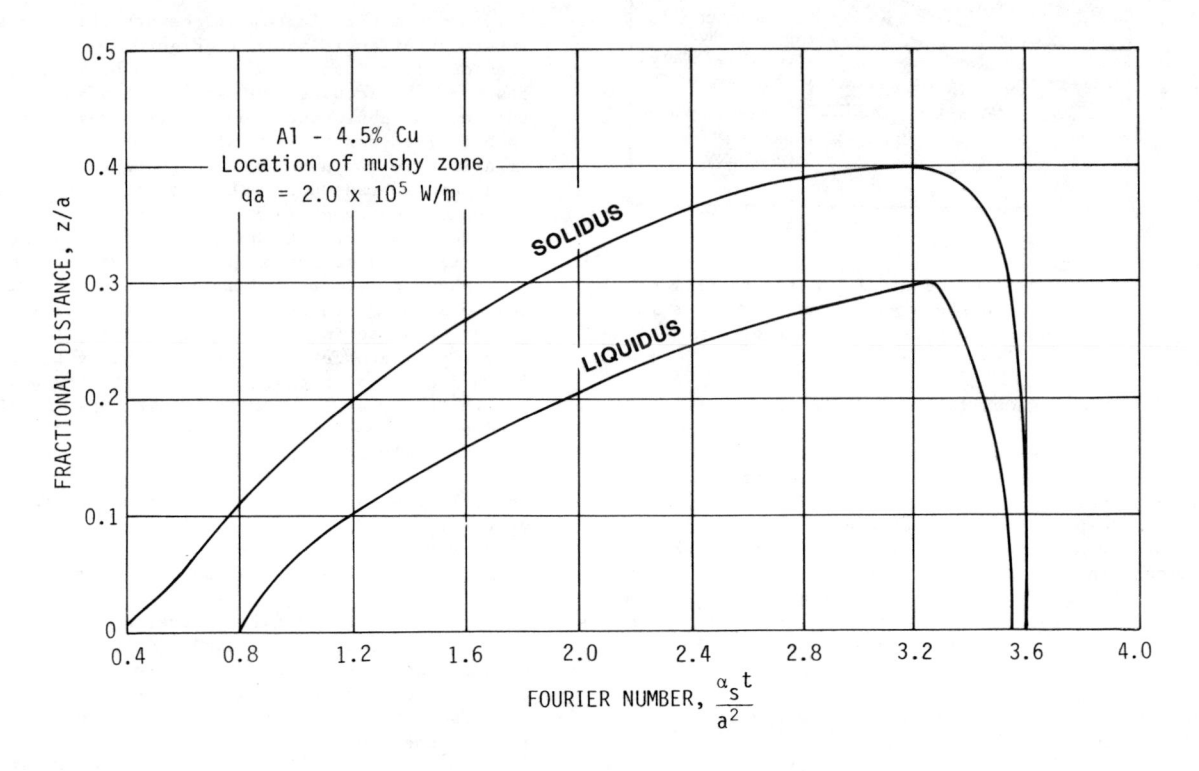

Figure 2. The dimensionless size and location of the "mushy" zone along the z-axis versus time for product $qa = 2.0 \times 10^5$ Wm^{-1}. T_{923} and T_{821} denote the locations of the liquidus and solidus isotherms, respectively [4].

The steady-state shapes and locations of the "mushy" zone, liquidus and solidus isotherms as a function of both dimensions r and z are readily obtained from the computer calculations. In Figure 3, we show the positions of the steady-state "mushy" zones for two values of the product qa. Both values are within the range noted above for which $T_L \leq T(0,0) \leq T_v$. While the theoretically calculated size of the "mushy" zone does not change significantly with increasing value of qa, from 2×10^5 to 2.7×10^5 Wm^{-1}, Figure 3 shows that the melt depth almost doubles. Experimental melt profiles of the 2219 aluminum alloy are plotted on the same figure for comparison. Detailed experimental procedures are described in a later section. Agreement between the two is reasonably good considering the various assumptions of the calculations including a slightly different alloy composition, and constant but different thermophysical properties for the solid, the "mushy" zone, and the liquid. The calculations are based on complete absorption of the incident energy by the substrate. This is a reasonable assumption for electron beam surface melting.

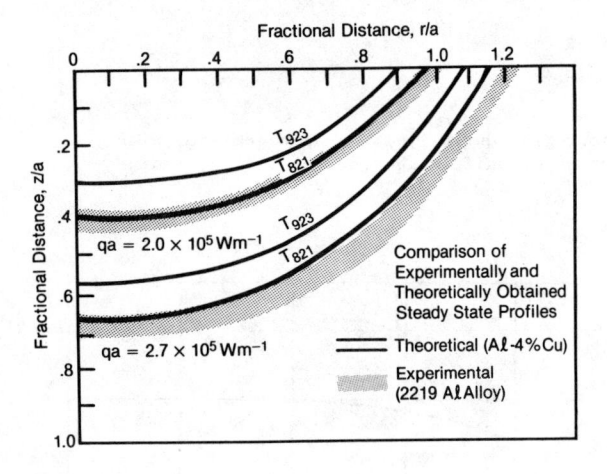

Figure 3. Comparison of predicted and experimentally measured steady-state melt profiles for aluminum-copper alloy at two values of qa.

ACOUSTIC EMISSION THEORY

Acoustic emission is the elastic radiation from changes of stress in a body. Such changes may be either positive (e.g., elastically constrained thermal expansion during absorption of a laser pulse) or negative (e.g., the growth of a crack through an already stressed region). The propagation of elastic waves through an observation point within/on the body results in transient elastic displacements of the point. These are the basic physically observable quantities in an acoustic emission experiment.

Calculations [6-8] using dynamic elasticity are able to predict acoustic emission waveforms. To demonstrate the calculations, the transient displacement vertically above four point sources, a vertical dipole, a horizontal mode I loaded crack, a thermal expansion (dilatation), and an inclined shear loop are shown in Figure 4 [9]. If such waveforms due to actual sources are measured, the dynamics and other source properties can be deduced [11].

EXPERIMENTAL

Materials

In the present investigation the acoustic emission from two aluminum alloys has been simultaneously studied. One alloy was 1100 aluminum, a 99.0 percent commercial purity aluminum. The second alloy studied was 2219 aluminum containing 6.6 wt.% Cu, 0.08 wt.% Si, 0.3 wt.% Mn, and 0.104 wt.% Zr. The 2219 aluminum alloy was in the -T87 condition. Rectangular plate samples were cut from each alloy with dimensions of 4.0 x 1.5 x 0.4 cm.

Electron Beam Facility

Samples were mounted in the facility shown in Figure 5 and placed in an electron beam heating unit. To insure attainment of steady-state conditions, the samples were heated with a 100 ms duration stationary electron beam pulse with a circular cross section having a 1 mm radius a. The electron beam was accelerated through 25 kV and the beam current deduced from the voltage developed across a resistor placed in series with the line to ground. The heat flux was calculated for each pulse, and the current varied so that heat flux could be investigated as an experimental variable. The product qa was systematically varied between $\sim 1.6 \times 10^5$ and 3.4×10^5 W/m to produce steady-state melt profiles at the long times $t \sim 100$ ms noted above.

Acoustic Emission Measurements

A schematic diagram of the acoustic emission instrumentation is shown in Figure 6. A piezoelectric transducer was used to convert the elastic displacements of the acoustic emission into a voltage signal that could subsequently be electronically processed. To optimize sensitivity (signal-to-noise ratio) signals were bandpass filtered between 0.02 and 1 MHz and a transducer with high sensitivity in this passband (see inset in Fig. 6) was used to measure emission. The gradual build-up of stress could activate rapid stress drops of ≤ 1 μs risetime accompanying deformation and fracture events. The coarse sampling interval required to follow the more gradual contraction stress results in effectively only random discrete sampling of the fast risetime (high frequency) signals. In effect, the bandwidth is limited by a 5 μs sampling rate to about 100 kHz. Studies showed, by varying the low pass filter, that approximately half the typical signal power was below 100 kHz. Much of this can be attributed to the filtering effect of the transducer; its sensitivity (Fig. 6) drops about 40 dB between 100 kHz and 1 MHz. Signals were amplified 50 dB for the 2219 aluminum alloys and 70 dB for 1100 aluminum.

The acoustic emission signal and the beam current were digitally recorded on separate channels of a 2-channel 8-bit transient recorder. For the examination of emission throughout the entire pulse, a coarse sampling interval of 200 μs (giving a record length of 400 ms) was used for a few pulses on each alloy. In this study our interest was focused on the acoustic emission during solidification and for the majority of measurements, digitization was performed at 5 μs intervals (10 ms record) from just prior to the moment the electron beam was cut off. The digitized data was temporarily stored on floppy diskettes at the remote electron beam

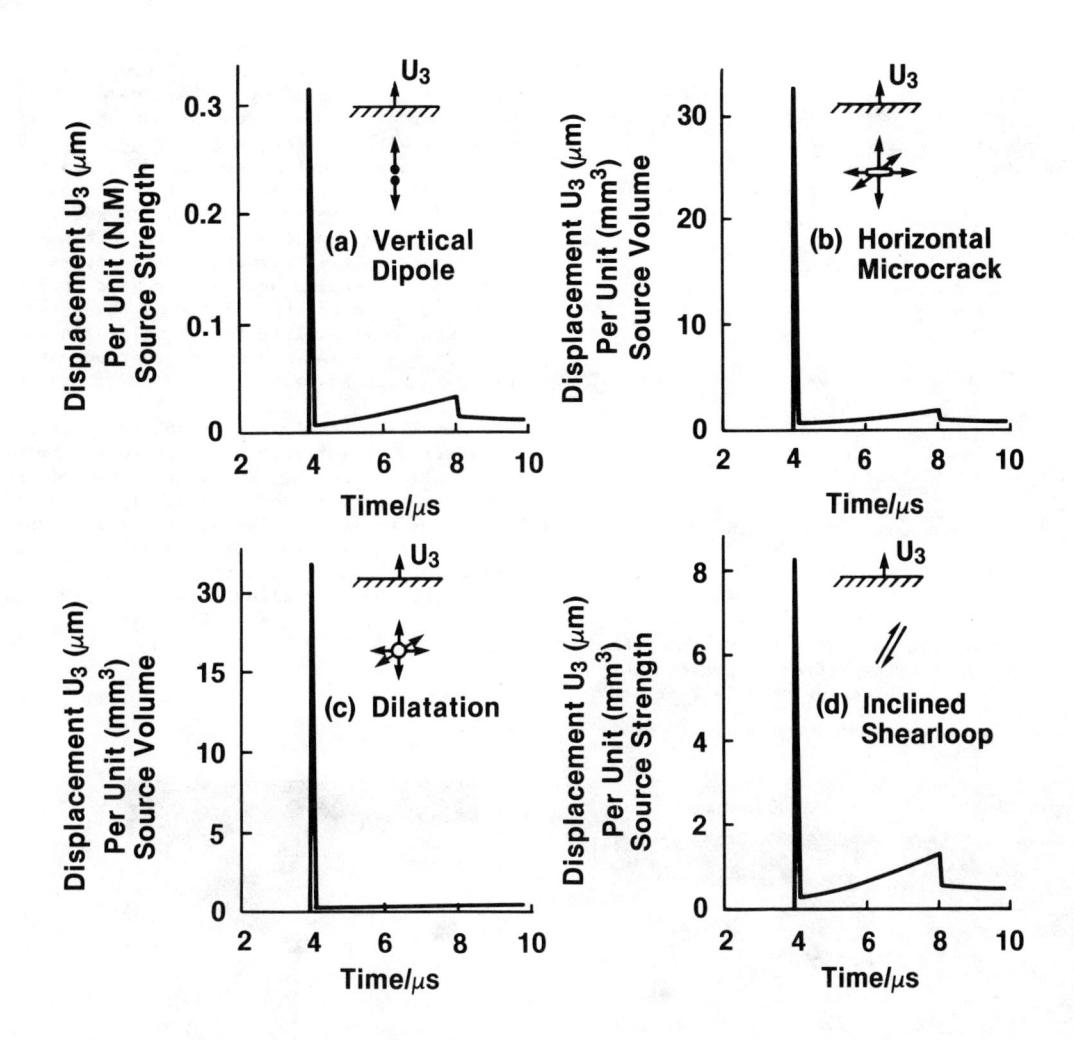

Figure 4. Transient epicenter vertical displacements above four point sources: (a) Vertical dipole; (b) Horizontal microcrack; (c) Dilatation; and (d) Inclined shear loop [9].

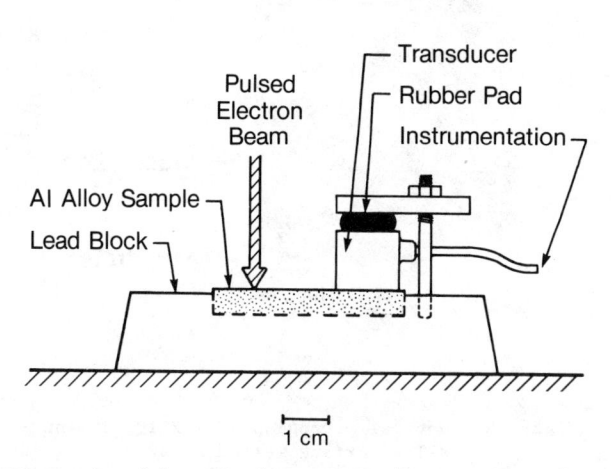

Figure 5. Schematic diagram of the experimental facility to measure acoustic emission during the rapid solidification of electron beam surface melted aluminum alloys.

installation and transferred, off-line, to a 1.5M Byte minicomputer for data analysis. In these experiments we are not attempting to deduce dynamic source information. Wave propagation within the sample is too complex to allow this at present. Rather, we are attempting to follow the kinetics of solidification through the temporal variation of acoustic emission activity. This "activity" was arbitrarily (but in a rigorously standardized manner) measured by computing the mean square voltage of the emission as a function of time and then integrating this over time until the signal became lost in noise. This provided a useful single number (related to acoustic emission energy) with which to describe the relative "activity" of the different electron beam pulses/samples.

RESULTS

Microstructure

The effect of increasing the product qa, absorbed heat flux $q(Wm^{-2})$ times radius of the circular region a, upon melt size and microstructure is shown in Figure 7 for both the 1100 aluminum and the 2219

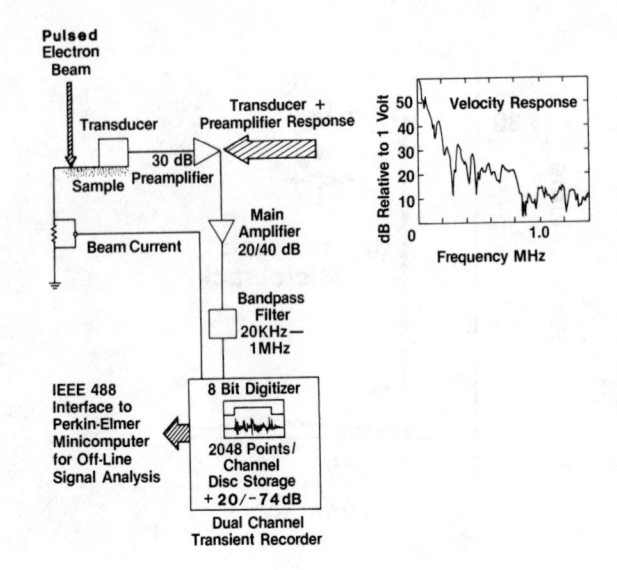

Figure 6. Schematic diagram of acoustic emission instrumentation together with the transducer plus preamplifier response determined by the breaking capillary method.

aluminum alloy. Figure 3 shows a comparison between the calculated and experimentally measured depth profiles for the 2219 aluminum alloy. Agreement is good considering various assumptions of the heat flow model including constant thermophysical properties for the alloy. The size of the melted region for qa values in the steady-state regime increased with qa. For the same qa, the alloy, with its lower liquidus temperature, had a slightly deeper melt depth. The 2219 alloy also had a less clear interface between the rapidly solidified region and the unmelted substrate. This is marked the "mushy" zone in Figure 8. It denotes a partially melted region between the melt pool and the unmelted substrate. Figure 9 shows the effect of qa upon melt profile for qa values in the vaporization regime. Deep penetration can be observed in both alloys.

Detailed metallographic observations indicate considerable plastic deformation to accompany melting and resolidification, and, in the case of the 2219 alloy, considerable hot tearing also occurred (see Fig. 8 for example). The effect of qa upon melt depth along the z-axis for both alloys is shown in Figure 10. Melting occurred at slightly lower qa values for the 2219 alloy and, for a given qa, melt depths were always greater for 2219.

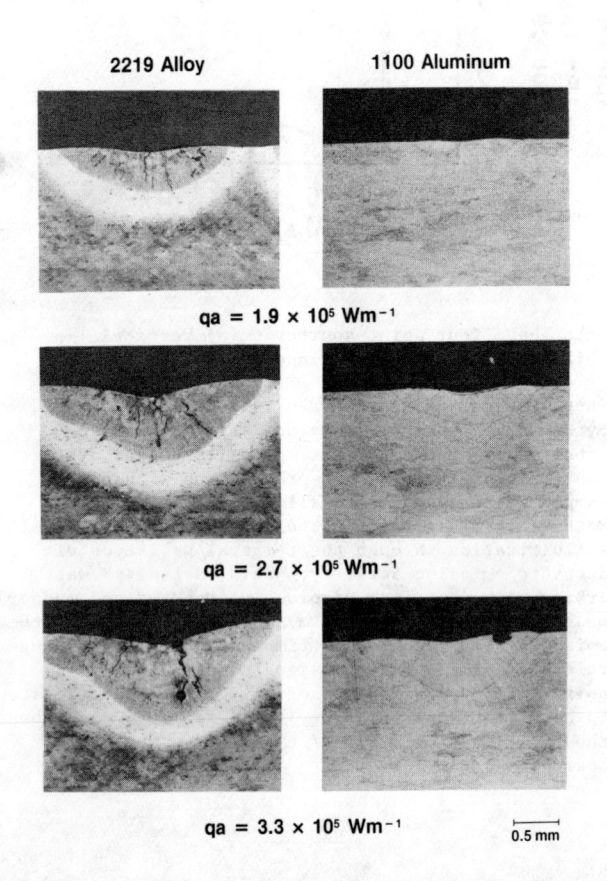

Figure 7. Optical micrographs of 1100 and 2219 aluminum alloys showing the effect of qa upon melt profile and ensuing microstructures in the "steady state" region.

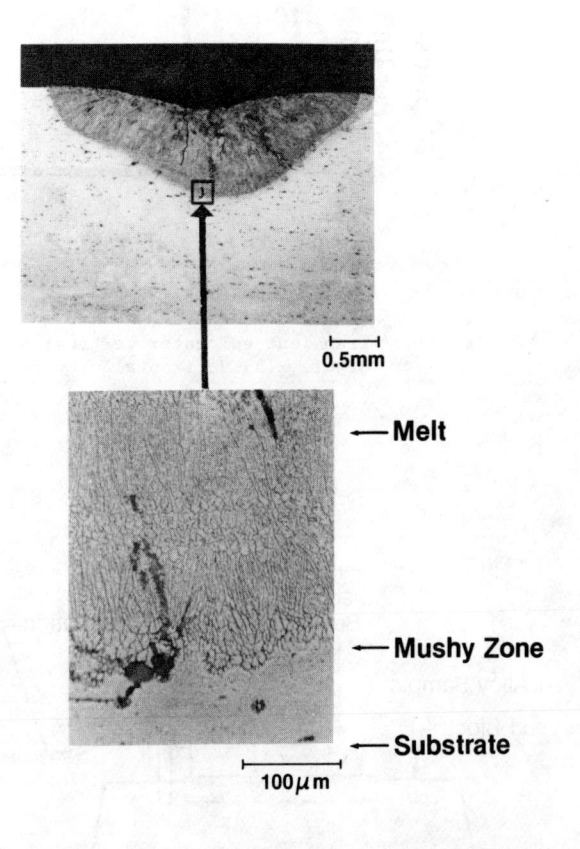

Figure 8. Optical micrographs of 2219 aluminum alloy surface melted at $qa = 2.7 \times 10^5$ Wm^{-1} showing detailed microstructure of the region of the "mushy" zone at the bottom of the melt pool.

Acoustic Emission

The temporal behavior of the acoustic emission voltage, together with absorbed heat flux, are shown in Figure 11 for $q_a = 2.7 \times 10^5$ Wm^{-1} for both metals over the entire beam pulse. It can be seen that very much more acoustic emission is observed from the 2219 alloy than from 1100 aluminum both while melting was occurring and during rapid solidification after beam cut-off.

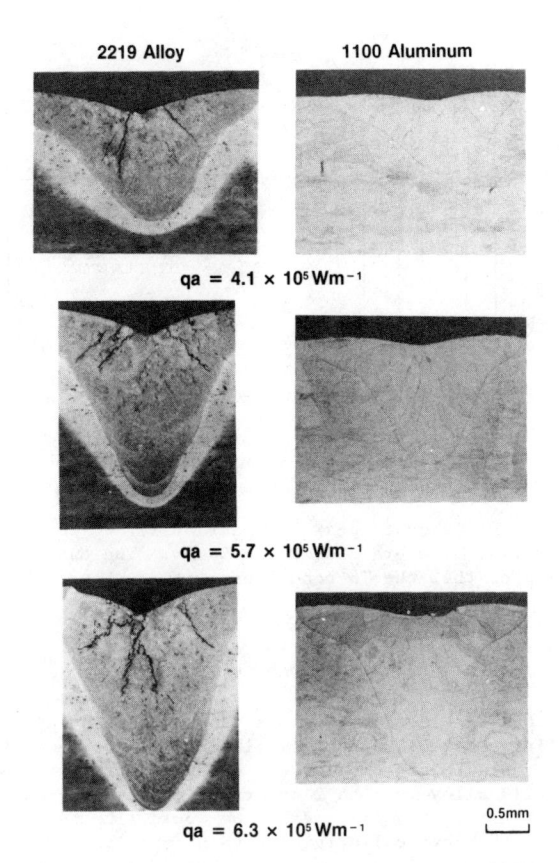

Figure 9. Optical micrographs of 1100 and 2219 aluminum alloys showing the effect of q_a upon melt profile and ensuing microstructure of q_a such that $T[0,0] > T_v$.

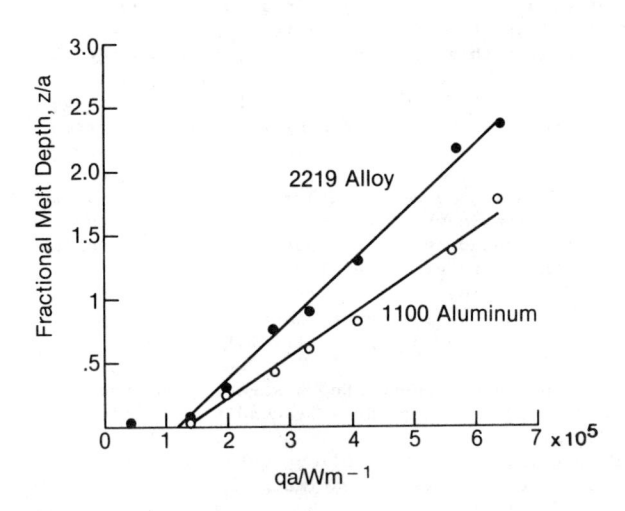

Figure 10. The effect of q_a upon melt depth for 1100 and 2219 aluminum alloys. Melt depth is measured from the liquidus isotherm to the substrate surface for the 2219 alloy.

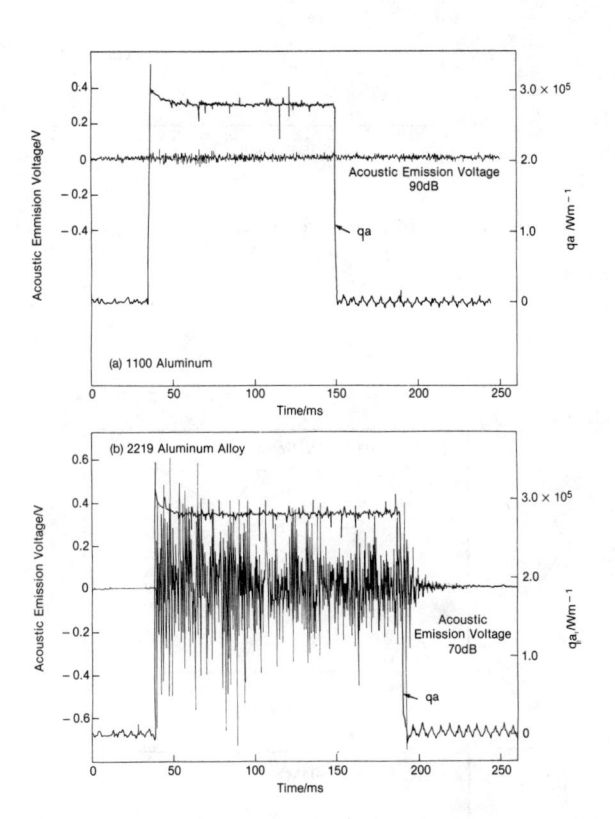

Figure 11. Acoustic emission generation during pulsed electron beam heating ($q_a = 2.7 \times 10^5$ W/m) and subsequent cooling of 1100 and 2219 aluminum alloys. Note that 1100 aluminum measurements were made at a 10x higher gain, but even so generated orders of magnitude less emission.

The acoustic emission energy during the period of solidification (during and after beam cut-off) is shown, as a function of q_a, in Figure 12. Figure 12(a) and (b) focuses on the AE behavior of 2219 material in a -T87 condition in the thermoelastic region, where no melting occurs, (12(a), $q_a = 1.36 \times 10^5$ W/m) and the melt region (12(b), $q_a = 2.70 \times 10^5$ W/m), showing the beam current (dotted line) and AE response (solid line). Note that melting in this case often produces an extra pulse of AE during solidification. In Figure 13 the cumulative energy of the AE signals during solidification is shown for both the 2219 and 1100 materials. Each of the data points are the average of 15 separate electron pulses on three different, but nominally similar samples. It can be seen that the acoustic emission is typically a hundred times more energetic in the 2219 aluminum alloy.

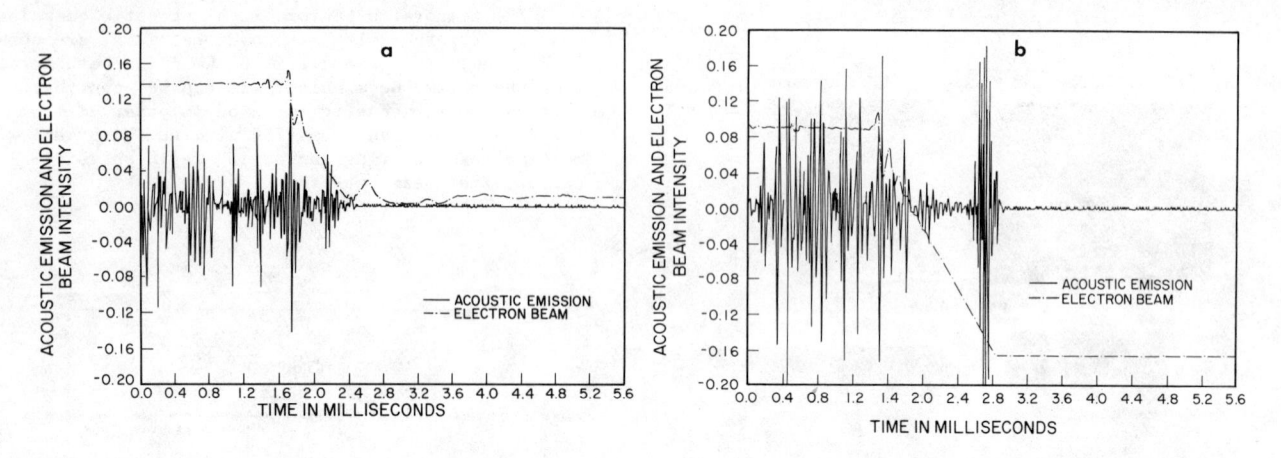

Figure 12. Effect of qa level on the type and quantity of AE during rapid solidification of 2219 aluminum. (a) qa = 1.36 x 10^5 W/m. (b) qa = 2.70 x 10^5 W/m.

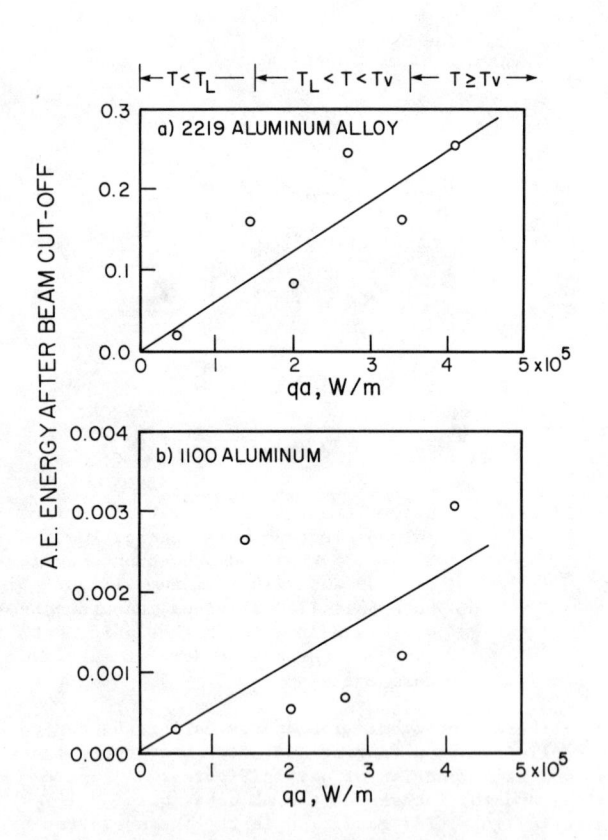

Figure 13. The effect of qa upon acoustic emission after electron beam cut-off for steady-state conditions in (a) 2219-T87 and (b) 1100 aluminum alloys. Note the different scales for the two alloys.

Acoustic emission during solidification was measured from samples with different substrate microstructures. Table II shows that the amount of AE during rapid solidification is dependent on the microstructure of the substrate. In this experiment, 2219 aluminum was tested in the solutionized condition (75 min. at 535 °C, ice water quench), and -T6 conditions (aged to peak hardness at 177 °C for 18 hours). The grain size was 50 μm. The table indicates that the T6 condition emitted the greater AE.

DISCUSSION

During electron beam heating and subsequent cooling, both 1100 and 2219 aluminum alloys generate detectable levels of acoustic emission but that from the 2219 alloy is 1 to 2 orders of magnitude greater in intensity (Fig.11). Here, we would like to deduce the physical origin of this emission. To do this we shall use our understanding of the differences in the behavior of the two alloys during solidification, in conjunction with elementary elastodynamic calculations, to deduce the stress change mechanisms that generate detectable signals.

Every change in stress radiates elastic waves. Due to noise and to sensitivity limitations not every stress change generates detectable elastic waves. Many of the stress changes occurring during solidification of the 2219 aluminum alloy are clearly detectable, while very few in 1100 aluminum are of sufficient strength to satisfy detectability criteria. To develop this argument further, it is essential to elucidate the detectability criterion for this system. That is, what is the relation between the measured acoustic emission energy and the source stress change?

We have devised a rather simple method for this calibration in which we use the predictable thermal stress induced by the absorption of a short duration laser pulse to generate elastic waves at the natural source position. For the experiments we used a 25 ns duration 12 mJ energy Nd:YAG pulse from a Q-switched laser as the artificial source.

This source generated a signal whose energy, when measured using the same conditions as for the 2219 aluminum alloy experiments, was 0.30 units. From this, since the smallest detectable energy was about 0.0005 units, we estimate the smallest detectable dipole source to be about 1 μNm [7]. Differences in the radiation patterns of various sources are likely to be reduced because of the multiple paths of reflected rays that contributed to measurements.

44

Table II

Acoustic Emission After Beam Cut-off as
Affected by Substrate Material Condition

Material Condition	Rockwell B Hardness	Energy (Relative Units)		
		$qa \leq 1.5 \times 10^5$ W/m	$qa > 1.5 \times 10^5$ W/m	Total
2219-Solutionized	39	0.008	0.013	0.021
2219-T6	77	0.056	0.117	0.173

Origin of Acoustic Emission During Cooling

From above, it is clear that only those stress changes for which $D_{ij} \gtrsim 1$ µNm will give detectable acoustic emission in these experiments. During the electron beam treatments on 1100 aluminum we have observed very weak acoustic emission signals: presumably, the stress changes accompanying melting/resolidification here have dipole components $\lesssim 1$ µNm. The 2219 aluminum alloy has generated intense acoustic emission signals: the addition of 6.6 wt.% Cu has resulted in many stress changes during processing for which $D_{ij} \gg 1$ µNm. We can use these differences to aid our deduction of the acoustic emission sources.

Since the solidification interface velocity, cooling rate, and thermal constants do not vary appreciably between the two alloys we would expect the development of thermoelastic contraction stress to be similar in both alloys. Since little or no acoustic emission was observed from the 1100 alloy it is reasonable to assume this potential acoustic emission is undetectable in these experiments. Very high contraction stresses undoubtedly develop, but in solidification from a steady state condition the cooling rate is presumably too slow to generate detectable elastic waves above 20 kHz frequency. The more rapid cooling of short duration, high flux density laser pulses might be measurable however.

The maximum potential thermal stress that could develop upon cooling can be estimated from the relation:

$$\Delta\sigma = (\lambda + 2/3\mu)\beta\Delta T$$

where ΔT is the temperature drop, about 600 °C in aluminum. This results in a stress of ~ 900 MPa, many times the yield strength of both alloys. Thus, during cooling, we expect plastic deformation to relax the thermal stress. Each dislocation that slips will radiate elastic waves and could be a detectable emission source.

Suppose a unit of slip consists of the simultaneous expansion of n dislocation loops each expanding over an area dA on the same slip plane. Then the dipole matrix [8] is given by

$$D_{ij} = n\, C_{ijk\ell}\, b_k\, dA_\ell$$
$$\cong n\mu\, b\, dA$$

For detectability this must exceed 1 µNm leading to the criterion $ndA \geq 0.13$ mm^2.

Previous studies reviewed in reference [10] have indicated the fraction of slip events during uniaxial plastic deformation able to generate detectable signals is highly sensitive to composition and microstructure. For grain sizes below about 100 µm it is unlikely the 1100 aluminum alloy could generate detectable signals because n would have to be unrealistically large to satisfy the detectability criterion. In the alloy, during solidification a spectrum of microstructures exist ranging from hot solid solution to the peak hardness precipitate distribution. Previous studies have indicated that deformation of aluminum alloys containing small ($\lesssim 100$ Å) diameter shearable precipitates generates intense acoustic emission due to cooperative slip phenomena. Variation of the substrate microstructure, which we expect to have only a small effect upon processes other than plastic deformation, resulted in large changes in acoustic emission. The most active sample condition (T6) was the one containing a mixture of θ" and θ' precipitates. Thus, for $qa \geq 1.56 \times 10^5$ Wm^{-1} a part, at least, of the extra acoustic emission from the 2219 aluminum could be due to plastic deformation. While for $qa < 1.56 \times 10^5$ Wm^{-1} (for which no melting occurred) probably all the emission is from plastic deformations.

For $qa \geq 1.56 \times 10^5$ Wm^{-1} melting resulted, and upon subsequent resolidification hot tearing occurred raising the possibility of an extra emission source. The dipole strength for formation of a circular microcrack under mode I stress σ is approximately given by

$$D_{ij} = (\lambda + 2\mu)\, 2(1-\nu^2)\, \frac{\sigma_r^3}{E}\, \pi$$

where ν = Poisson's ratio, r = crack radius, and E = Young's modulus. Substituting physical property values for aluminum and assuming $\sigma = 100$ MPa yields the detectability criterion $r \geq 1$ µm. Many cracks clearly satisfied this criterion. For these cracks to generate detectable signals they should also propagate at speeds greater than 1 ms^{-1}. Data on the micromechanisms of crack extension during hot tearing are sparse. Metallographic evidence indicates cracks have propagated beyond the prior melt boundary into unmelted material suggesting fast fracture may indeed have occurred. However, such evidence is far from conclusive and, because of the pronounced substrate effect, for the moment, we may only speculate that a part of the emission from samples for which $qa \geq 1.56 \times 10^5$ Wm^{-1} was from cracking.

ACKNOWLEDGMENTS

We are grateful for the advice and assistance of R. Schaefer, C. Brady, J. Chang, and C. Turner and the financial support of the Defense Advanced Research Projects Agency under DARPA Order No. 4275.

REFERENCES

1. R. Mehrabian, Int. Metals Rev., 1982. 27(4), 185.

2. M. C. Flemings, Solidification Processing (McGraw Hill, New York, NY, p. 142, 1974).

3. S. C. Hsu, S. Kou, and R. Mehrabian, Metall. Trans., 1980, 11B, 29.

4. J. A. Sekhar, S. Kou, and R. Mehrabian, accepted for publication Metall. Trans.

5. S. C. Hsu, S. Chakravorty, and R. Mehrabian, Metall. Trans., 1978, 9B, 221.

6. J. A. Simmons and R. B. Clough, "Theory of Acoustic Emission," Dislocation Modeling of Physical Systems, J. Hirth and M. Ashby, Eds., Acta/Scripta Metallurgica, Pergamon Press (1981).

7. H. Wadley, C. Stockton, J. Simmons, M. Rosen, S. Ridder, and R. Mehrabian, "Quantitative Acoustic Emission Studies for Materials Processing," DARPA/AFML Annual Review of Quantitative NDE, held in 1982.

8. H. N. G. Wadley, C. B. Scruby, and J. E. Sinclair, Phil. Mag., 1981, 44, pp. 249-274.

9. C. B. Scruby and H. N. G. Wadley, Accepted for publication in J. Phys. D, 1983

10. H. N. G. Wadley and R. Mehrabian, to be published.

11. H. N. G. Wadley, C. B. Scruby, and G. Shrimpton, Acta Met. 1981, 29, 399.

LASER MARKING TECHNIQUES

Burton Bernard
General Photonics Corporation

Almost from the beginning, individuals have had a need to mark and identify their belongings. Now, more than ever, individuals and organizations require a convenient way to permanently mark items for safety information, batch/date coding, inventory control, quality assurance, government requirements and theft protection as well as personal identification.

The first marking system probably consisted of a charred tree branch used to draw simple images on large rock formations. After hundreds of years it was discovered that color could be added to the drawings by using juices from plants and blood from animals. Eventually, the printing press was invented allowing one to repetitively mark the same message at comparitively high speeds.

More recently, improvements of reliable and practical lasers coincidedent with the advancement of computer technology and the evolution of optical components have led to the development of Laser Marking Systems. These new and unique non-contact systems may be used to mark or engrave an almost endless list of materials including metals, plastics, ceramics, glass, wood and leather as well as painted or printed surfaces and photographic emulsions.

There are two basic methods for laser marking: they are Mask Marking and Beam Deflected Marking. Both methods take advantage of the relatively high powers generated by lasers. Both methods also use optical techniques to enhance the power densities to levels sufficient to etch the surface of the material to be marked.

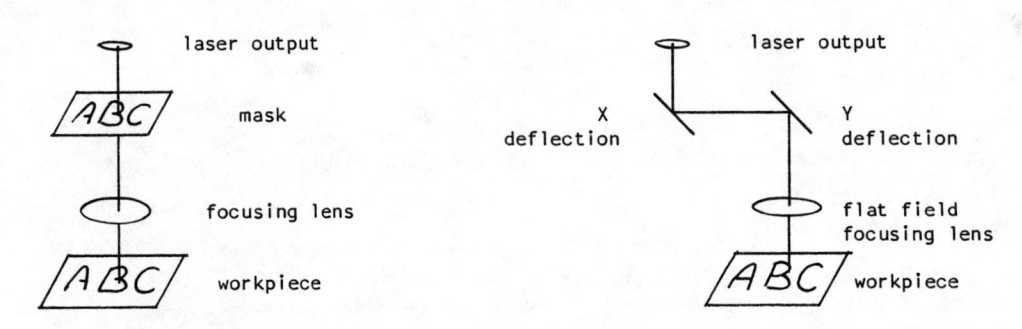

Figure 1 MASK

Figure 2 BEAM DEFLECTED

The mask marking system images the laser beam through a special mask, or series of masks, containing the data to be marked and then reimages that pattern on the workpiece. Using commercially available pulsed gas lasers it is possible to reproduce the mask image on to the surface of a variety of materials at rates up to 25 marks per second.

The beam deflected marking system uses a mini or microcomputer to deflect the laser beam in a preprogrammed manner to generate characters or patterns. The computer also turns the laser on and off as required by the message being marked.

The deflected beam is directed through precision optical components that focus the laser output to spot sizes of about 25 microns (.001 inches) creating extremely high power densities. Typical marking rates are 5 to 15 characters per second.

	MASK	DEFLECTED BEAM
Marking Area		+
Throughput		
repeat message	+	
varying message		+
Character Formation		+
Moving Objects	+	
Depth Penetration		+
Initial Cost	+	

Figure 3 COMPARISONS

Mask marking systems are normally limited to smaller marked areas (generally less than 35 square millimeters for uncoated surfaces and up to 600 square millimeters for coated surfaces) due to the decrease in power density as image size is increased. Beam deflected marking systems are typically available with marking windows of at least than 2500 square millimeters (2 inches x 2 inches). Larger marking areas can be obtained by mechanically moving the material being marked.

Either system can produce characters as small as 0.25 millimeters (0.010 inches). The largest character size is limited by the marking window available.

Mask marking systems have a higher throughput than beam deflected marking systems provided that each marked image stays the same and consists of more than one or two characters. When each marked message is to be different, computer updating of the deflected beam marking systems is faster than manual or electro-mechanical changing or modification of the mask or stencil.

Character formation using computer control provides an instant and almost endless list of possibilities including different character fonts for alpha-numerics, foriegn letters, logos, bar codes, line drawings and special patterns.

A typical mask marking system uses a Pulsed Carbon Dioxide Laser that produces about 5 megawatts per square centimeter at the workpiece. Depth penetration is usually less than 25 microns (0.001 inches).

A typical beam deflected marking system uses a CW Pumped, Q-Switched Nd:YAG Laser that produces peak powers of more than 100 megawatts per square centimeter at the workpiece. Depth penetration is about 25 to 100 microns (0.001 to .005 inches). Reducing the laser output power or increasing the writing speed will reduce the amount of penetration.

Laser marking systems are considerably more expensive than most conventional type marking systems. However for many applications the advantages of non-contact, the omition of inks and fluids, the elimination of typesetting, the ability to mark irregular shaped surfaces, the ease of initial setup and cleanup, the avoidance of drying time and the flexibility of varying character fonts can prove to be extremely cost effective.

Some objects cannot be marked by conventional printing techniques. For example, the Food and Drug Administration requires that all objects transplanted in the human body be permanently identified with name of manufacturer, date of manufacture and batch number. Mechanical contact with these delicate parts can create thermal shock causing the part to shatter or seriously reduce reliability. Also, inks can wash off causing contamination to the body. Laser marking appears to be the only feasable alternative.

The initial cost of a mask marking system is about $30,000 to 35,000. The beam deflected systems are about $65,000 to 75,000. Custom material handling systems can add up to $100,000 or more to either system.

When selecting a laser marking system for a particular application there are many factors to consider.

POWER DENSITY
TIME
THERMAL
    Thermal Conductivity
    Heat Capacity
    Melting Point
    Heat of Vaporization
REFLECTIVITY
    Material
    Wavelength
    Temperature

Figure 4 CONSIDERATIONS

Power density at the workpiece is more important than the power generated by the laser alone.

POWER DENSITY = PEAK POWER / AREA

Power density is determined by the amount of peak power generated by the laser divided by the area of the focused beam. Wavelength, beam divergence and quality of optics become important factors in determining how small a beam can be produced.

For mask marking systems the focused beam area is about 0.36 square centimeters. Spot sizes for the beam deflected system are about 25 x 10(-6) square centimeters. The higher the power density, the easier it will be to mark the material and the deeper the penetration.

The amount of time the laser power is focused onto the material also has impact on the ease of marking and depth of penetration. The pulsed Carbon Dioxide Laser produces a pulse of about one microsecond duration at rates up to 25 pulses per second.

CW pumped, Q-switched Nd:YAG Lasers generate 100 nanosecond pulses at rates up to five kilohertz. Flash pumped Nd:YAG Lasers provide 100 microsecond pulses at rates up to 100 pulses per second.

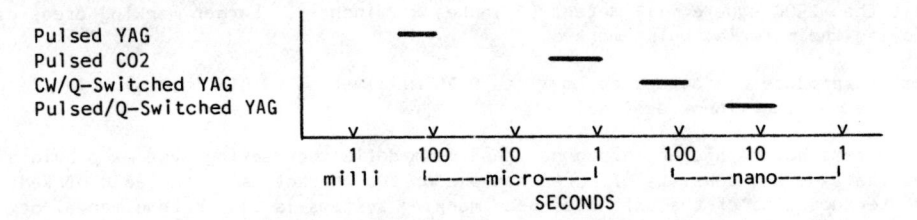

Figure 5   TYPICAL LASER PULSEWIDTHS

The marks are created by vaporization, melting, thermally induced chemical reactions or removal of an overcoat of ink or paint to expose a contrasting substrate. Most metals are quickly heated to the point of vaporization by the focused laser beam. Acrylic and some other materials exhibit a change in color due to a chemical reaction caused by the heat generated by the laser beam.

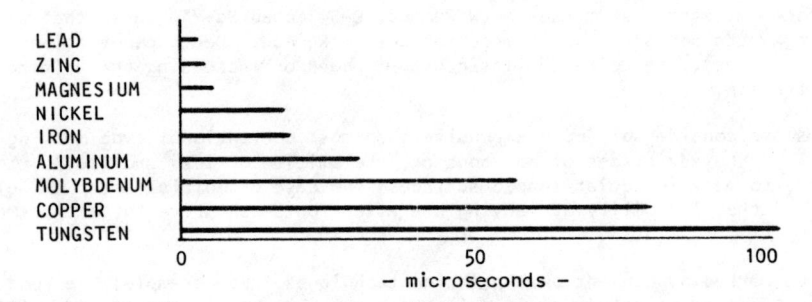

Figure 6
TYPICAL TIME TO REACH VAPORIZATION
(one megawatt/square centimeter)

Highly refective materials will not absorb all of the laser energy focused onto it. The absorptivity of a given material depends upon its reflection and transmission characteristics. For any given material, the reflectivity plus the transmisivity plus the absorptivity must equal one (100%).

% ABSORPTION + % REFLECTION + % TRANSMISSION = 100 %

Absorptivity is effected by the type of material, surface condition (ie; smooth or rough, polished or oxidized), wavelength and surface temperature.

The Nd:YAG Lasers used in the Beam Deflected systems emit energy at 1.06 micrometers (near infrared). CO2 Lasers used in the Mask systems emit energy at 10.6 micrometers (infrared). In general, metals absorb a greater percentage of Nd:YAG Laser energy than that of Carbon Dioxide Lasers. On the other hand, white paper and most transparent materials absorb a greater amount of Carbon Dioxide laser energy. Some materials such as silicon absorb the same percentage of energy from either laser.

	YAG	CO2
ALUMINUM	25	3
COPPER	9	2
GLASS	5	94
IRON	35	6
MOLYBDENUM	41	5
NICKEL	28	3
PAPER (white)	25	95
SILICON	72	72
TUNGSTEN	31	4
ZINC	50	2

Figure 7 % ABSORPTIVITY (typical)

At least 1000 laser marking systems are now in daily operation throughout the world. There are hundreds of different uses. A few examples of some of the more common application are shown in the following photographs.

EXAMPLES OF LASER MARKED MATERIALS

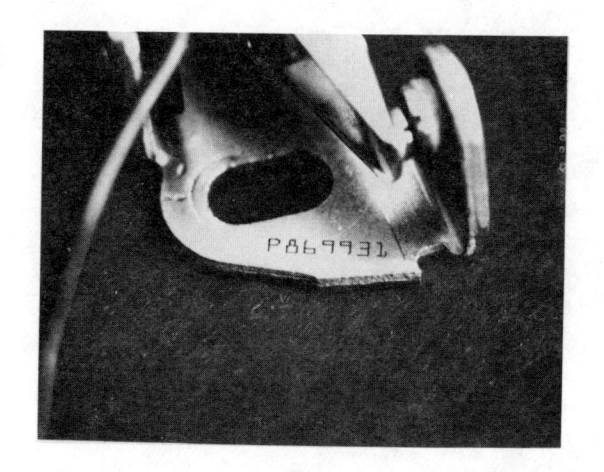

Figure 8 Automobile Component

Figure 9 Plastic Retainer

Figure 10 Silicon Wafer

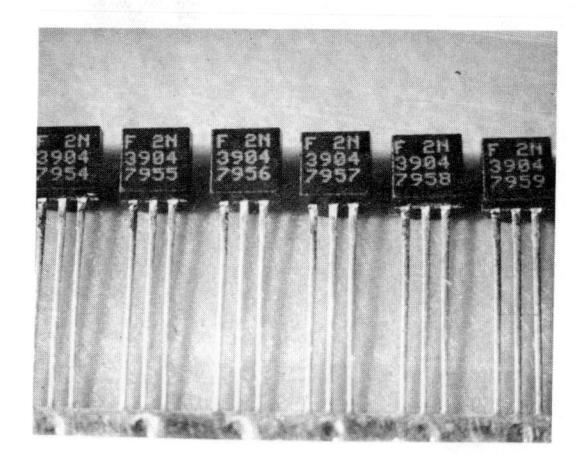

Figure 11 Transistor Cases
Note serialization

Figure 12 Anodized Aluminum

Figure 13 Hand Tool

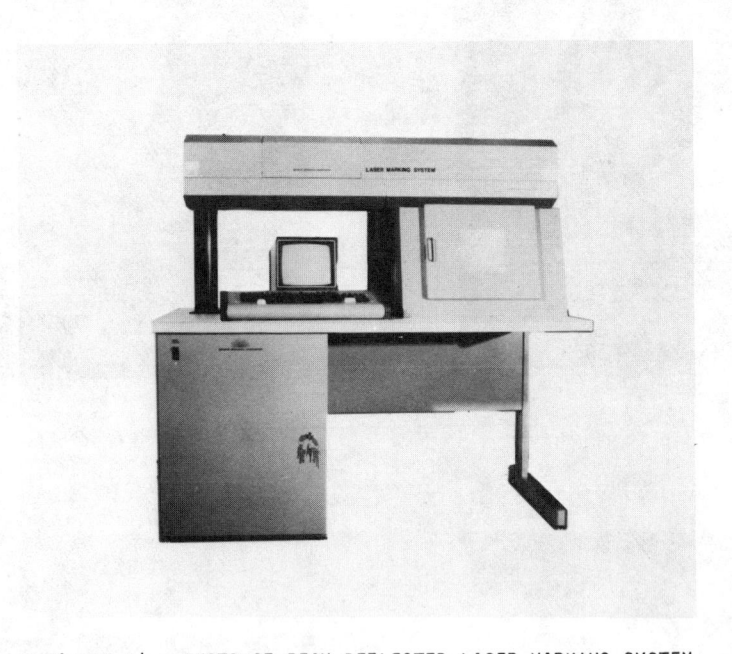

Figure 14 PHOTO OF BEAM DEFLECTED LASER MARKING SYSTEM

8301-006

LASER MARKING OF COMPONENT PARTS

Albert V. Gress, Jr.
Monsanto Research Corporation
Mound*
Miamisburg, Ohio 45342

INTRODUCTION

As a part of the U.S. Department of Energy's weapons complex, Mound has long been required to indelibly mark components. Permanent identification of components and subassemblies for traceability and historical purposes is essential for assemblies subject to long term storage. Marking requirements run the gamut from simple functional alphanumerics for terminal or wire numbers to complex component identification involving program nomenclature, part number, manufacturer's code, serial number, date code, and lot or batch number. The wide range of opaque materials marked includes both ferrous and nonferrous metals, plastics, composites, and ceramics.

MARKING METHODS

The variety of marking requirements at Mound necessitated a number of general marking methods. These are defined in the applicable standard shown in Figure 1. The specific class and method, along with character size, are usually called out on the component drawing. If the marking method has not been specified on the drawing, it is selected, based on the following factors:

- Piecepart size
- Piecepart material
- Piecepart geometry
- Piecepart strength
- Compatibility
- Physical constraints
- Chemical constraints
- Temperature constraints
- Quantity
- Cost

One of these factors, quantity, requires an explanation. Production at Mound is run in small lots or batches. A typical lot could be fewer than 200 assemblies. Obviously, this is a very important consideration. For a variety of reasons, a number of the components were marked using engraving and offset ink printing. Both processes have a number of disadvantages.

Engraving

Engraving with a pantograph is a slow process requiring operator skill both in setup and operation. A typical setup is shown in Figure 2. On some plastic parts, depth must be closely controlled to a maximum of 0.127 mm (0.005 in.). The resulting groove can become a stress riser and adversely affect the strength of the part. In addition, changing templates for serializing and replacing worn tools combine to increase production time and cost. Another cost element is the unforgiving nature of the process. For the most part, the scrap produced cannot be salvaged. The process at Mound was also limited to flat surfaces.

Offset Ink Printing

Offset ink printing was used to avoid part damage and to mark convex parts such as plastic-covered cables. With offset printing the ink itself becomes a problem. First, the ink must be compatible with the material being marked; second, it must cure to a permanent indelible mark. In the ink marking industry, inks are not stocked because of their vast number and limited shelf life. Rather, inks are compounded for specific applications, and shelf life length starts as soon as the ink is made. At Mound, modified epoxy inks containing no metals are required. Because cure times can be as much as 24 hr, storage facilities are necessary. Cure time can be reduced to 3 or 4 hr by the application of heat, provided the component can tolerate it. But the application of heat obviously presents a problem for explosive components. The offset type used is standard, but manufacturers do not stock it because of the great number of sizes and styles. The user must therefore stock an inventory of spare type and must anticipate the lead time required for replacements or changes in requirements.

Ink marking has been used in several production areas at Mound, necessitating a number of dedicated presses. The cost of the Markem 200 AD presses was significant, but tooling cost could often equal or exceed the press costs. The tooling expense was made up of three elements: the part holding fixture, a special holder for the indexing serializing head and type insert, and a possible part indexing means for spatial considerations. A typical example is shown in Figure 3. Another disadvantage is that the ink cleaner compound is a highly volatile, toxic chemical requiring hooding and venting. Finally, the cost of labor must also be considered. Ink adjustment is difficult and the operation is, for the most part, manual.

LASER MARKING INVESTIGATION

Early in 1977, an investigation was launched into the laser process of marking parts. At that time, Hadron's Korad laser system offered the greatest promise. Based on tests and demonstrations conducted at the Korad plant, Mound acquired a Korad laser system.

The Korad package can accurately be described as a group of modules that must be packaged and assembled to make up an operational system. For the initial tests

---

*Mound is operated by Monsanto Research Corporation for the U.S. Department of Energy under Contract No. DE-AC04-76-DP00053.

Class	Description	Method
A-1	Ink w/o covercoat	A. Stamp B. Stencil C. Silk Screen
A-2	Ink with covercoat	A. Stamp B. Stencil C. Silk Screen D. Mechanical Guide
A-3	Ink w/o covercoat	A. Freehand
A-4	Ink with covercoat	A. Freehand
B-1	Paint w/o covercoat	A. Stencil B. Silk Screen
B-2	Paint with covercoat	A. Stencil B. Silk Screen
C-1	Impression	A. Die Impact
C-2	Impression	A. Freehand
C-3	Impression	A. Sand Blast
C-4	Impression	A. Roll Die
D-1	Hot Stamp w/o covercoat	A. Heated Die and Foil
D-2	Hot Stamp with covercoat	A. Heated Die and Foil
E-1	Engrave w/o fill	A. Chemical or Electro-Chemical Etch B. Mechanical
E-2	Engrave with fill	A. Chemical B. Mechanical
F-1	Surface Conversion	A. Electro-Chemical
G-1	Integral	A. Embossed (Raised) B. Recessed
H-1	Indirect	A. Bag B. Tag C. Tape D. Heat-Shrinkable Sleeve E. Embossing Tape F. Box
J-1	Dry Transfer with Covercoat	A. Manufacturer's Instructions
J-2	Dry Transfer w/o Covercoat	
K-1	Laser	A. Laser

Figure 1. General Marking Methods

Figure 2. Typical Engraving Setup

Figure 3. Ink Marking Press and Tooling

at Korad, some representative parts and other pieces of metal were marked. It was clear, considering the small number of pieces tested, that the control parameters were not critical and that satisfactory marking could be achieved with uncomplicated work holding devices. Some of the test pieces were hand held, though this is definitely not a recommended procedure. Other parts were supported very simply.

In these initial tests, small round steel bars ranging from 3.18 mm (0.125 in.) to 9.53 mm (0.375 in.) in diameter were marked both axially and radially to see whether curved surfaces could be engraved successfully. Two important points were noted. First, the depth of field or focus was large enough to mark "around the horn." Second, as the deviation of the beam from normal to the surface increased, enough power remained to mark the material. (Bear in mind that the part is stationary, and the beam is moving. This configuration is standard in the industry.)

In a second trip to the Korad plant, a designed experiment was conducted for process evaluation and the preparation of preliminary laser marking specifications. The experiment used specially machined steel blanks, configured to facilitate metallurgical examination. A total of 131 samples was marked in the engrave mode using various control parameters. The results ranged from too heavy to too light. In the absence of an objective testing medium, six samples were selected by a panel. Their selection was based subjectively on aesthetic appearance. At that point, readability was the chief criterion. Five samples of each of the six selected control parameters were marked to provide a basis for more detailed investigation and testing. At Mound, the testing included light sectioning microscope measurements, 50X photographs, and metallurgical sectioning for depth analysis.

LASER MARKING PROCESS AND MODES

Laser marking has two main operational modes: engraving and dot matrix. Laser engraving is similar to mechanical engraving in that the beam actually vaporizes the subject material at the point of focus. The moving spot of coherent light traces out each alphanumeric character individually, scribing to the precise depth desired. In the dot matrix mode, characters are formed by a series of shallow craters or tiny holes in a standardized format such as 5 x 7 or 7 x 9. The dot matrix mode can be used anywhere a minimal marking is required, and is often used on plastic parts to reduce heat buildup. An example of dot matrix mode is shown in Figure 4.

Figure 4. Dot Matrix Mode Examples

All of the marking in the designed experiment was done in the engrave mode. The results of the testing narrowed the number of acceptable groups to three. A typical cross section of the laser groove in the engrave mode shows a shallow trench and a rise above the part surface on either side. A cross section is shown in Figure 5. Two of the tested groups exhibited a double trench and were deemed unsatisfactory because of the possibility of loose debris contaminating a product. The double trench phenomenon is not rare or peculiar to one particular system. A cross section of one is shown in Figure 6. Initially, the double trench was attributed to a high pulse repetition rate. A third group was also eliminated because satisfactory marking could be achieved at a lower power setting, thereby conserving the use of the excitation lamp.

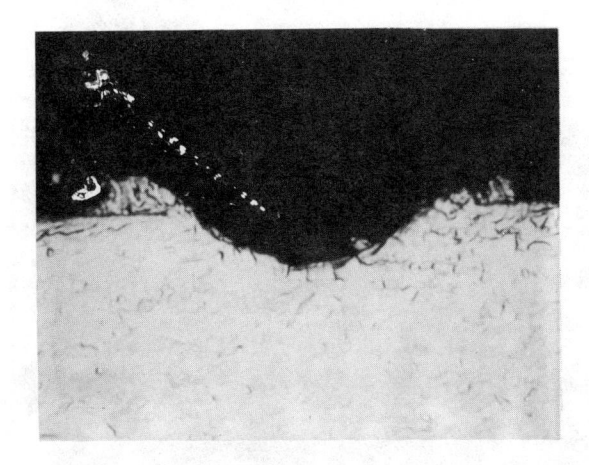

Figure 5. Typical Engrave Mode Cross Section

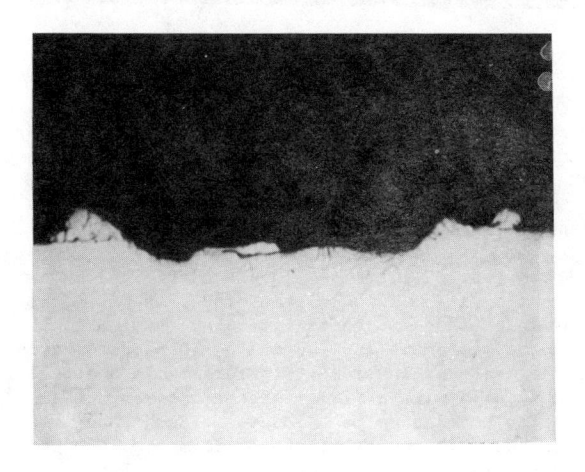

Figure 6. Double Trench Cross Section

The measured results of the tests on the three selected groups confirmed that parameter settings were not critical. Acceptable markings were achieved over quite a range of settings. Measurements of the rise ranged from 0.0102 mm (0.0004 in.) to 0.0127 mm (0.0005 in.). Depth varied between 0.0076 mm (0.0003 in.) and 0.0089 mm (0.00035 in.). In some cases, the rise is of more importance than the depth because it affects dimensional integrity, particularly on close-toleranced parts. Metallographic examination of the grooves at 500X showed the heat affected zone to be very shallow, approximately 0.00254 mm (0.0001 in.) This fact indicated that thin sections, as small as 0.127 mm (0.005 in.), could be marked with little degrading effect on their mechanical properties.

KORAD MARKING SYSTEM

In late 1977, a Korad KLM-1 system was ordered. The standard package consisted of the laser head module, power cabinet, control cabinet, and Teletype. The standard computer software did not handle serial incrementing; therefore, this feature was added to the software. No viewing or aligning means was incorporated in the KLM-1. A television camera and monitor were procured to provide the operator a means of viewing the marking area. This accessory has proven to be a useful addition.

The laser and optics assembly came mounted on a wide-flanged beam; however, it was necessary to design and fabricate functional packaging. This included the support table, tooling adaptor, and safety hooding. A view of the completed system is shown in Figure 7. The modules shown from left to right are as follows: power cabinet, control cabinet, Teletype, TV monitor, laser unit, hooding and table. The module on top of the laser is a fixture indexing control unit, the purpose of which will be explained later.

Figure 7. Korad Marking System

The Korad system is controlled by minicomputer and uses a YAG laser with Acousto-Optic-Q switch. After the laser beam exits the laser head, it passes through an up collimator which enlarges the beam diameter. The laser beam is focused on the part by mirrors and focusing lenses. High speed galvanometer beam positioners move two mirrors to control the laser beam in the X and Y axes. The scanning laser beam then passes through the fixed flat field focus lens onto a 50.8 mm x 50.8 mm square (2 x 2 in.) image area. For most purposes, it can be considered to have a 1.27 mm (0.050 in.) depth of field.

The marking pattern is controlled by the computer program; however, the actual depth of engraving is controlled by three adjustable parameters: beam power, the pulse repetition rate, and the scan or write speed. Power is adjustable up to 100%. The sytem is rated at 50 W peak power. Voltage and amperage, have a maximum of 400 VDC and 50A, respectively. The repetition rate can be varied from 1 to 100 KHz per second. Writing speeds can range from 0.0254 mm (0.001 in.) to 762 mm (30 in.) per second. Power and repetition rate are adjusted by the operator with the power panel controls. Writing speed is entered via programming.

Safety Considerations

In 1977, the applicable safety regulation was the "American National Standard for the Safe Use of LASERS, ANSI Z136.1-1976." The Korad laser unit, rated at 50 W, was classified as a Class IV laser. This is the only class capable of producing hazardous diffuse reflections. For use on production at Mound, it was necessary to declassify the system to Class V by totally enclosing the laser path and work area. The interlocked aluminum safety hood and operator access door are clearly shown in Figure 7. The hood is pivoted at the right end to provide easy access for changing part holding fixtures and for maintenance.

Table Assembly

The support table on which the laser assembly, TV monitor, and hooding are mounted was fabricated using plywood and Unistrut construction. To maintain laser path alignment and a fixed focal plane, the fixture holder or adapter was mounted on an auxilliary cantilever beam fastened to the main laser head support beam. The resulting assembly provided a rigid C-frame and insured fixture interchangeability.

Tooling

The tooling concept was based on interchangeable fixturing so that changing from one part to another would not require adjustment in any of the three axes. The fixture adaptor consisted of two fixed locating rails forming a right angle and two screws so that the tooling can be urged into corner registry. The fixture height is controlled so that the surface of the part to be marked will lie in the laser focal plane.

A further improvement or simplification of tooling was achieved by the use of a master fixture and inserts. This fixture along with the typical insert part holder is shown in Figure 8. For comparison, a one piece integral fixture is also shown. For a number of applications, this greatly reduced fixturing size and cost. An added advantage was that less tool storage space is required.

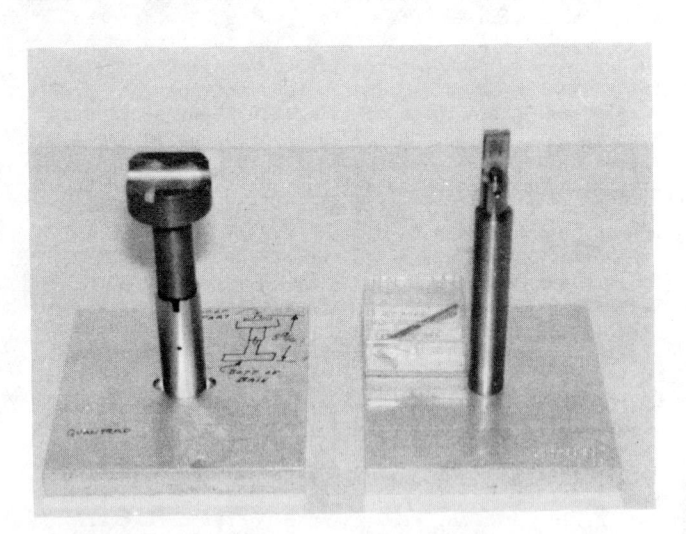

Figure 8. Fixturing and Tooling

A third type of fixture was required for parts that must be marked on more than one surface. If the part could be moved in one plane for multiple markings, that was done with a Slosyn motor driven indexing fixture. A view of an indexing fixture and Slosyn drive is shown in Figure 9. The stepping motor, which is premanently fastened to the table, drives by means of a flexible shaft, a 3:1 speed reducer which is a part of the fixture. With this configuration, there are 600 steps per revolution. The construction is a cost effective means of reducing tooling expense by using one permanently

Figure 9. Index Fixture and Drive

installed and connected motor. The flexible shaft drive accommodates fixtures with rotation in various axes. Although any amount of index which is a multiple of 0.6° can be programmed, the design intent was to implement marking on 2, 3, 4, 5, 6, or 8 sides of a component.

The Korad control system could not handle the indexing drive and additional logic; therefore, it was necessary to design an indexing control module. A view of the index panel is shown in Figure 10. The control consisted of a Texas Instruments programmable controller in combination with a Slosyn preset indexer. A selector switch permits up to 9 indexes. Each index must contain the same number of counts as set on the preset indexer dials. The indexing control is integrated into the Korad system so that several sides of a piece can be engraved with different markings in a single cycle.

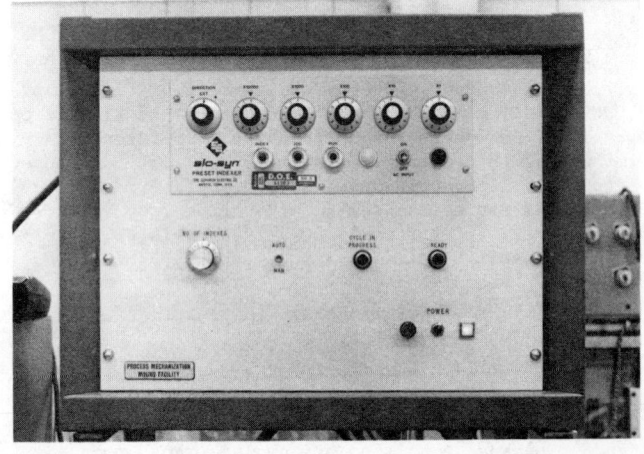

Figure 10. Indexer Panel

Programming

The use of a minicomputer to control marking operations results in a flexible, easy to use system. Programming may be entered via the high speed paper tape reader or Teletype. The latter was adopted as standard procedure because of its simplicity and ease of programing.

Serial numbering was an additional consideration. Only a few parameters need to be entered via the keyboard. The programming consisted of the character format specifications, writing speed, messages, serial incrementing, indexing, and marking mode.

The marking format is variable. Alphanumeric characters are controlled by three parameters: height, width, and slant. With the standard software, all letters are block capital and numbers are standard unless controlled by special programming. Character height may be varied from 0.254 mm (0.010 in.) to 50.8 mm (2.0 in.). Width specification is optional. If it is not programmed, the standard aspect ratio contained in the software will apply. This results in a character having a width of 2/3 the height. Slant in radians may also be specified.

Writing speed is one of the three factors controlling depth of marking. It can be varied as previously described. The remaining program parameters have been previously discussed or are self explanatory.

For production, the programming instructions for an individual component are contained in the Appendix of the Mound Laser Marking technical manual. A typical example of a laser marking instruction is shown in Figure 11. With the exception of the initialization of the five digit serial number and write speed, all of the information necessary to enter the program is contained on this sheet. Four sides of this part are to be marked as shown in the illustration box. The instructions are self explanatory with the possible exception of "Indexer Settings." The dual entry 3/150 defines the number of indexes as three and the setting of the preset indexer dials at 150 steps. This setting gives 90° indexing because 600 steps constitute one complete revolution of the indexing fixture. The number of indexes is one less than the number of sides to be marked. After the last side is marked, the fixture returns to start position by means of the Texas Instruments programmable controller logic.

Marking Experience and Growth

The Korad-Mound system became operational in mid-1978. Production was inaugurated with tooling for six components. The number quickly grew as the laser process proved its utility and value by effectively marking a variety of parts and materials without undesirable effects such as contamination, mechanical or thermal stress. When the system's simplicity and marking speed became known, product engineers began to specify more laser marking than ever before. With the ease of programming and the feasibility of improvised fixturing, there were numerous requests for custom marking on individual pieces. Typical examples of production marking are shown in Figure 12. The brass nameplate in this illustration was not laser engraved, though it could have been. Plaques have been routinely engraved with the system.

All this is not to say there have been no problems. Because this system was one of the first six sold by Hadron, we expected some problems would occur with the technology. The principal problem we encountered was the two flash/lamp laser head itself. Several modifications were made to the laser to increase lamp life and reliability. The factors contributing to downtime were lamp life, lamp installation, mirror alignment, service, and the procurement of replacement parts.

LASER MARKING INSTRUCTION

Product MC3355 TRAINER   Drawing No. AY317194

Fixture No. AYD780200   Adapter _____

Type Material to be marked Aluminum/Nickel Plated

Marking sample Yes X No ___ Sample ID# _____

I. FORMAT

  A. Letter Height 40 (approx., in mils)

  B. Letter Width   30 (normal = 2/3 Height, approx.)

  C. Letter Slant None , approx.

    Engrave Mode X   Dot Matrix Mode _____

    Default case is the engrave mode.

II. MESSAGE

  A. Index Yes X No ___ Indexer Settings 3/150

  B. Message

    1. (Serial No.) XXXXX
    2. (Program) MC3355 TRA
    3. (Part No.) 317194
    4. (Suffix)   -A
    5. (Notation) Inert

III. REWORK AND SPECIAL INSTRUCTIONS

  A. Rework Permitted Yes ___ No X

  B. Instructions:

Figure 11. Laser Marking Instruction

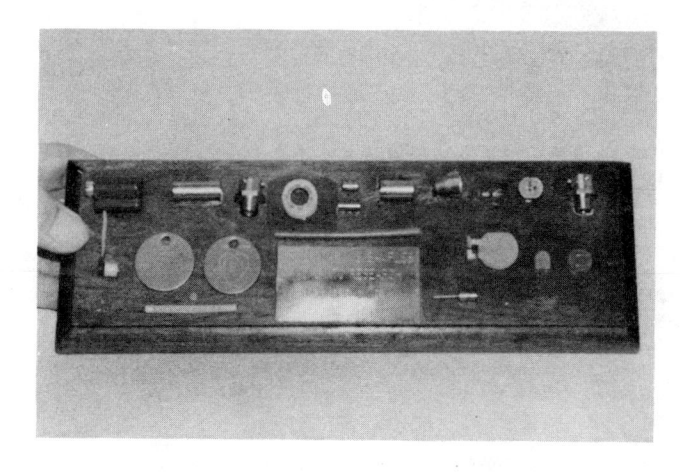

Figure 12. Laser Marked Examples

NEW REQUIREMENTS

By late 1980, the need for an additional laser marking system at Mound was recognized. The four principal reason for this need were:

1. The reliability of the system
2. The greatly increased demand for laser marking
3. Plant logistics
4. The need for greater depth control

The first two factors have already been discussed. The third, logistics, concerns component movement or transportation. At Mound, the fabrication and assembly areas are located in a number of different buildings having different environmental conditions. It is this factor that justified the need for two new systems. The additional system was needed in the machining area so that pieceparts could be marked as produced without disrupting production. This marking requirement resulted in part from the need to identify components for quality control measurements used in variation engineering control. Yet another justification for a second new system was the need to cost effectively mark in situ tooling and miscellaneous other items.

The fourth factor was the need for greater depth control. The laser marking process worked satisfactorily for most of the products, but for certain components, the process could not be used. There were two reasons for this. On alphanumeric characters having intersection points where the laser beam must cross over a point already traversed, the crossover was deeper than the product specifications permitted. This same condition also occurred at the beginning and end of a character path.

LASER MARKER SYSTEMS EVALUATION

Our needs at Mound led to an evaluation of the available laser marker systems. Our survey concentrated on four suppliers. In the final analysis, the six factors of our criteria were compared to choose the best system for Mound. The comparison focused on the following considerations:

- Laser Head
- Computer
- Program Storage
- Ease of Operation
- Manufacturer's Performance Record
- Price

Of course, other factors were considered as well. But two of the important ones, character variation capability and marking speed range, were nearly identical in all four systems, and this similarity removed these control parameters from consideration and focused attention on other system differences. Some of the available features or options might be of importance to other buyers, but were not considered important at Mound. Examples of these are logo generation capability and wobble mode. In wobble mode, a system produces broader scribed lines, by moving the laser beam in small circles at high speeds while engraving.

QUANTRAD BLAZER SPECIFICATIONS

In March 1981, at the conclusion of our evaluation and investigation of the available applicable laser marking systems, we decided to purchase two Quantrad Blazers. The purchase specifications for the systems were as follows:

- Model ID-2000
- 100-watt CW multimode Laser (Model 1514)
- Z80-based micro-computer with PROM-based operating software
- Standard font on tape
- Laser power supply and cooling system
- RS-232C interface to CRT keyboard or remote programming station
- Manual parts handler
- 16-key operator console with 18-digit display
- Software to provide "fill-in-the-blanks" programming capabilities
- Diagnostic display panel for system parameter monitoring
- Welded tubular steel frame enclosed with fiberglass laminate covers
- Beam conditioning and focusing optics
- Galvanometer beam positioner
- CRT keyboard and tape storage (Hewlett Packard Model HP-9875A)
- 15-in. closed circuit television viewing system with internal illumintion
- Special font digitized and stored on EPROM to eliminte line intersections and crossovers
- First pulse suppression capability
- Nozzle for inert gas cover during marking operation
- Data entry terminal for writing and entering new marking programs (Hazeltine 1510)
- Tape cassette drive for data storage.

A view of the Quantrad unit is shown in Figure 13. The Blazer is a complete turn key system and requires only tooling to make it operational. Tooling presented no problem or additional expense. The existing Korad fixtures are used with the new markers.

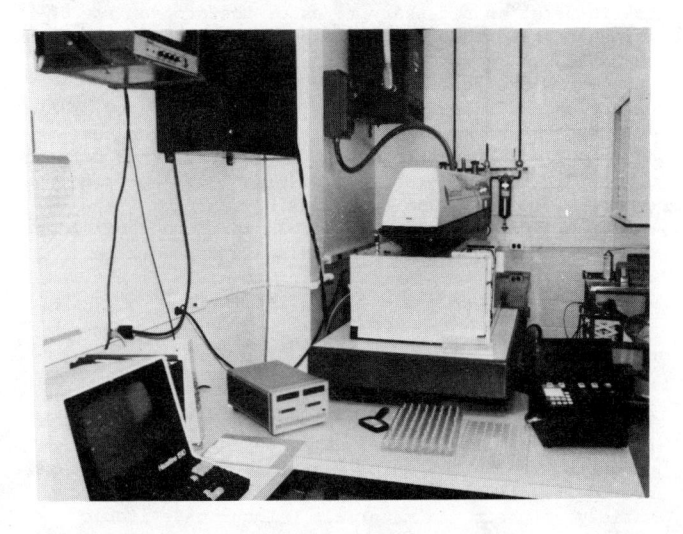

Figure 13. Quantrad Blazer System

Laser Head

The system uses a laser head manufactured by Control Laser Corporation. This is a cavity which has a reputation for being reliable and easy to maintain. It uses only one flashlamp, which can be changed in 15 min without requiring any realignment of the laser mirrors. This means minimal downtime and lower lamp replacement cost. The laser produces 100 watts of power - double that of the Korad unit. This level of available power will not normally be needed for marking operations, but will provide standby capability if required for some new operation. In the event the laser mirrors were slightly misaligned, there would still be sufficient power available to mark most materials without having to immediately realign the mirrors. Production could continue uninterrupted and downtime would be minimized.

Computer

The Quantrad uses a micro-computer of their own design utilizing standard and widely available parts. It has 16K Random Access Memory (RAM), and no special computer language is required to operate the system. 16K memory is adequate for Mound's purpose because magnetic tape cassettes are used for external program storage. The system can engrave logos, emblems, or other special symbols, but the programming requires the use of an optional digitizer or the purchase of the program in the form of an EPROM or tape. This option was not purchased because Mound rarely has a need for such capability, and the Korad marker does have it.

Ease of Operation

As has been stated, the Blazer uses no special computer language such as Fortran. This makes the system very easy to use. With the Quantrad programming system, commands such as character height, width and spacing, serialization, the message to be marked and other commands are entered into the computer memory on a "fill-in-the-blanks" basis. The programming format, or "Screen Fields" as Quantrad calls it, are displayed on the operator's CRT as shown in Figure 14. The programmer simply enters the desired command information into each applicable blank, enters the message to be engraved in the ten lines below the character parameters, and enters the program into memory. This is a much simpler process than that used on the Korad.

Programming Capabilities

It is beyond the scope of this paper to discuss each of the entry or program fields; however, some of them will be briefly described to explain and/or clarify system's capabilities or features. The format display is divided into four parts: command line, page data, lot/serial data and message area. The command line gives the file name and enables the operator to communicate with peripherals. The page data contains the marking or engraving parameters. Lot and serial data are self explanatory as are the messages or text to be marked.

The X and Y coordinate fields are located in the upper left hand corner of the page data. Three modes are provided for coordiate start location of the text to be marked: absolute, auto center, and relative. The X and Y coordinate fields are used to position the start within the 50.8 mm (2.0 in.) x 50.8 mm (2.0 in.) marking window in both the absolute and relative modes. Programming uses the conventional coordinate axes nomenclature including directions and signs.

The marking radius field allows marked text to be engraved in a circular pattern of a specified radius. All the characters are rotated so that they follow the circumference of the circle. The bottom of the characters rests on the circumference. The range of entry is 1.0 mm (0.040 in.) to 254 mm (10.0 in.). The text is evenly spaced on both sides of the specified X,Y coordinate location.

Laser speed and spot overlap must be considered together because these parameters interact to affect the appearance of the marked parts. Laser speed sets the rate of travel of the laser spot across the marking surface. It ranges from a minimum of 0.5 mm (0.020 in.)/sec to a maximum of 254 mm (10.0 in.)/sec. Spot overlap sets the number of times the laser is fired for every 0.0254 mm (0.001 in.) of distance the laser spot travels. The product of the laser speed and spot overlap equals the laser firing frequency. The allowable limits of the latter, which was termed repetition rate by Korad, are 10 to 30,000 Hz.

Three of the page data fields are not used because the associated optional equipment was not purchased by Mound. Laser focus and laser position are active only when the optional moving optics package is installed. The third inoperative field is wobble. This optional function was previously described.

The load program field is associated with the loading of a stored graphic such as a logo. Entering a file number and source of load allows the operator to load a graphic program from either the permanent memory in the Remote Control Unit (RCU), a tape cassette, or a digitizer. The latter, an option previously discussed, was not purchased by Mound.

The marking mode/font field is used to select the operational mode, either engrave or 5 x 7 dot matrix. The font selection is either the standard normally supplied or the special font required for close depth control. Quantrad, using a hardware approach, incorporated into their standard system a circuit which prevents the deeper marks at the starting and ending points of each character. The special font is contained in the software. Crossover or overlap points are eliminated by leaving small gaps at the points where the laser beam would normally touch a second or even a third time during the course of character generation. The special font permits the laser marking process to be used on a greater variety of components than before.

file: command:/ : : : : : : :   page:

x coord x	y coord y	marking radius r	laser speed s	spot overlap o	laser focus fo	laser position po	load program l
marking mode/font	wobble x/y	char height h	char width w	char tilt t	char spacing cs	line spacing ls	graphic size(%) p
lot #(/)		serial # (:)		increment size		cy total remain	

0
1
2
3
4
5
6
7
8
9

Figure 14. Screen Fields

Character height and width fields are used to specify in mils these parameters. Height and width are independent and must be specified. Both parameters range from 0.508 mm (0.020 in.) to 50.8 mm (2.0 in.).

The character tilt field is used to rotate the characters or graphics controlled by the page. Rotation is about the bottom left corner of the character or graphic in a counter clockwise direction. Tilt is controllable from 0 to 359° in 1° increments.

Character and line spacing fields are used to specify the distance between characters and lines. Both functions must be specified in mils. Both parameters range from 0 to 50.8 mm (2.0 in.).

Graphic size field sets the scale of the graphics in percent. The graphics fill the 50.8 mm (2.0 in.) window when 100% is selected. Graphic size can be varied from 1 to 100% in 1% increments.

LASER MARKING GROWTH AND STATUS

In the fall of 1981, the two Quantrad Blazers were received and installed. The assembly production unit became operational in November and quickly proved its worth. The Korad unit was relegated to backup or standby status. The increased work load has easily been handled by the two new units. The systems have performed up to expectations, and downtime has been decreased.

One aspect of the laser marking process that can be improved on is part handling. The importance of this in planning and implementing tooling and fixturing cannot be overemphasized. It constitutes the only real speed limitation to the process. Robotizing or automating the part handling will result in more throughput and decreased labor costs. This is not to say that operational costs have been high. As a matter of fact, experience has shown that one worker is now required for the laser process where six workers were formerly required with other methods.

Not only has there been a labor saving, but there has been an improvement in the physical marking itself. The quality of the engraved mark will make it possible to automatically read data such as the serial number into a data bank or automated system. Character recognition equipment for this purpose is currently available from companies such as General Electric, Cognex, Key Images, and Data Copy.

The success and usefulness of the Korad system and our experience with the Quantrad Blazers has clearly demonstrated the advantages of laser marking. The principle advantages are these:

- adaptable to a variety of materials
- noncontact
- permanent
- minimal effect on materials
- versatile
- flexible
- programmable
- operating simplicity
- inexpensive tooling
- high speed
- low cost

Obviously, no single process can answer all marking needs; but at Mound, laser marking has proved to be a most cost effective process.

INVESTIGATIONS IN OPTIMIZING THE LASER CUTTING PROCESS

F. O. Olsen

1 INTRODUCTION

Since the appearence of CO_2-lasers in the power range over 200 watt, laser cutting of sheet metals has been investigated in many laboratories and laser-companies.

The first major step to increase laser cutting in sheet metals was the introduction of gas assisted cutting about 1970 /1/. By using a coaxial oxygen jet, cutting rates and cutting quality were increased.

Through the next 10 years, the CO_2-laser was used for some special cutting applications. However a real breakthrough for contour cutting with CO_2-lasers didn't occur.

In 1979 the influence of the polarization of the laserbeam when cutting in sheet metal was discovered /2/. This result quickly lead to the introduction of systems to control the polarization of the laserbeam, so that the cutting was performed with a circular polarized laserbeam. By introduction of these polarization systems the cutting has been more stable, the cutting rate has been enhanced sligthly, and the quality of the cut has been improved.

However, calculations show that the laser cutting rates still are far below the theoretical maximum, and that further optimizing of the laser cutting process will be possible.

In this paper some of the investigations carried out at the Technical University of Denmark aiming at an optimization of the laser cutting process will be described. These investigations were initiated in 1977, when the laser processing laboratory at the University was established. Since then several projects in laser cutting have been carried out and the major results of these investigations will be described in this paper.

The investigations include cutting experiments with a plane-polarized medium power CO_2-laser in sheet metals, mainly mild steel. The investigations include cutting in different directions with respect to the plane of polarization. Furthermore, experiments with different lens/nozzle systems are included. Both the lens/nozzle system delivered from the laser manufacturer and a system designed at the laboratory have been tested, as well as the geometry of the nozzle tip. The influence of the gas pressure on laser cutting has been investigated in the pressure range from 1 to 7 bar.

The striations of the cut are examined, and the striation formation mechanisms are discussed.

2 THEORY

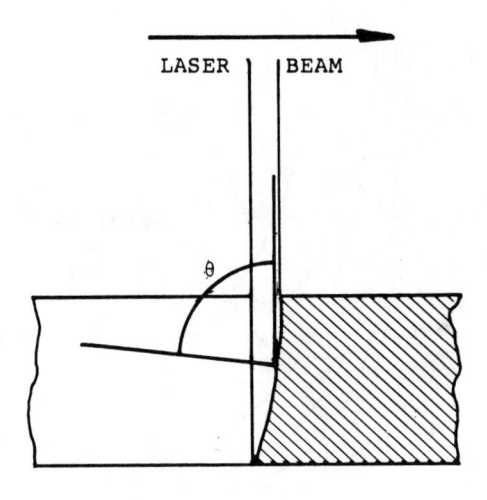

Figur 2-1: The cutting front

In figure 2-1 is shown the formation of the cutting front, when laser cutting in sheet metal. Typical, the laser beam diameter in the cut kerf is about 0.1 to 0.2 mm, when using a medium power CO_2-laser. Therefore, when cutting in sheet metal with a thickness larger than 0.7-1.0 mm, the laser light is absorbed under incident angles close to tangential. From the surface of the cut front, melted and evaporated material is ejected, and the assisting coaxially gas jet, normally oxygen, are blowing the melted and the evaporated material out of the cut kerf.

The main factors which affect the cutting process are:

- laser beam parameters: power, mode, polarization and focusing.

- gas jet parameters: type of gas, gas pressure and the direction of the gas jet.

- material parameters: thermal properties, reflectivity, viscosity of the melted material, optical properties of melted and evaporated material and the chemical and mechanical properties of the melted material.

To determine the maximum cutting rates, obtainable for a given laser power in a material, a rough eastimate of the heat balance can be used.

To achieve these maximum cutting rates, the temperature gradients in the front of the cut kerf shall be very large and the thickness of the melted and the evaporated film in the cut front shall be negligible.

2.1 HEAT BALANCE

When cutting with an active gas the energy balance in the cutting proces can be described as:

$$\frac{dH_e}{dt} + \frac{dH_s}{dt} + \frac{dH_l}{dt} + \frac{dH_r}{dt} + \frac{dH_p}{dt} = 0 \qquad (2.1)$$

where H_e is the energy of the material removed, H_s is the energy tranferred to the specimen, H_l is the energy of the laser beam and H_r is the energy from the chemical reactions between the gas and the material and H_p is the energy of the gas jet.

If the gas is inactive to the material, $H_r=0$.

H_e is a sum of thermal energy and the potential and kinetic energy of the outcoming gas jet.

If we assume that:

-the potential and kinetic energy of the outcoming gas jet is negligible,

-the energy, H_g, of the gas jet from the nozzle is negligible,

-the energy, H_r, of the chemical reactions is negligible,

-the energy transferred to the specimen, H_s, is negligible, which can be assumed when cutting at high speed,

-all of the energy of the laser beam is absorbed in the front of the cut kerf

-and the temperature of the material removed from the cut kerf is equal to the melting temperature,

we can estimate the maximum cutting speed by:

$$V_{max} = \frac{W}{t\,b\,m\,(c(T_m-T_0) + H_m)} \quad (2-2)$$

where W is the laser power, t is the thickness of the specimen, b is the width of the cut kerf, m is the specific mass of the material, c is the specific heat of the material, T_m is the melting temperature of the material, T_0 is the room temperature and and H_m is the melting energy of the material.

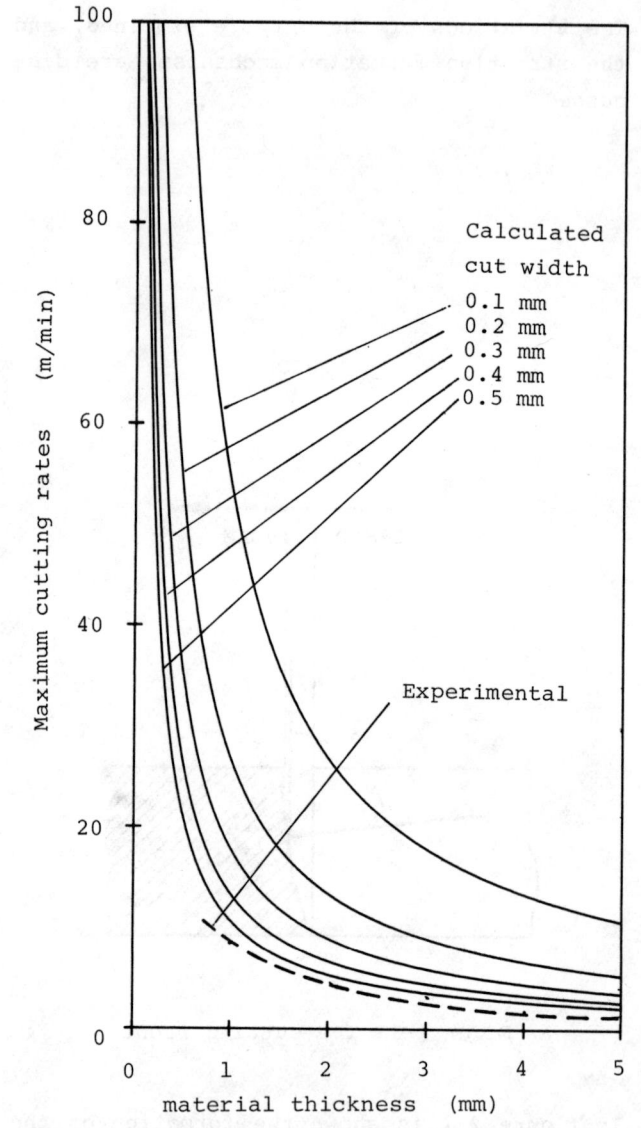

Figur 2-2: Maximum cutting rates versus material thickness

In figure 2-2 the maximum cutting rates calculated by eq. 2-2 are shown versus material thickness for different cut widths. The laser power is 500 watt and the material is mild steel.

In figure 2-2 is also shown typical experimental obtained maximum cutting rates /3/.

Generally experimental results show that the cut width is between 0.3 mm and 0.5 mm for a normal 500 watt CO_2-laser, however the beam diameter in the focal point generally is about 0.1 mm.

These calculations show that there is a considerable gap between theoretical cutting rates and experimental cutting rates obtainable with the technology of today.

2.2 REFLECTIVITY

The reflectivity of a material surface, irradiated by a laser beam can be calculated from the Fresnel formulaes:

$$R_p = \frac{\tan^2|\phi_1-\phi_2|}{\tan^2|\phi_1+\phi_2|} \quad (2-3)$$

$$R_s = \frac{\sin^2|\phi_1-\phi_2|}{\sin^2|\phi_1+\phi_2|} \quad (2-4)$$

where R_p is the reflectivity for light polarized in a plane parallel to the plane of incidence, and R_s is the reflectivity for light polarized perpendicular to the plane of incidence. The angles ϕ_1 and ϕ_2 are the angle of incidence and the angle of transmission respectively, where the angle of transmission can be calculated from the Snell theorem:

$$\sin \phi_2 = \frac{\sin \phi_1}{\underline{n}} \quad (2-5)$$

where $\underline{n}$ is the complex index of refraction of the material.

The optical properties of a material can be expressed by the permittivity, $\underline{\varepsilon}_r = \varepsilon_1 - i\varepsilon_2$, which is a complex material constant. The index of refraction and the permittivity is related by:

$$\underline{n} = \sqrt{\underline{\varepsilon}_r} \quad (2-6)$$

The permittivity of a material is a function of the wavelength (or frequency) of the incident light. In addition the permittivity is varying when the material surface is heated by a high power CO_2-laser. These changes in the optical properties of a material, when heated by a high power CO_2-laser, is generally known from both laser heat treatment and laser welding, where the coupling of light energy in steel raises dramatically, when the surface temperature reaches the melting temperature.

In figure 2-3 is shown the reflectivity of normal incident light, R_0, numerically plotted as a function of the complex permittivity for negative values of the real part of the permittivity, e_1.

Experimental values /4/ of material constants at room temperature indicates that the permittivity of steel at the wavelength of CO_2-lasers (10.6 micrometers) is in the region about $\varepsilon_1=-1000$ and $\varepsilon_2=1000$.

Using equations 2-3, 2-4, 2-5 and 2-6, the reflectivities R_p and R_s of a steel surface is shown versus angle of incidence in figure 2-4. These curves show that there is a significant difference between R_p and R_s at large angles of incidence.

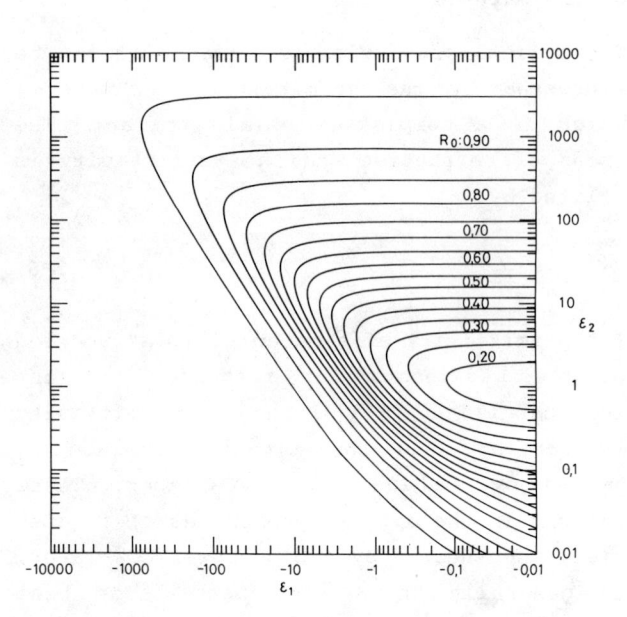

Figur 2-3: R_0 versus the complex permittivity

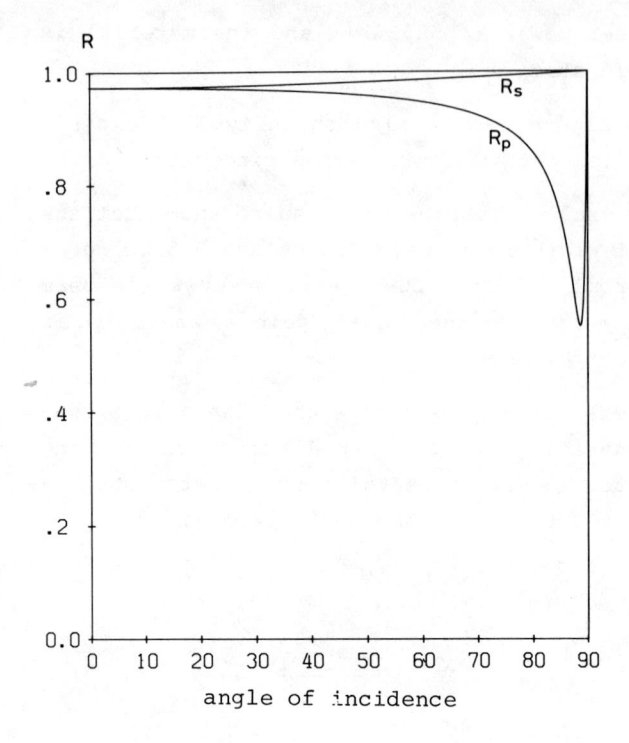

Figur 2-4: R_p and R_s versus angle of incidence of the light

For metals irradiated by a CO_2-laser, the coupling coefficients A_p and A_s are:

$$A_p = 1 - R_p \qquad (2-7)$$

$$A_s = 1 - R_s \qquad (2-8)$$

which means that all the light passing the surface is absorbed.

As stated above, the reflectivity of normal incident light drops dramatically when the temperature rises. Although the exact temperature variation of the permittivity cannot be determined, the influence of the temperature on the ratio A_p/A_s can be estimated. In figure 2-5 the ratio A_p/A_s is shown versus angle of incidence for different values of R_0, where $\varepsilon_2 = -\varepsilon_1$. These curves show that the ratio A_p/A_s for a given angle of incidence

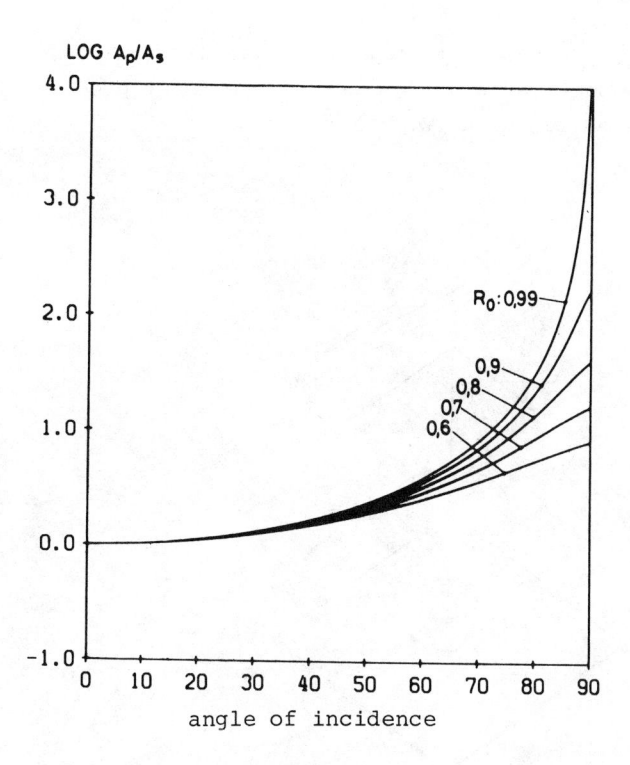

Figur 2-5: The ratio A_p/A_s versus angle of incidence for different values of R_0

is decreasing when the normal reflectivity is decreasing, i.e. when the temperature is increasing.

Therefore, it is difficult exactly to determine theoretically the ratio A_p/A_s for laser cutting, where the light is absorbed partially of melted material, vapor and plasma, and partially in solid material. However, the absorption of laser light under angles close to tangential in metals is depending on the angle between the plane of polarization and the plane of incidence.

3 INVESTIGATIONS IN CUTTING HEAD DESIGN

Through our initial experimental work, we found that it was difficult to achieve reproducable results with the standard lens/nozzle system delivered from the laser manufacturer. Therefore we started optimizing the design of the cutting head.

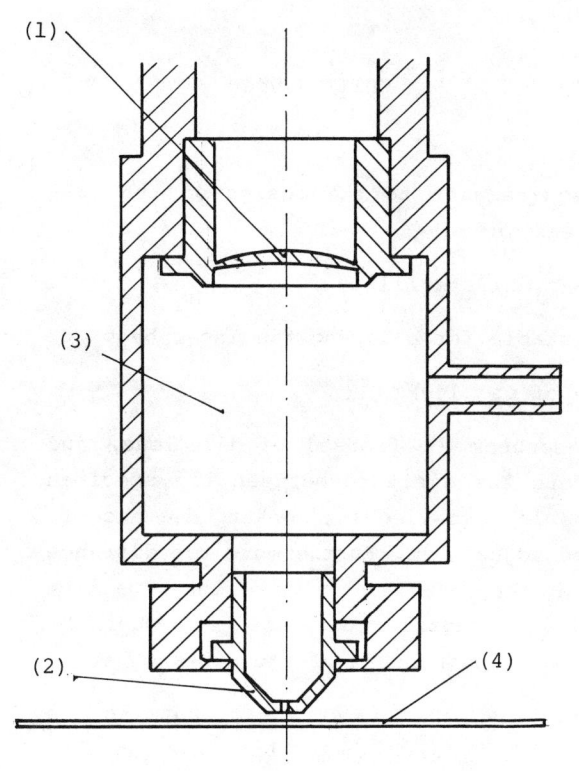

Figur 3-1: Cutting head

Figure 3-1 shows generally a cutting head for medium power cutting (based upon transmitting optics). The cutting head include (1) a focusing lens in a lens mount, (2) a nozzle tip. Between the lens and the nozzle tip is formed a pressure chamber (3), and from this pressure chamber the cutting gas is forming a gas jet coaxially with the laser beam towards the specimen (4), which is moving relatively to the cutting head.

These investigations were concentrated upon:

- the design of the cutting head
- the design of the nozzle tip and
- the design of the lens mount

3.1 CUTTING HEAD

The requirements to the design of the cutting head is:

- mechanical stability
- adjustable for aligning the laser beam
- free of gas leaks

The laserbeam is focused of the lens, and therefore the distance between the specimen and the lens, called the working distance f, must be adjustable. Furthermore the distance between the nozzle tip and the specimen called the nozzle distance h, must be adjustable independent of the distance f.

In practice the laser beam has to be adjusted to pass through lens and nozzle, which means that the cutting head must include some possibilities of adjustment of lens or/and nozzle in a plane perpendicular to the laser beam propagation direction.

This adjustment of the laser beam through the cutting head is a adjustment of 4 variables: the position of the beam arriving to the lens (x and y) and the angle of the beam arriving to the lens (x´ and y´).

To achieve the nessesary adjustments of these 4 variables, several adjustment procedures can be used.

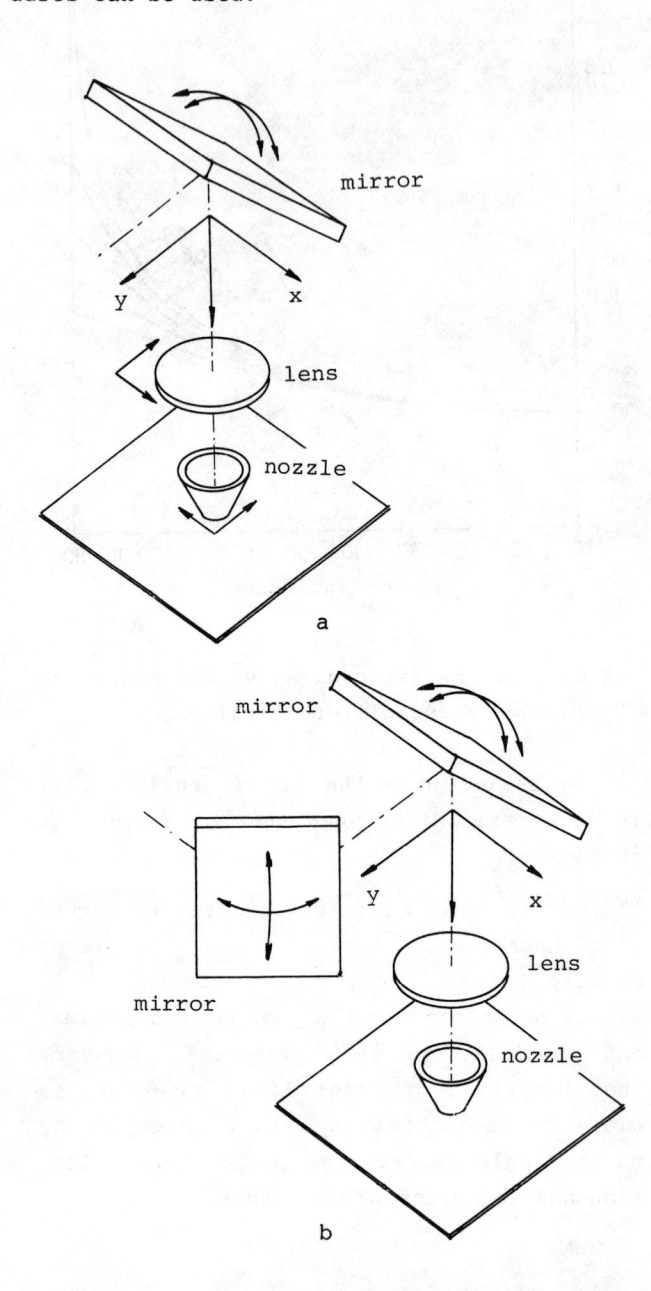

Figur 3-2: Adjustment possibilities for cutting head

In figure 3-2 the different methods for adjusting the laser beam through the cutting head is shown. In figure 3-2a is shown the possibility, where the adjustment of two of the four variables is performed by adjusting one bending mirror, which is placed before the lens, while the adjusting of the two remaining variables is performed by moving either the lens or the nozzle tip or both. In figure 3-2b is shown the possibility, where 2 bending mirrors are placed before the cutting head. Here the adjustment of all 4 variables are performed before the cutting head.

The cutting process here require that:

- the laser beam and the gas jet are coaxially

- the center line of the laser beam is perpendicular to the cutting direction.

The most common method to align the beam is the method, shown in figure 3-2a. However, these methods can only be used for small corrections, otherwise they will cause errors in respect to the two requirements above.

Therefore the alignment procedure, shown in figure 3-2b, where all alignment of the laser beam is conducted with mirrors before the cutting head, should be most suitable. This alignment procedure has the advantage that the design of the cutting head is simple and stable.

However, two independent adjustments are neccesary: the adjustment of the distances f and h.

In figure 3-3 is shown the different methods for adjusting the distances f and h. In figure 3-3a is shown the most common method:

a

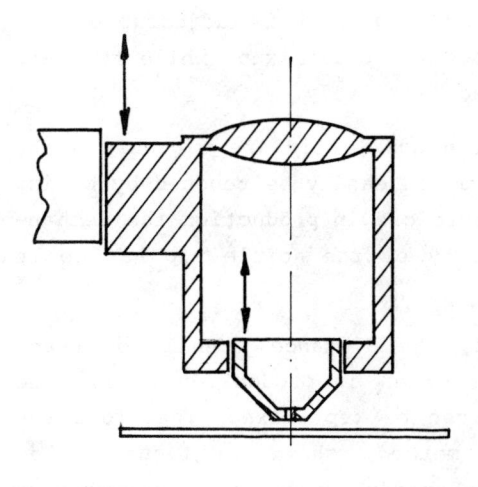

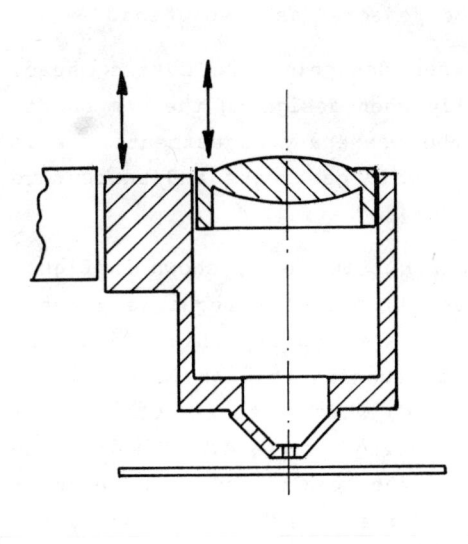

b

Figur 3-3: Adjustment of the distances f and h

the cutting head is adjustable. The lens is fixed in the cutting head, and the nozzle tip can be moved relatively to the cutting head. In figure 3-3b is shown the method, where the cutting head is adjustable, but here the nozzle tip is fixed, while the lens can be moved relatively.

Generally the nozzle tip are closest to the specimen and can easily be touched by moving objects. Therefore in production the mechanical stability of the nozzle tip has to be very good.

Furthermore, the distance f shall be readjusted quit often in production, while the distance h can be kept fixed. Therefore the adjustment method, shown in figure 3-3a, where each readjustment of f requires a readjustment of h, and where the mechanical stability of the nozzle tip generally is poor, must be rejected as a solution.

Generally, when designing the cutting head, and especially when designing the components to perform the nessecary adjustments, it is important to avoid gas leaks and unaccurate movements under adjusting.

In the typical cutting head, shown in figure 3-3a., the adjustment of f requires a readjustment of h, and typical this adjustment is achieved by turning the nozzle tip up or down in a thread, which acts as both adjustment and bearing surface. Hereby the positioning of the nozzle tip is quite unstable, and very often the leaks in the system are considerable.

3.2 NOZZLE TIP

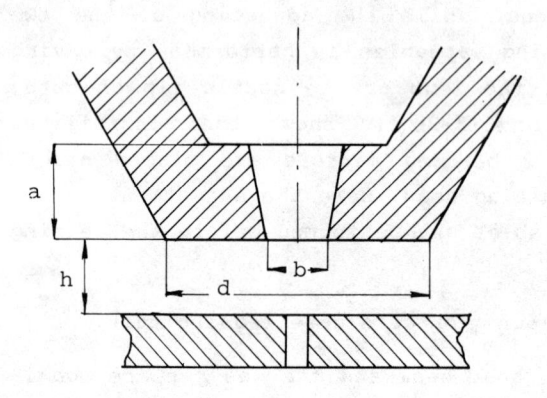

Figur 3-4: Nozzle tip geometry

In figure 3-4 is shown the major parameters in nozzle tip design:

- nozzle hole geometry, i.e. cylindrical, conical, convergent-divergent

- thickness of nozzle, a.

- nozzle aperture, b.

- nozzle tip outer diameter, d.

- nozzle distance, h.

A theoretical analysis of the flow conditions in the nozzle tip and the cut kerf is complicated. However, two situations can be evaluated qualitatively:

-the nozzle aperture b is small compared with the nozzle distance h, as shown in figure 3-5a.

- the nozzle aperture b is large compared with the nozzle distance h, as shown in figure 3-5b.

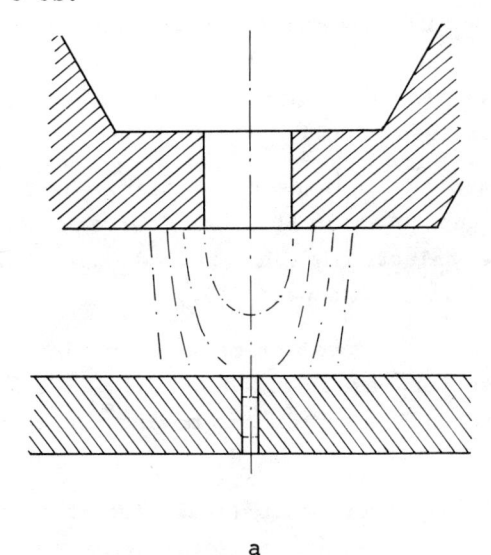

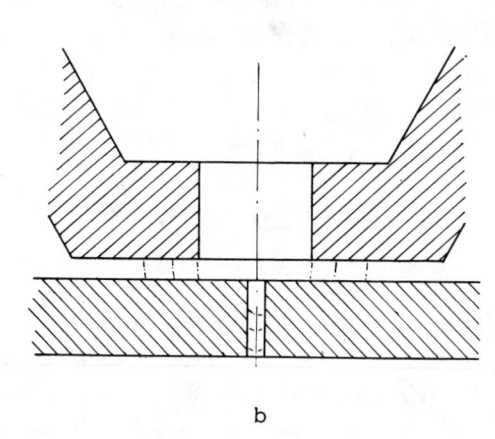

Figur 3-5: Gas flow through nozzle tip and cut kerf

In figure 3-5 is schetched how the gas flow through nozzle tip and cut kerf is depending upon the ratio b/h. If b/h is small, as shown in figure 3-5a, the gas pressure will decrease from the nozzle tip down to the cut kerf. Furthermore, pressure variations perpendicular to the propagation direction of the laser beam can cause lensing effects, which can disturb the laser beam. This lensing effect might be unstable, while the moving specimen can disturb the flow conditions currently.

In figure 3-5b is shown the situation, where the ratio b/h is large. Generally the nozzle aperture, b, is more than 2 times the cut width, and therefore the pressure decrease between nozzle tip and specimen is negligible. The pressure decrease occurs in the cut kerf and in the gap between nozzle tip and the specimen. Therefore, when using a large ratio b/h, lensing effects can be avoided and the gas pressure in the cut kerf are increased for a given pressure in the cutting head.

When the ratio b/h is large, the nozzle tip outer diameter, d, affects the gas flow. When d increases for a given h, the gas flow between the nozzle tip and the specimen decreases.

3.3 LENS MOUNT

Generally in medium power laser cutting, a standard lens mount is used.

In figure 3-6 is shown this standard lens mount This mount might cause unstability in laser cutting due to poor design. The lens is kept in position by a ring which acts as a spring. This ring is placed over the lens and when the pressure in the cutting head is

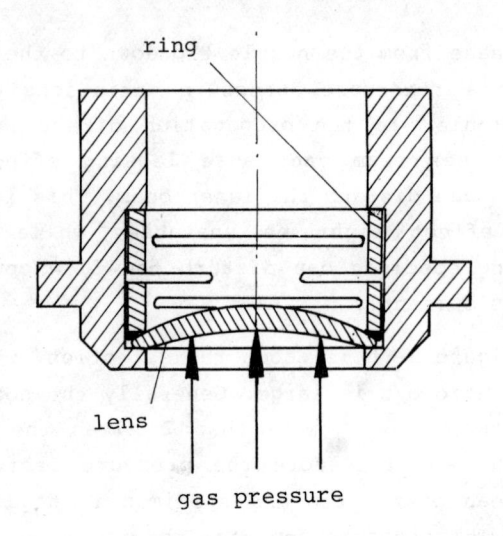

Figur 3-6: Standard lens mount

increased, this ring must withstand the pressure. Very often the user of a laser cutting system observes that this ring has been deformed.

Furthermore studies have shown that there can be considerable gas leaks between lens mount and nozzle head through the outer thread of the lens mount.

4 EXPERIMENTAL STUDIES

Our experimental studies have been concentrated upon cutting with a plane polarized CO_2-laser in sheet metal, mainly mild steel and stainless steel.

The laser is an industrial laser, which can deliver up to about 600 watt (cw), and which can be electronically pulsed (including enhanced pulsing capability).

The relatively movement between cutting head and specimen was obtained by using a simple one-axis moving system for moving the specimen.

The initial experiments (including the studies of polarization effects) were carried out with the cutting head delivered from the laser manufacturer.

For the latest experimental studies, a cutting head was designed according to the discussion in section 3.

4.1 POLARIZATION EFFECTS

The initial studies of the polarization effect were conducted in 1979. These studies included very few experiments with a pulsed laser beam in sheet metal and with a continuos wave laser beam in 7 mm acryl /2/. These experiments showed:

- when cutting in sheet metal, the polarization affects the cutting rates obtainable in different directions.

- when cutting in sheet metal, the polarization causes variations in the geometry of the cut kerf, typical demonstrated by the cutting of small holes with a linear polarized laser, where the hole on the back side of the specimen formes en ellipse.

- when cutting in materials with a low reflectivity for normal incident light, the polarization effects cannot be observed.

In 1980 further experimental studies were conducted. Here cutting were performed in 0.7 mm mild steel with a laser power of 400 watt (cw), and the cuts were performed with a linear polarized laser beam in two directions: parallel to the plane of polarization (v=0°) and perpendicular to the plane of polarization (v=90°) /5/.

In figure 4-1 is shown the cutting rates, where high quality cuts were obtained in the two directions versus the oxygen flow through the nozzle.

Here the standard lens/nozzle system from the laser manufactor were used. Because of considerable gas leaks in the lens/nozzle system, absolute values of gas pressure cannot be supplied. The nozzle tip was a cylindrical type with an aperture of 1 mm diameter.

These results show that there is a factor 2 between maximum cutting rates in the two directions parallel respectively perpendicular to the plane of polarization.

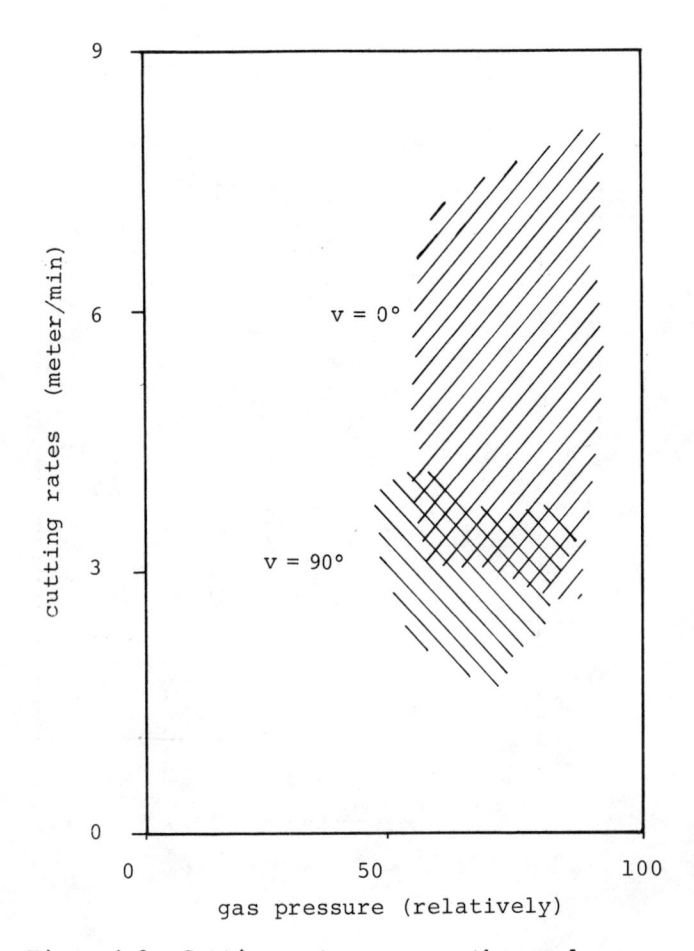

Figur 4-1: Cutting rates versus the angle v

4.2 EFFECTS OF GAS PRESSURE

The results shown in figure 4-1 demonstrated, that the gas pressure influences on the cutting results.

After these results, a new lens/nozzle system was developed, and further experiments were performed.

Figur 4-2: Cutting rates versus gas pressure

In figure 4-2 is shown high quality cutting rates versus gas pressure for cutting in 2 mm mild steel with a nozzle aperture of 1 mm, nozzle distance about 0.5 mm and for a laser power of 600 watt (cw).

These results show:

- at pressure levels lower than 1-2 bar, high quality cuts cannot be expected.

- in the pressure range 2-4 bar, the maximum cutting rates are increasing with increasing pressure.

- in the pressure range 4-6 bar, the maximum cutting rates are not affected of the gas pressure.

- at higher gas pressures, burning effects in the bottom of the cut impede a good cut quality.

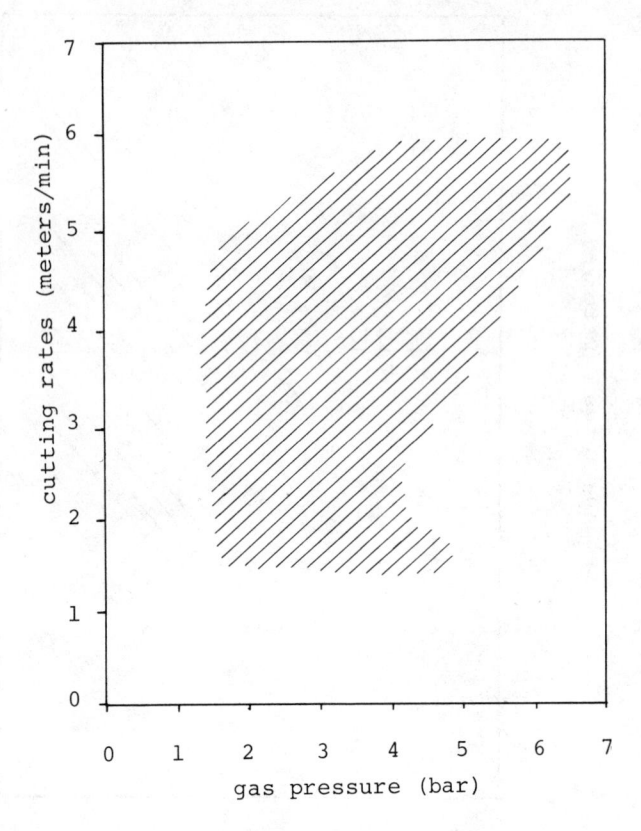

The experimental results in figure 4-2 were obtained with a nozzle tip with a converging hole. Corresponding results were obtained with a converging-diverging nozzle tip.

Further a few experiments with different outer diameter of the nozzle tip were carried out. These experiments show that a larger outer diameter of the nozzle tip is equal to a higher gas pressure in the cutting zone. For these experiments a b/h ratio about 2-3 was used.

Generally the experiments show, that the nozzle distance h has to be small, i.e. the ratio b/h must be larger than 1.

4.3 STRIATIONS

The dynamic behavior of the laser cutting causes the formation of the striations.

Attempts have been made to describe this dynamic behavior /6/.

These authors explain the striation formation as a phenomenon which occurs at low cutting speeds because of a higher velocity of the moving melting front from oxidation of the metal, than the velocity of the moving laser beam.

However, this model cannot describe the dynamic behavior of high speed laser cutting where the cutting velocity is higher than the velocity of the oxidation front, and where the striations are fine and have a more complex geometry.

Recent experiments have been performed to examinate the striation formation.

Samples from the cutting experiments were examined and different geometries of the striations were observed, as shown in figure 4-3.

At low pressure (about 2 bar) and medium cutting rates, the striations typically are straight lines, sometimes with a line parallel to the cutting direction, as shown in figure 4-3a.

At medium gas pressure (4 bar) and higher cutting rates the striations is curved, and often there are lines starting from the top of the cut kerf running to the buttom of the kerf. Often lines in between starts a distance from the top also running to the buttom, as shown in figure 4-3b.

When the pressure is in the range of 6 bar, and the cutting rates are high, cuts can be observed, where there is a region in the middle of the cutting kerf, where no lines are observed.

The curvature of the striations can be more or less complex, i.e. S-formed lines have often been observed.

Furthermore we have made two types of observations:

- At some special parameter sets (cutting rates, pressure, working distance) the process is unstable, i.e. when cutting, then suddenly the striations change characteristics from fine striations to rough striations. These changes occurs suddenly, as jumping between two types of cut.

- Very often when examining the two opposite sides of the cut, different striations

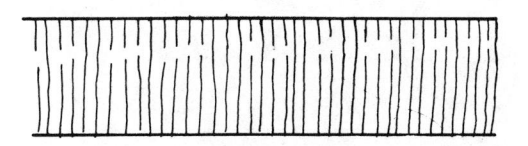

low pressure, medium speed
a

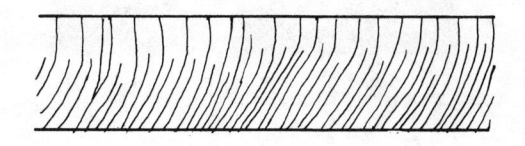

medium pressure, high speed
b

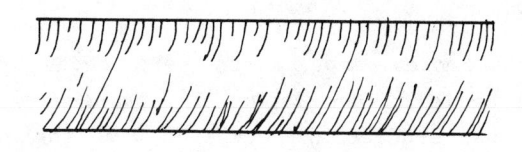

high pressure, high speed
c

Figur 4-3: Striations in 2 mm mild steel

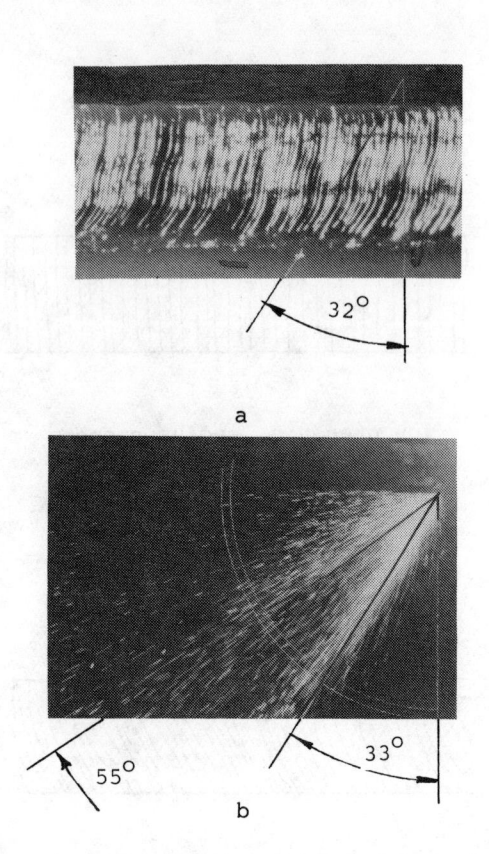

Figur 4-4: Striations and spark-showers, 9 cm/sec, 4 bar

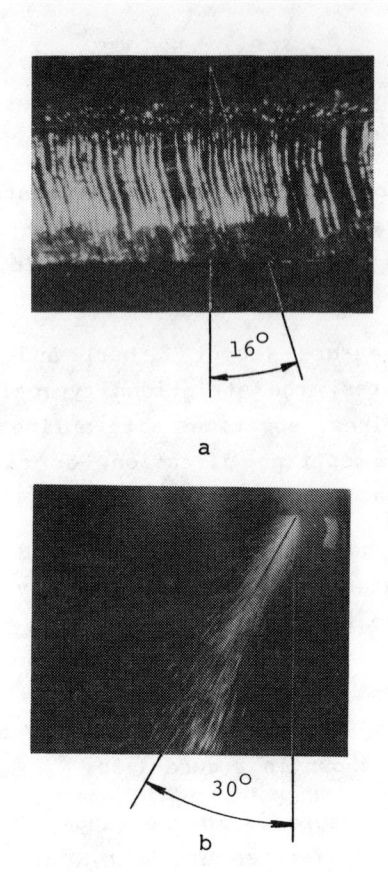

Figur 4-5: Striations and spark-showers, 8.5 cm/sec, 4 bar

are observed on the two sides. Two different directions of the striations and two different striation frequencies can be observed.

Furthermore, experiments were made, where the angle of the shower of sparks from the cut was studied and compared with the angle of striations in the buttom of the cut. In the figures 4-4, 4-5 and 4-6 is shown photographs of spark-showers and the striations from some experiments conducted in 2 mm mild steel.

In figure 4-4 is shown an experiment, where the main angle of the spark-showers corre-

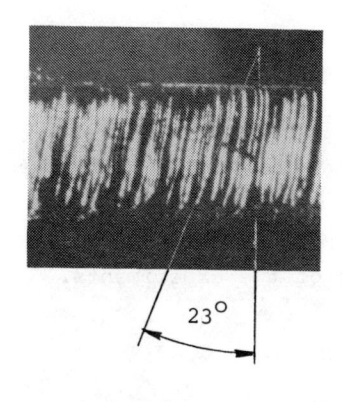

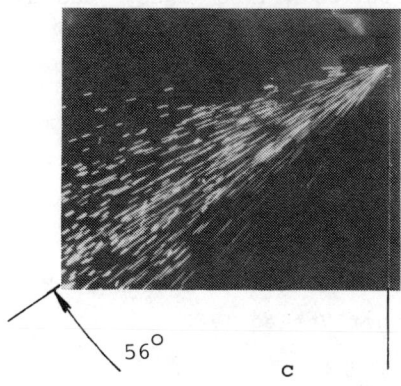

Figur 4-6: Striations and spark-showers, 7.4 cm/sec, 3 bar

In figure 4-5 is shown an experiment, where the main angle of the sparks differs from the angle of the striations. Here the cutting rate was 8.5 cm/sec and the oxygen pressure was 4 bar.

In figure 4-6 is shown two experiments, which gave the same pattern of striations, but where the angles of the spark-showers are very different. Both experiments were made with a cutting rate of 7.4 cm/sec and a gas pressure of 3 bar. The only difference between the two experiments was a change in working distance of 0.2 mm.

5 CONCLUSIONS

Considering the purpose to optimize the laser cutting process following conclusions from the described experiments can be drawn:

- The mode of the laserbeam has to be good, and the focal spot has to be as small as possible, so that the width of the cut kerf can be minimized, which means that a higher cutting rate and a better cutting quality can be achieved with a given laser power.

- The polarization of the laserbeam must be plane, and in contour cutting the plane of polarization must be turned, so that the plane of polarization is always parallel to the actual cutting direction.

- The lens/nozzle system must be improved to meet the requirements of mechanical stability under operation, and permit easy

sponding to the angle of striations in the buttom of the cut. Here the cutting rate was 9 cm/sec and the pressure of the oxygen was 4 bar.

adjustment of the laserbeam through lens and nozzle, controllable gas pressure in the nozzle and proper geometry of the nozzle tip.

- Further investigations in cutting at high gas pressures are necessary.

The studies of striation formation mechanisms indicates, that the formation of striation is caused not only by blowing out of droplets from the front of the cut, or by differences in the velocity of the moving laser beam and the velocity of the oxidation front. Both the fact, that the striations on the two opposite sides of the cut can differ very much, the fact that the striation formation in special situations can change characteristics currently under an experimnet and the fact that the angle of the spark-shower from the cut can differ from the angle of the striations, indicate that the striations can be formed on the sides of the cut kerf as waves in the thin film of molten material behind the moving laser beam.

6 ACKNOWLEDGMENT

The author would like to thank professor L. Alting for helpful discussions and F.B.Thomassen, M.Sc., for his assistance in conducting some of the experiments.

REFERENCES

(1) Sullivan, A.B.J. & Houldcroft, P.T.
Metal cutting by oxygen assisted laser, Brit. Weld. J. Aug. 1967 p 443-445

(2) Olsen, F.O.
Cutting with polarized laser beams, DVS-berichte 63 1980, p. 197-200

(3) Tamaschke W
Schneiden von Stahlbelech mit CO_2-laser, Zeitschr. f. ind. Fertigung, 72 (1982) p. 323-326 (german)

(4) Landolt-Bornstein
Phys.-Chem. Tabellen, 6. Auflage, Bd. II/8, Springer Verlag 1962

(5) Olsen F.O.
Studies of sheet metal cutting with plane-polarized CO_2-laser, Proc. 5.int. congress: Laser 81, p 227-231

(6) Arata Y., Maruo H., Miyamoto I & Takeuchi S.
Dynamic behavior in laser gas cutting of mild steel. Trans. JWSI v. 8 nr 2, 1979

LASER SHAPING OF MATERIALS

Stephen M. Copley
Department of Materials Science
University of Southern California

ABSTRACT

The carbon dioxide laser can be employed to shape both metallic and ceramic materials. Metallic materials are shaped by turning in a lathe with the laser beam focused on the workpiece shoulder directly in front of the cutting tool. The laser heats material lying on the shear plane causing a decrease in cutting force, which can be exploited to attain increased material removal rate. Ceramic materials are shaped by forming a groove or shallow hole at the surface by vaporization. Shaping operations such as turning, facing, thread cutting and milling have been carried out by overlapping the grooves or holes through laser beam or workpiece manipulation. This paper reviews recent results on laser shaping. In Ti-6Al-4V, material removal rate increases of 100% have been attained by laser heating without decreasing tool life. A theory has been developed that predicts the beam power required for a specific increase in material removal rate. The shape of grooves formed in Si_3N_4 has been shown to depend on the angle between the translation direction of the workpiece and the polarization direction of the incident beam. Four point bend tests on laser shaped Si_3N_4 samples show that laser shaping decreases the strength by about 30% relative to diamond ground samples but also decreases the scatter in strength.

INTRODUCTION

In the period since the First International Conference, considerable progress has been made in the understanding and development of techniques for laser shaping of materials (1). Two techniques will be discussed here: laser-assisted machining (LAM), in which the laser beam heats material to be removed by a single point cutting tool; and, laser machining (LM), in which the laser forms a groove in the material by vaporization and expulsion. An investigation of LAM of Inconel 718 and Ti-6Al-4V alloys has shown that laser heating can be employed to significantly increase the material removal rate in force limited semi-rough turning on a lathe without increasing tool wear (2-4). An investigation of LM of hot-pressed Si_3N_4 has demonstrated that high material removal rates, smooth surfaces and good mechanical properties can be attained by this technique (5-7). It also has demonstrated the effect of beam polarization on the shape of laser machined grooves. The purpose of this paper is to summarize the results of these recent investigations.

LASER ASSISTED MACHINING

An arrangement for LAM is shown in Fig. 1. A CW laser beam is directed by turning mirror (M1) along a path parallel to the turning axis of the lathe. The beam is then reflected from carriage mirror (M2), cross-slide mirror (M3) and workpiece mirror (M4), and finally focused onto the workpiece a distance δ in front of the edge of the cutting tool. Because the direction of the carriage motion is parallel to the turning axis and that of the cross-slide motion is perpendicular to the turning axis, the beam retains its alignment during operation of the lathe.

The absorption of the beam is maximized if it is oriented perpendicular to the surface it is heating. Thus in turning, if it is desired to heat the shoulder of the workpiece, the angle θ is set equal to $\pi/2 - \kappa_r$, where κ_r is the major cutting angle. On the other hand, in facing, θ is set equal to κ_r. If it is desired to heat the cylindrical surface of the workpiece, then θ is set equal to $\pi/2$.

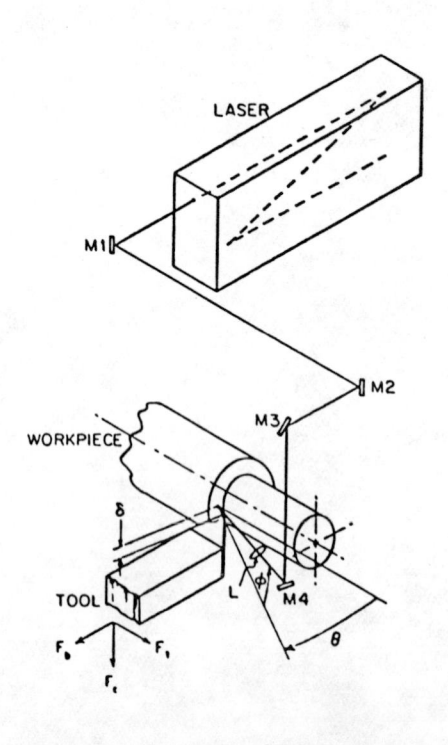

Figure 1 Experimental arrangement for laser-assisted machining.

In our experiments, the beam is produced by a 1400 W, CW-CO_2 laser operating in the TEM_{00} mode (Gaussian spatial distribution of intensity). The mirrors are highly polished, water-cooled copper flats and the lens is ZnSe. For very high intensity beams, the lens may be replaced by a focusing mirror. The three components of tool force (F_c, F_t and F_b), which are defined in Fig. 1, are measured by a piezoelectric dynamometer mounted in the base of the tool holder.

Potential Benefits

In LAM, the laser is employed as a heat source. It is used to heat the material on the shear plane as the chip is being formed. The advantage of the laser is its capability of heating material on most of the shear plane as the chip is being formed without heating significantly the material that contacts the edge or face of the cutting tool. This, of course, is a consequence of the heat flow associated with laser heating. Heating the material on the shear plane may result in benefits such as decreased cutting forces, increased material removal rate, increased tool life and improved surface conditions such as smoothness, residual stress or flaw distribution. However, heating the material that contacts the tool is likely to decrease tool life.

In LAM, the laser scan speed must equal the speed at which the cutting tool passes over the workpiece. Heating the material on the shear plane may change the mode of the chip formation from discontinuous to continuous or decrease the tendency to form a built-up-edge. Such changes affect the smoothness and flaw distribution of the machined surface. One of the most important reasons for heating material on the shear plane is to produce a decrease in cutting force. Such a decrease is anticipated in precipitation hardened alloys, if they can be heated by combined laser heating and shear plane heating into the temperature range where the yield stress decreases markedly with increasing temperature. We will present evidence for this effect in Inconel 718 and Ti-6Al-4V alloys. Such a decrease in machining forces may result in a direct benefit such as making possible the machining of a workpiece that would deflect elastically too much to maintain tolerances or, perhaps, even plastically deform under conventional machining conditions. Such a decrease in machining forces might also produce an increase in tool life, if the concomitant small increase in temperature of the material contacting the tool due to laser heating does not offset this effect. Finally, such a decrease may make possible an indirect but very important benefit; namely an increase in material removal rate and thus a decrease in machining cost. A discussion of this possibility will now be given.

Consider the cost of turning a unit volume of material. If C_o is the labor and overhead rate, C_t is the tool cost per edge, T is the tool life and t_m is the machining time, then the cost is given by the equation

$$U = C_o t_m + C_t \frac{t_m}{T}. \quad (1)$$

Nonproductive time such as loading and unloading time and tool change time have been omitted in deriving this equation because they depend on the specific workpiece and the shop conditions. For machining a unit volume, the machining time is given by

$$t_m = \frac{1}{Z} \quad (2)$$

where Z is the material removal rate.

Thus

$$U = \frac{1}{Z}\left(C_o + \frac{C_t}{T}\right). \quad (3)$$

One approach to decreasing the cost is to increase tool life; however, even for an infinite tool life the cost of machining a unit volume of material is C_o/Z. A more effective approach is to increase the material removal rate.

In single point turning, the material removal rate is given by the equation

$$Z = v a_c a_p \quad (4)$$

where v is the cutting speed; a_c is the undeformed chip thickness, which in single point turning is equal to the feed; and a_p is the back engagement, which is also known as the depth of cut. The magnitude of the cutting force is approximately proportional to the undeformed chip cross-section ($a_c a_p$). For roughing and semi-roughing operations, commercial practice dictates selecting as large an undeformed chip cross-section as possible taking into consideration the rigidity of the tool-workpiece system and the power of the lathe. If, for such force-limited cutting, it is possible by laser heating to decrease cutting force, then it is possible to increase the undeformed chip cross-section and thus the material removal rate. The use of a laser would, of course, increase the magnitude of C_o in Eq. (3). Also it might result in a decrease in tool life, T. Ultimately, the use of the laser would have to be justified on the basis of decreased cost.

Results

At this time, LAM is under active investigation in several laboratories including our own. Although there is data available supporting many of the benefits mentioned previously, insufficient data is available to draw firm conclusions regarding the technological feasibility of the LAM process. In this section, we give results for Inconel 713 and Ti-6Al-4V alloys where the shear plane was heated during chip formation.

Figure 2 shows the results of an experiment designed to illustrate the effect of velocity on force drop. A workpiece was machined in the facing configuration so that the cutting velocity continuously decreased as the cutting tool approached the turning axis. During a facing cut the laser was turned on and off several times and the force recorded continuously. Turning the laser on was observed to produce a force drop. The magnitude of the force drop increases with decreasing velocity. The experimental conditions for this experiment were 600 W incident beam power, beam diameter (1/e) = 212 μm, δ = 5 mm, a_c = 0.0075 cm and a_p = 0.0375 cm.

Figure 3 shows the effect of varying the distance δ between the impingement point of the laser and the cutting tool edge on the reduction in cutting force for two cutting speeds, 10 and 20 cm s^{-1}. The other conditions for the experiment were: incident beam power = 600 W; beam diameter (1/e) = 212 μm; a_c = 0.0075 cm; and, a_p = 0.050 cm. At both cutting speeds, the reduction cutting force first increases and then decreases with increasing distance, δ. The maximum in cutting force reduction occurs in the range δ = 3-4 mm. At distances less than this range, chip interference is thought to account for the decrease in force reduction. At distances greater

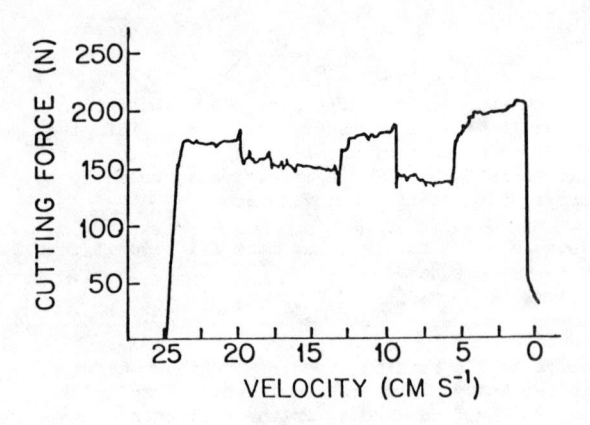

Figure 2 The effect of cutting speed on force drop in the laser-assisted machining of Inconel 718.

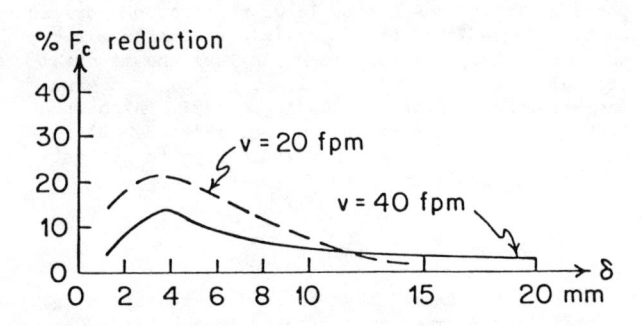

Figure 3 The effect of varying the distance, δ, between the impingement point of the laser and the cutting tool edge on per cent cutting force reduction.

than this range, the material has time to cool so the average temperature of the shear plane with the laser turned on decreases. At the 20 cm s^{-1} cutting speed, there is less time for the absorbed power of the laser beam to heat the material on the shear plane than at 10 cm s^{-1}. Thus, the reduction in force observed at highest velocity is less than that observed at the lowest velocity.

From the preceding discussion, one might conclude that by increasing the incident power it might be possible (1) to increase the magnitude of the force reduction and (2) to obtain a significant force reduction at increased material removal rates. It was found, however, that when the beam power was increased for cutting velocities in the 10-20 cm s^{-1} range, the observed force changes were not reproducible and, sometimes, when the laser was turned on the force actually increased. At 50 cm s^{-1}, it was not possible to obtain a significant force drop with the available incident beam power.

To obtain a better understanding of the changes in cutting force induced by heating the material on the shear plane with a laser, an investigation was carried out in which chip formation was videotaped and displayed on a split screen with simultaneously recorded tool forces. This investigation showed that at low cutting speeds (v < 30 cm s^{-1}), a stable built-up-edge forms on the cutting tool, see Fig. 4. Laser heating at low power does not disturb this built-up-edge. If the beam power is increased, however, the built-up-edge becomes unstable. Loss of the built-up-edge changes the effective rake angle from ~ 20° to 8°, and should result in an increase in cutting force. The force increase associated with such a loss could outweigh the force decrease due to heating the material on the shear plane. Thus, the loss of a built-up-edge provides a satisfactory explanation for the lack of reproducibility in the force change and the occasional increase in force induced by laser heating.

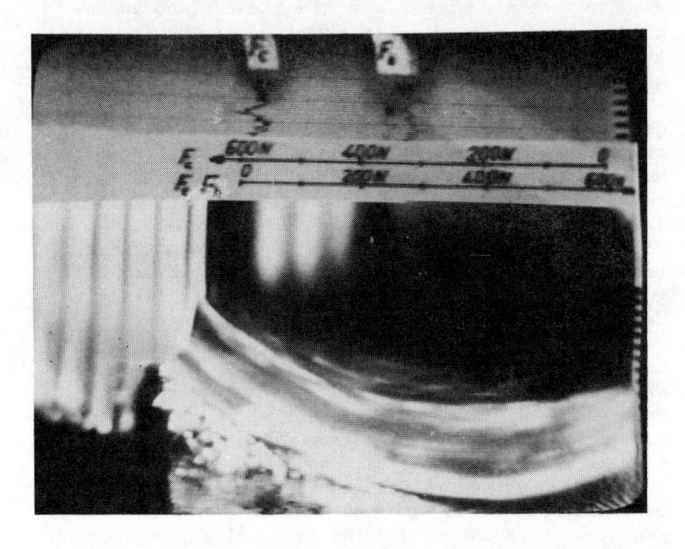

Figure 4 Stable built-up-edge formed on cutting tool in machining of Inconel 718 at v = 15 cm s^{-1}.

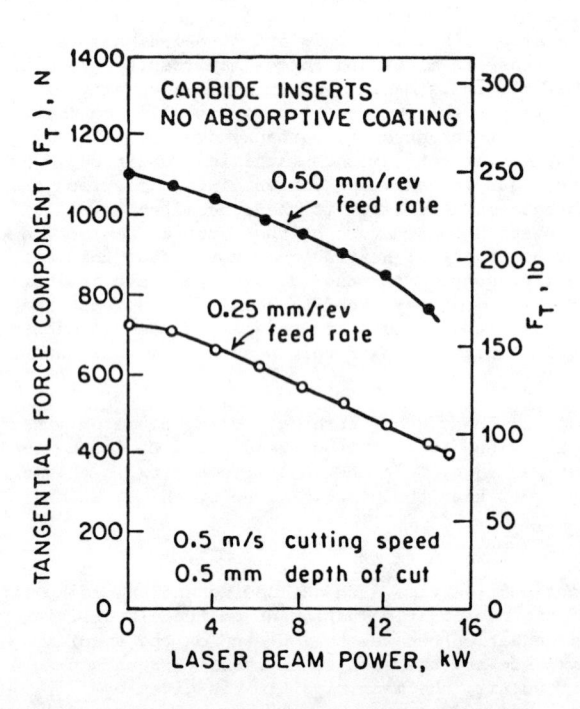

Figure 5 Cutting force versus incident beam power in machining of Inconel 718 at 50 cm s^{-1} (ref. 2).

At cutting speeds greater than 30 cm s^{-1}, no built-up-edge was observed. The failure to obtain a significant force drop at 50 cm s^{-1} appears to be due to insufficient incident power with the 1400 W laser.

Results reported by Rajagapal et al. (2) support the hypothesis that our failure to observe significant force drops at 50 cm s^{-1} was due to insufficient incident beam power. Their data is reproduced in Fig. 5, which shows a plot of cutting force versus incident beam power. The data was obtained at: constant cutting speed, 50 cm s^{-1}; constant feed, 0.025 cm rev^{-1}; and, constant depth of cut, 0.050 cm. After recording the cutting force with no laser heating, the laser was turned on at 2 KW and then increased in 2 KW steps up to 14 KW. The cutting time at each power step was 5 s. Tool wear effects were taken into account by repeating the experiment at decreasing power levels, again changing the power in steps of 2 KW. The value plotted at each power setting was found by averaging the values obtained for increasing power and for decreasing power. In this experiment, the distance δ was 12 mm and the focused beam was 0.1 cm x 0.4 cm ellipse. Clearly, sufficient shear plane heating occurred at the higher incident powers to produce a significant force drop. Unfortunately, with carbide tools excessive tool wear was observed suggesting that heating by the laser penetrated too deeply. In contrast, experiments with ceramic tools indicated that laser heating improves tool wear.

Analysis of the temperature distribution due to heating with an elliptical Gaussian beam, using the integral solution described by Arata et al. (8) suggested that more efficient heating with less penetration could be attained by changing the orientation of the beam with respect to the workpiece. Figure 6 shows the orientation of the beam used in the Inconel 718 experiments and the resulting temperature distribution on the shear plane. Figure 7 shows the improved beam orientation and the predicted temperature distribution.

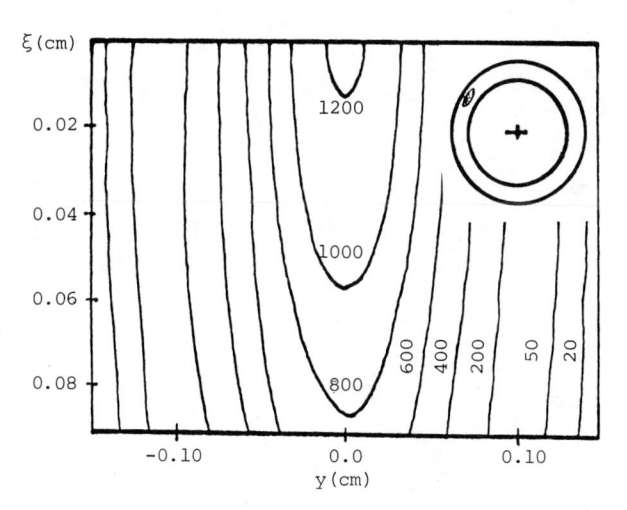

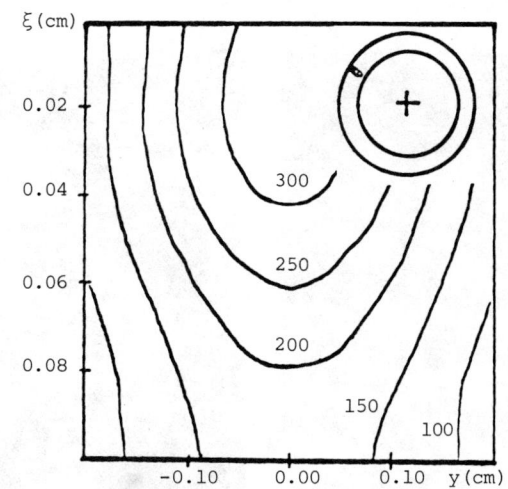

Figure 6   Temperature distribution in shear plane due to 2000 W, 0.1cm x 0.4cm elliptical beam (major axis tangent).

Rajagapal and Plankenhorn have recently reported the results of LAM of Ti-6Al-4V using the improved beam orientation in turning. They found that a two-fold increase in material removal rate was readily attainable by LAM at 12 KW without an increase in flank wear rate. For example, the feed was increased from 0.50 mm rev^{-1} to 1.00 mm rev^{-1} by laser heating for a cut taken at a speed of 0.7 m s^{-1}, and a depth of cut of 2.5 mm using a C-2 carbide insert while maintaining a 0.75 mm min^{-1} flank wear rate (9).

Figure 7   Temperature distribution in shear plane due to 2000 W, 0.1cm x 0.4cm elliptical beam (major axis radial)

An evaluation of the technological feasibility of LAM requires further data of the type obtained for the Ti-6Al-4V alloy. Also fatigue studies are needed. The economic benefits of increased material removal rate must be compared to the increased costs associated with laser heating.

LASER MACHINING

The arrangement for LM is similar to that employed in LAM except that no cutting tool is used, see Fig. 1. After reflection by the cross-slide mirrors the laser beam is focused by a ZnSe lens on the workpiece along a radial direction for turning or parallel to turning axis for facing.

Mechanism of Groove Formation

In early LM experiments, a high velocity N_2 gas jet was directed at the irradiated surface. It was observed that the substitution of O_2 for N_2 resulted in an increase in groove depth suggesting that oxidation might play an important role in the vaporization of Si_3N_4 laser machined in air (9). Subsequent experiments employing a slowly flowing cover gas have shown, however, no detectable difference in material removal rate for LM in O_2, N_2 and He. It appears that the previously observed increase was due to slight unintentional changes in nozzle orientation. It is now believed that the vaporization of grooves occurs by decomposition of the Si_3N_4 to form N_2 gas and Si liquid. The Si liquid is either expelled from the grooves as droplets or remains on the groove walls.

Effect of Beam Polarization

Polarization is known to affect metal cutting rates, however, the effect of polarization on groove cross-section and shape has not been investigated previously (10). Laser beams are often partially polarized due to the use of internal Brewster windows and turning mirrors.

Figure 8 shows the cross section of grooves laser machined at an incident power of 560W and a translation speed of 5 cm s^{-1}. The angle θ is the angle

θ = 0°   θ = 42°   θ = 72°   θ = 90°

.1mm   V_s

Material: NC-132 Si$_3$N$_4$
Power: 560 Watts CW
Translation Speed: 5cm/sec

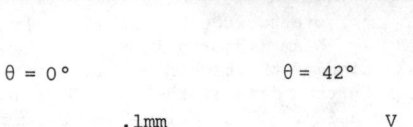

Figure 8 Shape of groove cross section as a function of θ.

between the electric vector of the polarized beam and the direction of translation of the specimen. When θ = 0°, the groove is straight, narrow and the deepest observed at any angle. When θ = 90°, the groove is also straight but is wider and shallower than the θ = 0° groove. For angles between 0° and 90°, the groove is curved.

The mechanism, which is believed to explain the polarization induced curvature, is illustrated in Fig. 9. The diagram shows the angular relationship between the translation direction and the electric vector of the incident beam. Also shown are two vectors designating the TM and TE reflections. The TM reflection is produced when the incident beam's electric vector is parallel to the plane defined by the incident and reflected beams; the TE, when it is perpendicular. For steady-state cutting represented in Fig. 9, the focused beam will interact with the slanted wall of the groove and be reflected. Although the whole front surface reflects the light, for simplicity only the two reflections mentioned previously are shown, and because the velocity vector is oriented 45° from the electric vector, the direction of the reflected beam is symmetric around the velocity vector as shown in Fig. 9. Although the direction of reflections is symmetric the magnitude of the TE and TM reflections are not equal. Fresnel's law predicts, for any angle of incidence larger than 0°, the reflectivity for the TE reflection will be the largest. For large angles of incidence, e.g. 80°, the reflectivity for the TM ray is 0.65 and that for the TE ray is 0.97. This larger magnitude of the TE reflection is represented schematically by the longer length of the TE vector in Fig. 9. After the initial reflections both the TE and TM rays are directed down and across to the opposing wall where the energy will be absorbed. But since the intensity of the light reflected as TE rays is larger than that reflected as TM rays more energy will be deposited where the TM rays are reflected and therefore more material removed in this area. The uneven reflection and subsequent non-symmetric energy deposition creates the curved cross section.

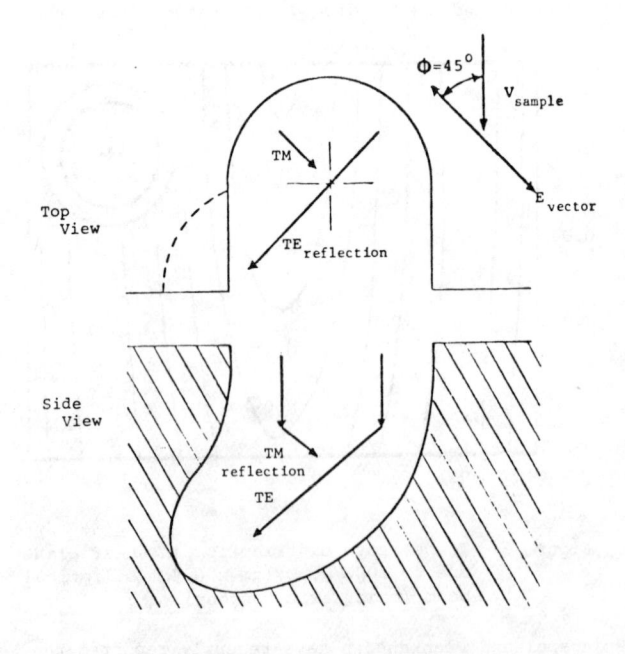

Figure 9 Effect of polarization on groove shape.

| 10.6 | 24.3 | 71.7 | 125.2 |

Velocity (cm/sec)

.1 mm

Material: NC-132 Si_3N_4
Power: 560 Watts CW
Gas Environment: O_2

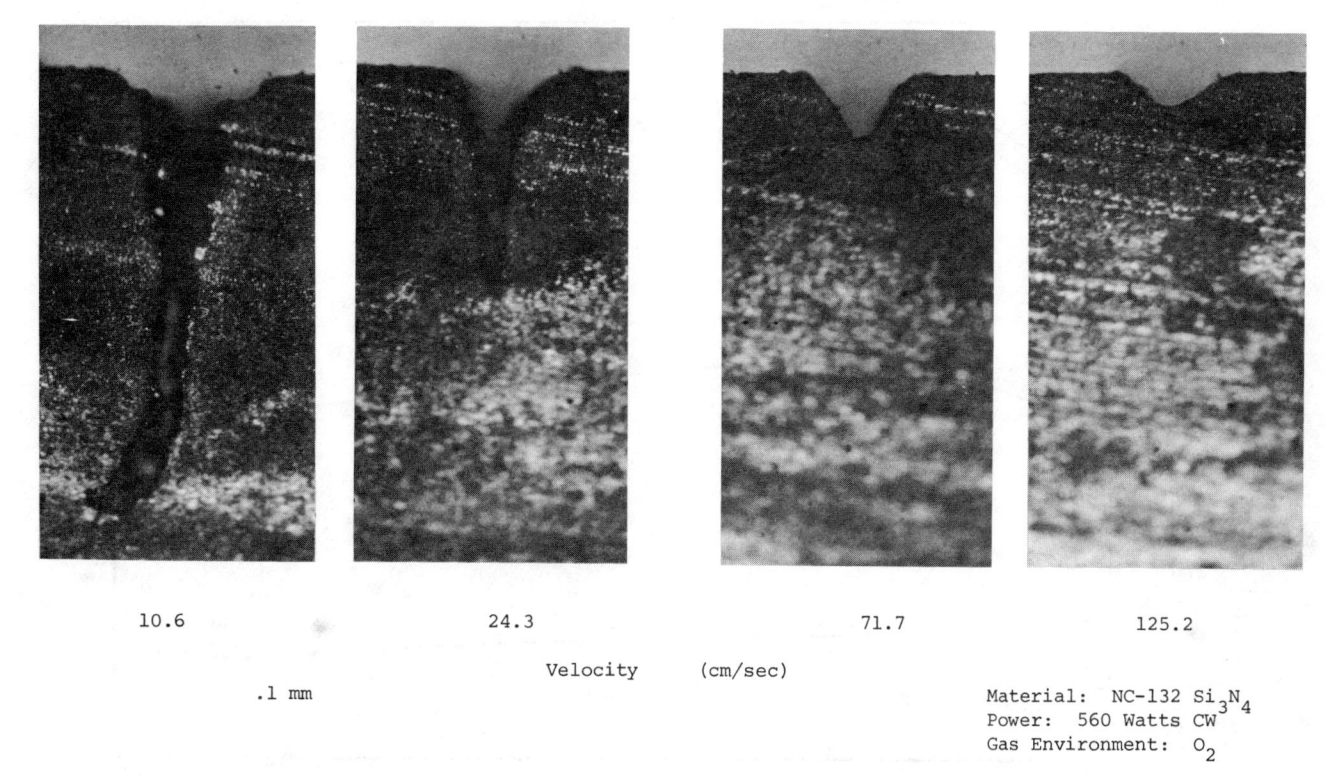

Figure 10 Cross section of laser machined grooves versus velocity at θ=0°

| 9.8 | 23.0 | 69.7 | 125.2 |

Velocity (cm/sec)

.1mm

Material: NC-132 Si_3N_4
Power: 560 Watts CW
Gas Environment: O_2

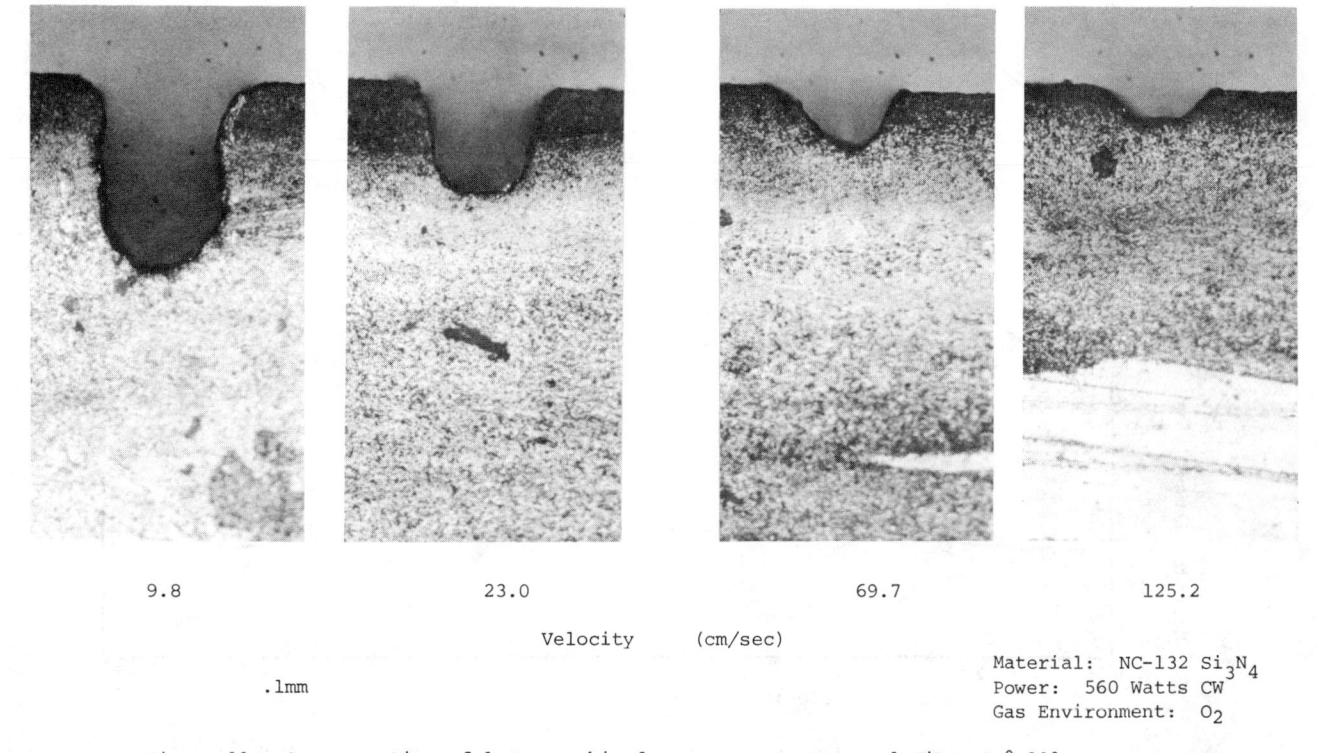

Figure 11 Cross section of laser machined grooves versus velocity at θ=90°

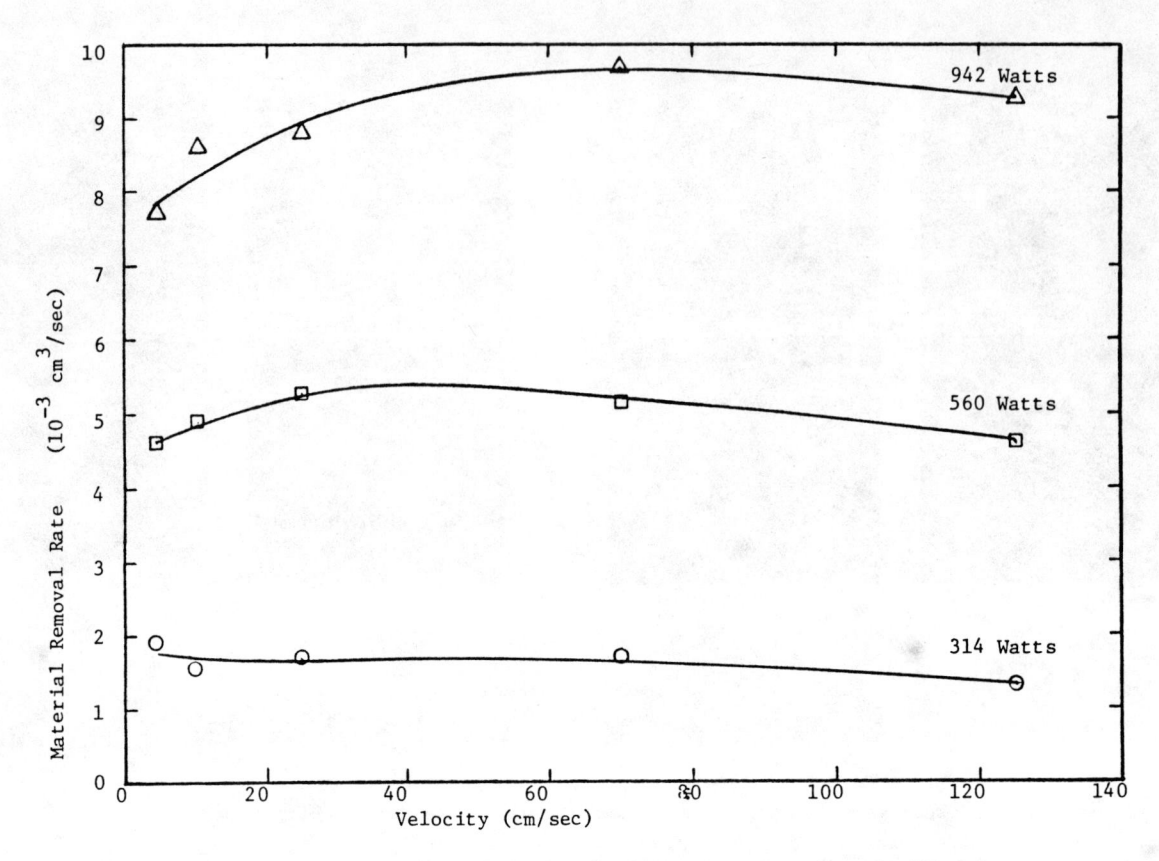

Figure 12 Material removal rate versus speed for θ=0° grooves at various incident beam powers.

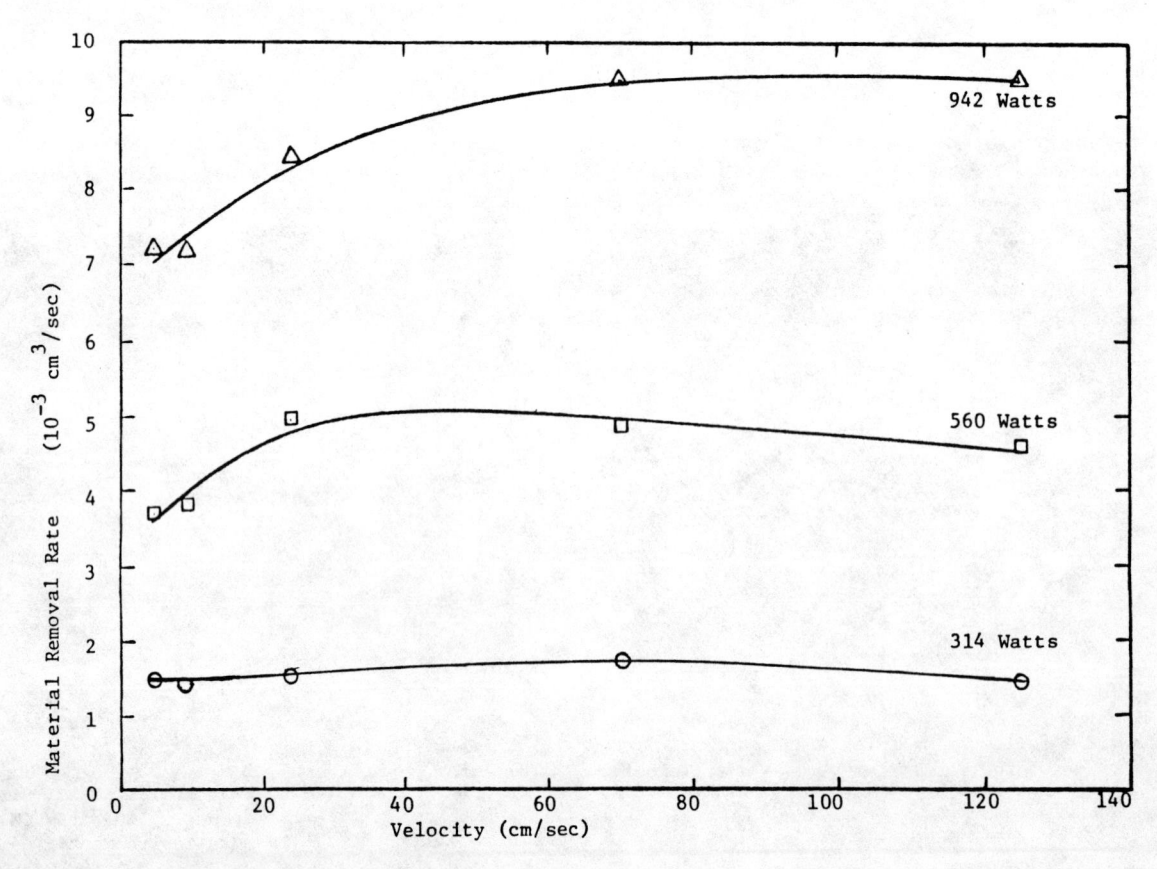

Figure 13 Material removal rate versus speed for θ=90° grooves at various incident beam powers.

Effect of Speed and Power

Figures 10 and 11 show the effect of varying scan speed on groove cross-section for 560 W incident beam power and for $\theta = 0°$ and $\theta = 90°$, respectively. At high speeds, the $\theta = 0°$ and $\theta = 90°$ groove cross-sections are similar although the $\theta = 0°$ groove has a more angular shape. By overlapping such grooves it is possible to remove layers of material as in turning on a lathe.

Figures 12 and 13 show material removal rate, which is the product of the cross-sectional area of the groove and the beam speed, as a function of beam speed for various incident beam powers, for the $\theta = 0°$ and $\theta = 90°$ orientations, respectively. Two trends should be noted: (1) at high speeds, the material removal rate decreases with increasing beam speed; (2) at low speeds, the material removal rate decreases with decreasing speed, particularly at high powers. The first trend can be explained on the basis of the total energy deposited per unit area, which decreases with beam speed. If the sample is translated fast enough, the irradiated volume near the surface will not absorb enough energy to raise its temperature to the point where the vaporization reaction becomes rapid enough for appreciable material to be removed. The second trend is thought to result from two other factors. The first factor is the blocking of the incoming beam by ejecta. As the groove becomes deeper at lower velocities, silicon droplets formed near the bottom of the groove are blown up into the incoming beam preventing the beam's energy from reaching the bottom of the groove. The deeper the groove, the longer will be the path that the incoming beam must penetrate. Consequently, less energy is absorbed by the sample resulting in a lower material removal rate. The second factor is conductive loss. As already discussed, the removal of material requires that the temperature be raised to a critical value. If the sample is translated slowly, there is time for energy to be conducted away from the volume being irradiated and because of this loss, more incident energy is needed to remove the same amount of material.

Effect of Feed

Figure 14 shows the arthmetical mean roughness and material removal rate versus feed calculated by overlapping the cross-sections of single grooves. The material removal rate increases with increasing feed approaching the single groove value because as the feed is increased less groove overlap occurs. The roughness increases because the spacing of groove bottoms becomes greater with increasing feed resulting in a more pronounced scalloping effect.

The observed behavior was not so straightforward and it was found that light guiding effects created two types of surfaces. The first was characterized by an increase in roughness with decreasing feed as predicted by the simple overlap theory. The second is characterized by deep initial grooves followed by a region of grooves of moderate depth and then a region where the grooves become shallow as the end of the cut is approached. Both these effects are illustrated in Fig. 15, which shows multiple pass groove cross-sections for a series of different feeds and an incident power of 560 W. The grooves, which were formed at a low speed (9.1 cm s^{-1}) and the $\phi = 0°$ orientation, are similar to those seen at higher velocities and also in the $\phi = 90°$ orientation.

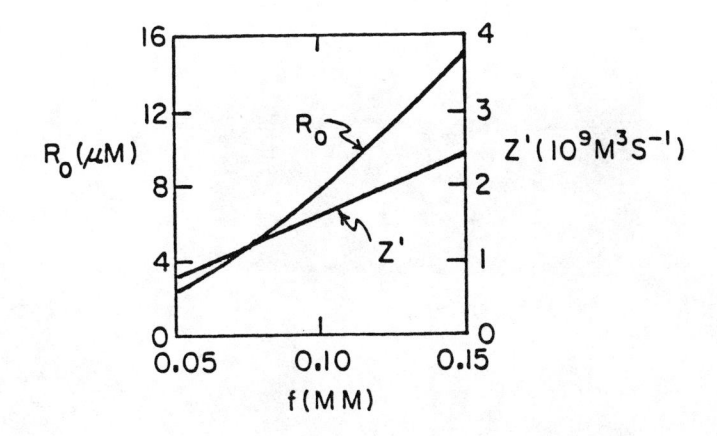

Figure 14 Roughness and effective material removal rate versus feed.

Figure 15a shows the bottom surface of a multiple pass, laser machined sample, machined at a feed of .022 cm. At this feed, a scalloped bottom surface as would be expected from the repetition of the single pass groove was observed. As the feed was reduced to .0178 cm, Fig. 15b, the scallops seen in Fig. 15a were almost completely removed and the surface became very smooth. Further reduction of the feed would be expected to further increase the degree of smoothness, however, this was not observed. Once the feed becomes significantly smaller than the width of a single pass groove the light guiding effect becomes important. Figure 15c shows an example of increasing surface roughness with decreasing feed. A feed reduction from .0178 cm in Fig. 15b is .015 cm in Fig. 15c, creates a large increase in roughness.

In Fig. 15c, the first laser machined groove is on the left. Only a small ridge from the first pass remains. During the second pass, the groove spacing is sufficiently small to cause a large portion of the laser light to be guided to the left by the walls of the first groove. During the third pass the effective groove spacing has been increased because the beam was guided to the left during the second pass. Thus the third pass groove is not guided by the previously machined surface and is similar to a single pass groove. The fourth pass is then guided similarly to the second pass and the process repeats giving the highly contoured profile cross-section shown in Fig. 15c. It should be noted that this gives a repetitive groove shape the spacing of which is not equal to the feed.

The second type of multiple pass surface obtained with a feed of .010 cm varied in depth as shown in Fig. 15d. This surface is typical of that produced by laser machining where the feed is reduced to the point that every groove after the first is always guided to the left by the previous grooves. This creates a depression of the type shown in Fig. 15d. For very small feeds, the depression can become very deep relative to a single groove and also can undercut the wall formed by the first pass. Such undercutting is only slightly evident in Fig. 15d. As one moves to the right from the initial depression, there is a region in which the depth remains constant. From this region the bottom surface gradually rises. This gradual rise occurs over a distance corresponding to many passes.

a (f=.022cm)

b (f=.0178cm)

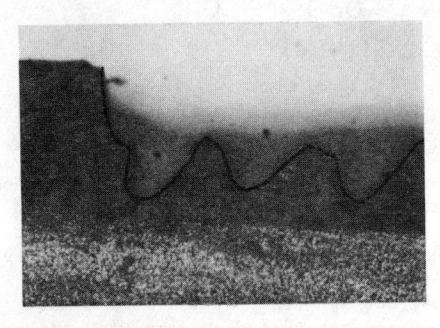

c (f=.015cm)

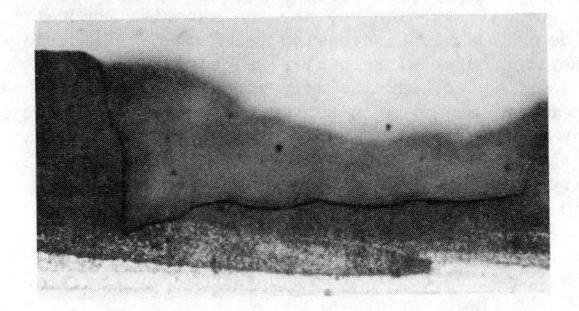

d (f=.010cm)

Material: NC-132
Power: 560 Watts CW
Velocity: 9.1 cm/s
O=90

Figure 15   Cross-section of overlapped grooves.

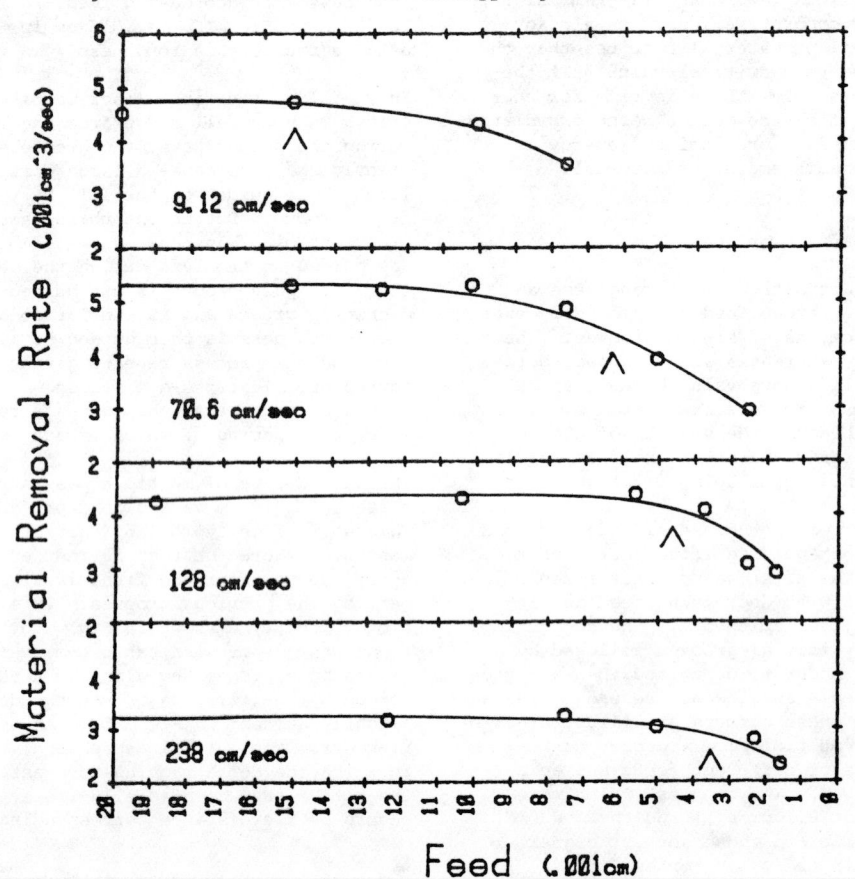

Figure 16   Material removal rate versus feed for various velocities and an incident power of 560 W.

The removal rate was previously calculated for single pass grooves. This was done by multiplying the area of the single groove, as measured from a photomicrograph with polar planimeter, by the sample translation speed. For multiple overlapping grooves, the average material removal rate equals the total area of material removed times the sample translation speed divided by the number of passes. The results are shown in Fig. 16, where the material removal rate is plotted for different feeds and velocities and an incident power of 560 W for the $\phi = 0°$ orientation. Both $\phi = 0°$ and $\phi = 90°$ orientations gave similar results.

In Fig. 16, one can see that laser machining of Si_3N_4 produces a constant material removal rate over a wide range of feeds. It was found that this constant rate was equal to the material removal rate for a single groove for the same power, speed and beam diameter. For each curve corresponding to a different velocity, there is a point beyond which the material removal rate begins to decrease with decreasing feed. The feed corresponding to the onset of this decrease decreases with increasing velocity. For the 9.12 cm/sec velocity curve, the onset is 11×10^{-3} cm. This value is reduced to approximately 3×10^{-3} cm for the 238 cm/sec translation speed curve. This decrease in material removal rate is not related to the decrease predicted by the machining analysis previously proposed which assumed the shape of a single pass groove would be repeated during multiple overlapping of light guiding. Because of light guiding by the already machined walls, the incident energy is deposited over a larger area at small feeds than at large feeds and, therefore, is not as effective in removing material.

Flexural Strength

Specimens with various laser machined surfaces were tested in 4-point bending. The test matrix consisted of seven sets of samples, A through G. The first set, A, was produced using only conventional diamond grinding and was used as a standard which could be compared to previously published results to verify the present experimental procedures. All sets were tested using a four point flexural strength testing fixture designed to conform to proposed military standards for structural ceramics. This included rounded steel knife edges rounded to a radius of 3.18 mm, provisions for the alignment of the bearing surfaces, and an outer and inner span of 2.54 cm and 1.27 cm, respectively. All tests were conducted at room temperature in air using a testing machine with a constant cross head speed of .0508 cm/min.

All bend specimens had the same dimension, .669 cm x .417 cm x 3.5 cm, with the long edges of the tension surface bevelled, .0794 cm at a 45° angle. The samples only had their tension surface laser machined with all other surfaces being ground employing a 520 grit resinoid bonded synthetic diamond wheel and an infeed of .000762 cm. The final grinding was done parallel to the long axis of the specimen. Because of the sensitivity of NC-132 to flaws and stress raisers, the smoothest possible laser machined surface was used. This was obtained with a scan speed of 237 cm/sec and an incident power of 560 watts. For this scan speed, a groove overlap of .00378 cm was chosen so that the laser machining would not be influenced by the light guiding effect previously discussed.

Sets B and C were laser machined using the $\phi = 90°$ orientation while sets D, E, F and G used the $\phi = 0°$ orientation. Sets C and D were longitudinally machined, i.e. their grooves were parallel to the long axis of the sample. Sets F and G were used to determine if any strength loss due to laser machining could be recovered with a finishing cut by diamond grinding. The laser machined surface of set F was diamond ground as previously described until the features of the laser machined surface became invisible to the unaided eye. Set G was machined in a similar manner but with one extra finishing cut of .0008 cm to remove microcracks which might have been created by laser machining.

A summary of the four point flexure strength measurements is presented in Table 1, which lists the number of samples tested, average bend strengths, standard deviations, and Weibull distribution characteristics.

Table 1 indicates the average strength of the laser machined sample as compared to that of set A, $694 MN/m^2$, was reduced by 30.6 to 41.9%. Sets B and D, which were laser machined in the transverse direction showed the biggest decrease. Set B ($\phi = 90°$) had an average flexure strength of $423 MN/m^2$ while set D ($\phi = 0°$) had an average strength of $403 MN/m^2$. Sets C and E were laser machined in the longitudinal direction. Set C was machined using the $\phi = 90°$ orientation while Set E had the $\phi = 0°$ orientation. Set C had a strength of $462 MN/m^2$ while Set E had a strength of $482 MN/m^2$. The closeness of these value suggests that at the speed of 237 cm/sec, used in this investigation, the resultant groove shapes of the two orientations were very similar. The average strength values of Sets F and G listed in Table 1 show that the reduction due to laser machining can be recovered by diamond grinding. Not only is the average strength increased but also the standard deviation is increased to the value observed in Set A and the previous investigators. The standard deviations of the laser machined sets in Table 1 show a marked decrease in comparison to the diamond machined sets. The previous investigations and Sets A, F and G have standard deviations of 9.4%, 8.0%, 8.0% and 10.3% of their average strengths, respectively while sets B, C, D and E have standard deviations that range from 2.7% to 3.3% with an average of 2.9%. Correspondingly the Weibull slope which is a measure of the variability of strength of the material with a larger value being associated with the least variability also varied significantly. Sets PI and A both had slopes of 12.6 while the laser machined sets B, C, D and E had values of 34.4, 33.0, 39.2 and 36.1, respectively. This along with the observed reduction in strength suggests that the laser machined samples had a narrow distribution of flaw sizes but that the flaws were larger than in the diamond ground samples.

Optical and SEM examinations were not able to reveal exact origins of fracture but did show that the fracture origins were associated with the surface and were not close to the chamfered edges. SEM photos also showed that the "wetted" silicon had very narrow cracks. Although these cracks were somewhat random in direction the majority were perpendicular to the groove direction and extended across the width of the "wetted" silicon.

The feasibility of laser machining ceramic materials by vaporization and expulsion of reaction products has been demonstrated. High material removal rates have been attained. The mechanical properties of laser machined articles are sufficiently good to encourage application of this technique.

Table 1

Four-Point Strength of NC-132 Si_3N_4

	PI*	A	B	C	D	E	F	G
				Sets				
Machining Tool								
D-Diamond (Grit)	D(320)	D(520)	L	L	L	L	L D(520)	L D(520)
L-Laser								
Machining Direction								
T-Transverse	L	L	T	L	T	L	T	T
L-Longitudinal								
O - Orientation			90	90	0	0	0	0
Number of Samples	50	10	10	10	10	10	9	9
Average Strength (MN/m^2)	632	694	423	462	403	482	643	683
Standard Deviation (MN/m^2)	59	56	12	15	11	14	52	70
Weibull Strength (MN/m^2)	658	720	430	469	408	489	668	715
Weibull Slope	12.6	12.6	34.4	33.0	39.0	36.0	12.5	9.8

*Previous Investigation

References

1. S. M. Copley and M. Bass, "Shaping Materials with a Continuous Wave Carbon Dioxide Laser," in <u>Applications of Lasers in Materials Processing</u>, E. A. Metzbower, Ed., American Society for Metals, Metals Park, Ohio (1979).

2. S. Rajagopal, V. L. Hill and D. J. Plankenhorn, "Laser Assited Machining of Inconel 718," in Annual Technical Report: Advanced Machining Research Program, D. G. Flom, Ed., DARPA Contract No. F 33615-79-C-5119, Ch. 18, August 16, 1980.

3. S. Rajagopal and D. J. Plankenhorn," Laser Assisted Machining of Ti-6Al-4V," in Annual Technical Report: Advanced Machining Research Program, D. G. Flom, Ed., DARPA Contract No. F 33615-79-C-5119, Ch. 9, August 17, 1981.

4. S. M. Copley, M. Bass and B. Jau, "Laser Assisted Machining of Inconel 718," in Annual Technical Report: Advanced Machining Research Program, D. G. Flom, Ed., DARPA Contract No. F 33615-79-C-5119, Ch. 9, August 17, 1981.

5. R. J. Wallace, M. Bass and S. M. Copley, "Effect of Beam Polarization on the Shape of Grooves in Si_3N_4 Produced by Laser Machining," submitted to J. Appl. Phys. (1983).

6. R. J. Wallace, M. Bass and S. M. Copley, "Laser Machining of Si_3N_4: I. Energetics," submitted to J. Am. Ceram. Soc. (1983).

7. R. J. Wallace, M. Bass and S. M. Copley, "Laser Machining of Si_3N_4: II. Shaping and Mechanical Properties," submitted to J. Am. Ceram. Soc. (1983).

8. H. Maruo, I. Miyamoto, T. Ishida and Y. Arata, "Investigation of Laser Hardening," J. JWS <u>50</u>, 208 (1981).

Acknowledgements

The author is pleased to acknowledge support of this research by the Defense Advance Research Projects Agency with the Air Force Wright Aeronautical Laboratories/MLTM, Air Force Systems Command, Wright-Patterson Air Force Base under Contract F33615-79-C-5119 and with the Department of the Army Defense Supply Service under Contract MDA 903-80-C-0436.

8301-009

PARAMETRIC EVALUATIONS OF LASER/CLAD INTERACTIONS FOR HARDFACING APPLICATIONS

J. I. Nurminen and J. E. Smith
Westinghouse R&D Center
Pittsburgh, PA 15235

ABSTRACT

A series of parametric studies have been conducted to characterize laser beam/clad coupling efficiencies for use in hardfacing applications. Laser beam/clad interaction efficiencies were determined for the cobalt-base Stellite 156 alloy in the form of a loose powder, a sintered preform and a fused preform. Process variables included laser beam power density, beam interaction time and shielding gas composition.

Test results show that laser beam energy absorption efficiencies range from 20 percent for fused preforms to nearly 80 percent for loose powder. Corresponding melting efficiencies ranged from 38 percent for fused preforms to nearly 60 percent for sintered preforms and loose powder. Argon gas shielding was determined to be more effective than helium for maintaining uniformity of the weld bead contour and was utilized as the shielding medium for the bulk of this study.

1.0 INTRODUCTION

Hardfacing or hardsurfacing treatments are applied to a variety of engineering alloys to reduce friction and to improve resistance to wear. The varied product applications can include items such as drill bits, feed screws and augers, chain-saw guides, bearing surfaces, steam turbine blades, power shovel guide teeth, and valve gate-discs and seat rings, to name but a few.

In principle, hardfacing involves the deposition of a hard, wear resistant metal cladding to a metal substrate in the presence of a heat source which causes metallurgical bonding (welding) of the hardfacing deposit to the substrate. The technique for hardfacing can range from metal-arc processes such as stick-electrode or submerged-arc hard-facing, to thermal-spray processes such as oxy-fuel gas or nontransferred plasma-arc processes (which generally require an additional heat source to fuse the hardfacing deposit to the substrate) or, more directly, a transferred plasma-arc process which simultaneously bonds the hardfacing deposit to the metal substrate. When the hardfacing alloy is available in wire or rod form, the metal-arc processes are the least expensive techniques for hardfacing. However, these processes result in deep penetration into the substrate causing excessive base-metal dilution of the hardfacing deposit. Moreover, many of the brittle cobalt-base hard-facing alloys are not available in wire or rod form, requiring the use of alloy powder with either thermal spray or transferred plasma-arc hardfacing techniques. Fortunately, recent technological developments have led to the production of high-power, continuous-wave, CO_2 lasers suited for industrial applications.[1] These lasers, capable of generating in excess of 20 KW of output power, are being employed in such materials processing areas as deep-penetration autogenous welding; cutting; drilling; transformation hardening; heat treating; surface alloying; cladding; and hardfacing.[2]

A unique feature of laser processing lies in the ability to produce high power densities equivalent to thermal sources of temperatures in excess of 20,000°K which facilitate melting of alloys and thermal treatments that are localized at the surface with negligible subsurface heating. The high cooling rates associated with laser processing minimize substrate dilution effects during surface alloying, cladding and hardfacing, and prevent degradation of the metallurgical properties of the substrate. Furthermore, laser processing can be accomplished in air, vacuum, or inert gas shielded environments, and does not require the use of vacuum chambers as are required for analogous electron-beam applications.

2.0 LASER OPERATING PARAMETERS

A series of parametric studies were conducted using an AVCO-HPL high-power, continuous-wave CO_2 laser capable of 20 KW of power output. This transverse flow, electron beam ionized CO_2 laser emits an invisible infrared high energy monochromatic beam of light at a characteristic wavelength of 10.6 μm. Figure 1 illustrates a schematic cut-away view of the AVCO laser power module, and Figure 2 illustrates a schematic view of the laser supply in relation to the welding and heat-treating work/modules.

The parametric studies of S-156 hardfacing alloy were performed in a heat-treating work module containing water-cooled copper reflective optics. This module was equipped with a pair of scanning optics which could be orthogonally oscillated via audio-amplifier controls to raster the profile of the impingement of the scanning beam on the work surface in either a square or rectangular configuration. The static laser beam itself was used in a defocused, soft-beam mode such that the spot diameter at the point of impingement with the work surface was approximately 0.64 cm (0.25 in.). This focal condition yielded a power distribution within the beam which was approximately Gaussian in form.

For the parametric studies, the laser beam was rastered to form a rectangle measuring either 0.89 x 2.67 cm or 0.89 x 1.65 cm with the long side of the rectangle normal to the direction of travel. These beam raster dimensions produced effective beam raster areas of 2.37 cm^2 amd 1.47 cm^2, respectively. Laser power levels utilized for the parametric studies ranged from 10 to 15 KW yielding laser power densities ranging from 5.05 KW/cm^2 to 5.45 KW/cm^2 for the two beam raster configurations.

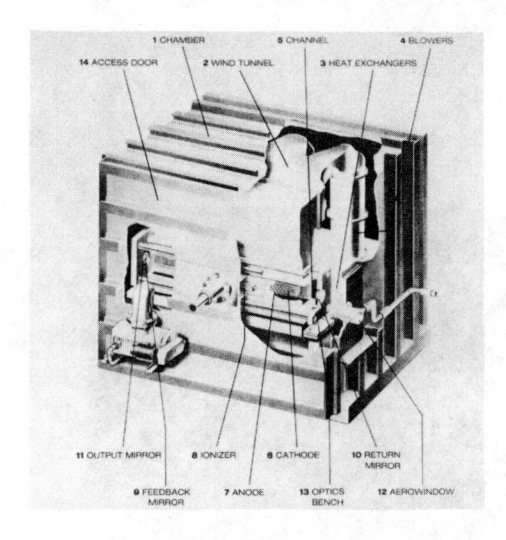

Figure 1--Schematic illustration of Avco HPL, continuous-wave CO$_2$ laser.

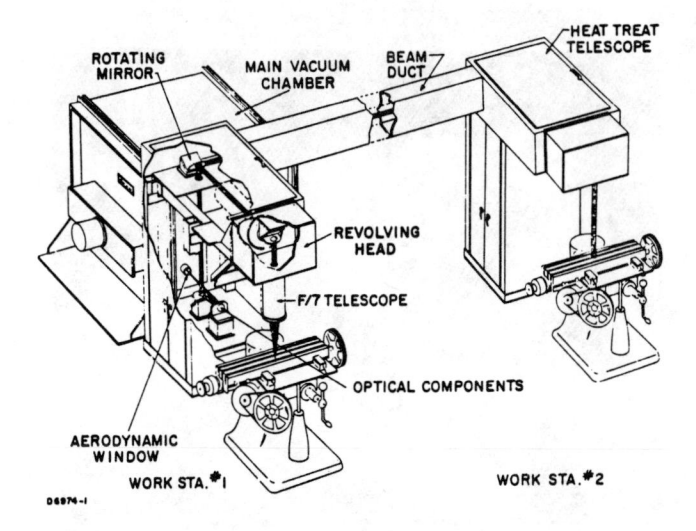

Figure 2--Schematic of Avco Laser Metalworker installed at the Avco Everett Research Laboratory. System includes two workstations, each with a separate control console. Workstations time share the laser on a first come-first served basis.

3.0 LASER PARAMETRIC STUDIES

A series of parametric studies have been conducted with Stellite 156 hardfacing alloy to characterize laser beam/clad coupling efficiencies for hardfacing applications. The purpose of these studies were twofold: first, to determine whether shielding gas species would affect the bead contour or uniformity of the hardfacing deposit; and second to determine whether coupling of the laser beam was affected by the physical nature of the preplaced Stellite 156 hardfacing alloy, i.e., loose powder, sintered power or fused preform.

Previous laser hardfacing studies had demonstrated the feasibility of laser hardfacing, but the results of these studies were very erratic, such that bead contour and uniformity would vary independently of laser power level and travel speed. Dilution levels for these hardfacing deposits were nominally in excess of 10%, and significant variation would occur along a given length of clad. A typical hardfacing overlay from these earlier studies is illustrated in Figure 3. Dilution levels corresponding to the metallographic sections cut from the same 25 cm length of hardfacing deposit vary from less than 5% to greater than 20% with notable variations in bead contour and uniformity. Similar results were obtained for other combinations of laser power, travel speed, and pre-placed powder depth which were applied to straight stainless steel bar segments and simulated stainless steel valve seat rings.

The inconsistent results of these earlier hardfacing trials led to speculation that inadequate gas shielding was somehow responsible for the variation in hardfacing uniformity. Therefore, high-speed (1000 fps), telephoto color motion picture photography was employed to study the mechanism of the hardfacing process. Notably, examination of the high-speed photography revealed that a drossy oxide film (slag layer) was forming on the surface of the molten S-156 pool during hardfacing, causing a reduction in the absorption of the laser beam. Further examination showed that thermally induced turbulence and eruptions within the molten puddle would occasionally break up the drossy film, thereby increasing the absorption of the laser beam; these variations of slag coverage and break-up corresponded with variations in hardfacing bead contour and uniformity resulting from changes in laser beam coupling efficiency during the course of a specific run. These observations confirmed the necessity for an evaluation of gas shielding practices during laser hardfacing.

3.1 Specimen Preparation

Specimens of Stellite 156 alloy (similar to those illustrated in Figure 4) were prepared for the parametric evaluations and gas shielding evaluations. Five hundred gram samples of Stellite 156 hardfacing alloy powder were placed in mild steel boats measuring 2.25 cm in depth by 3.75 cm in width by 30 cm in length. For loose powder/laser beam coupling evaluations, the powder was simply poured into the individual boats and leveled to a uniform depth. Sintered specimens were prepared by annealing the powder-filled boats in a hydrogen atmosphere at 1150°C for two hours to produce preforms of 88% theoretical density, i.e., 7.32 g/cm^3. Fused specimens of Stellite 156 were prepared by treating the powder filled boats in a hydrogen atmosphere at 1275°C for one hour. The density of the fused preforms was determined to equal 8.32 g/cm^3. Subsequent to preparation, all of the Stellite 156 specimens were stored in an argon atmosphere with dessicant to avoid moisture contamination prior to laser processing.

3.2 Shielding Gas Evaluations

Sintered specimens of Stellite 156 alloy were used to evaluate the possible effects of shielding gas species in relation to laser beam coupling efficiency. A stationary gas shielding hood (Figure 5) equipped with an open laser beam window and both off-axis and longitudinal gas jets was fabricated for use in these

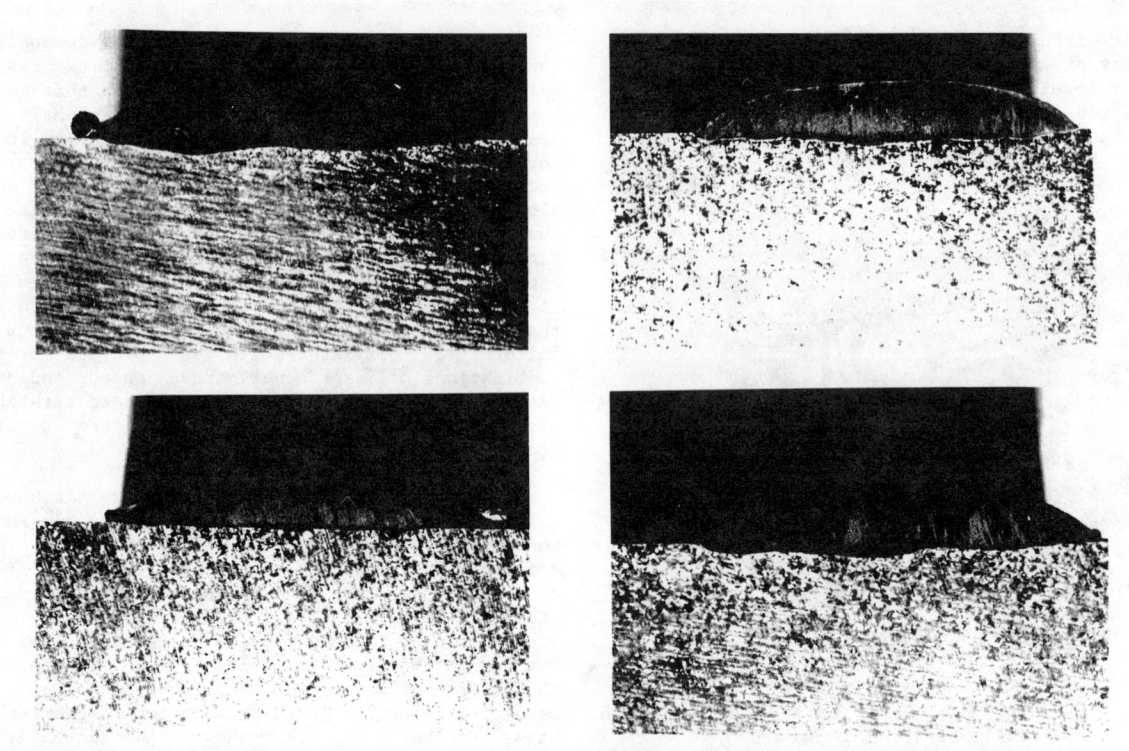

Figure 3--Metallographic sections of S-156/316SS hardfacing deposit. Segments illustrated were cut from same length of hardfacing deposit. 18 KW, 14.2 KJ/cm^2. Magnification - 4X.

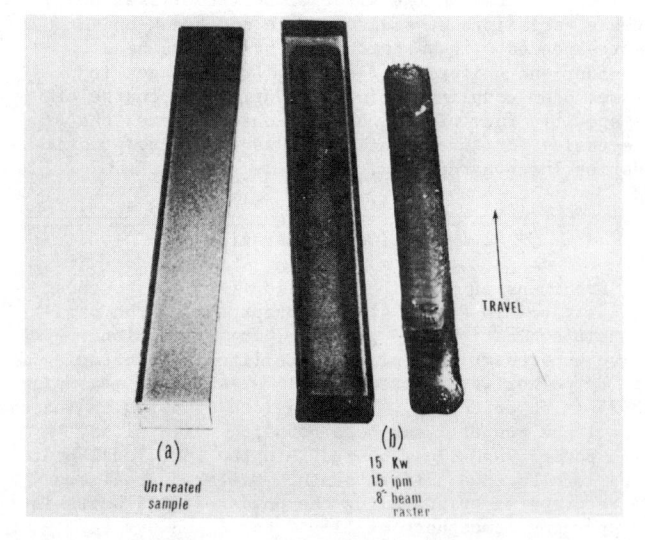

Figure 4--Sintered S-156 boat samples utilized for laser parametric evaluations. (a) Sintered specimen; (b) Laser processed specimen; 15 KW, 5.05 KW/cm^2; fusion envelope removed for illustration.

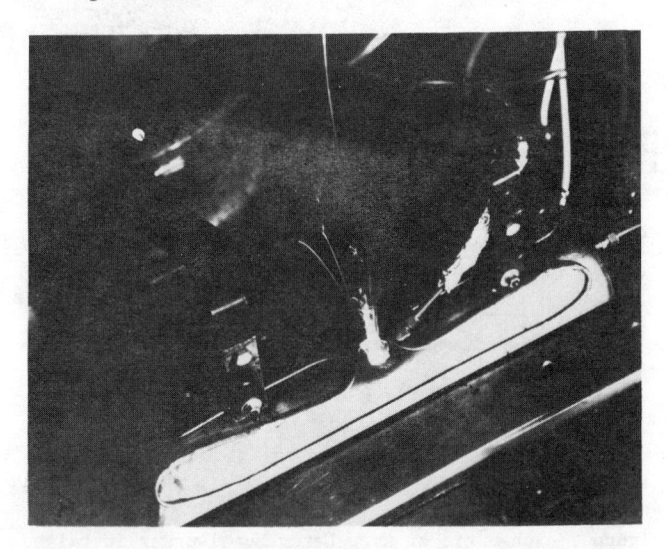

Figure 5--Shielding hood configuration utilized for laser parametric studies. Laser beam enters hood from above through window provided at center.

evaluations. In practice, the sintered boat specimen was placed upon an aluminum base plate mounted on a motor driven milling table and raised into position within the gas shielding hood by translating the milling table platform upward until contact was made between the aluminum base plate and the periphery of the shielding hood. Gas flow was initiated both in the off-axis and longitudinal gas jets and the shielding hood was purged for 5 minutes (75 volume exchanges) prior to each run. Gas flow rates of 30 CFH for the off-axis jet and 60 CFH for the longitudinal jet were determined to be the highest rates compatible with these trials.

After purging, the laser beam was initiated and ramped to full power and the milling table was remotely actuated to translate the sintered boat specimen beneath the laser beam at the preset travel speed. Figure 5 illustrates the shielding hood/milling table platform at the beginning of a laser parametric run. In this Figure, the laser beam is directed downward to impinge upon the sintered boat specimen located beneath the laser beam window at the center of the shielding hood; the milling table is simultaneously translating from the left toward the right. Subsequent to each run, gas purging was continued for an additional five minutes prior to removing the specimen from the shielding hood. The fused sample generated by each run was similar to that shown to the right of Figure 4.

TABLE 1
LASER OPERATING PARAMETERS
BEAM COUPLING CHARACTERISTICS VS SHIELDING GAS

Run No.	Material	Laser Power (KW)	Beam Raster Dimensions (cm)	Travel Speed (cm/sec)	Power Density* (W/cm²)	Energy/Unit* Length (J/cm)	Shielding Gas	Gas Flow Rate (CFH)	Dwell Time At Start-Up (sec)
1	Stellite 156 (Sintered)	15	0.89 x 2.67	0.64	5.05 x 10³	18.898 x 10³	He	30 Off-Axis/ 60 Longitud.	0
2	"	15	"	"	"	"	He	40 Off-Axis/ 60 Longitud.	0
3	"	15	"	"	"	"	He	50 Off-Axis/ 60 Longitud.	0
4	Stellite 156 (Sintered)	15	0.89 x 2.66	0.64	5.05 x 10³	18.898 x 10³	Ar	30 Off-Axis/ 60 Longitud.	0
5	"	15	"	"	"	"	Ar	"	1
6	"	15	"	"	"	"	Ar	"	2
7	"	15	"	"	"	"	Ar	"	4
8	Stellite 156 (Sintered)	15	0.89 x 2.67	0.51	5.05 x 10³	23.622 x 10³	Ar	30 Off-Axis/ 60 Longitud.	0
9	"	15	"	"	"	"	Ar	"	1
10	"	15	"	"	"	"	Ar	"	4
11**	"	15	"	"	"	"	Ar	"	2
12**	"	15	"	"	"	"	Ar	"	0

*Corrected for mirror absorption power losses.
**Direction of translation of boat specimen reversed to run counter-current to direction of shielding gas flow, i.e., left to right, Figure 5.

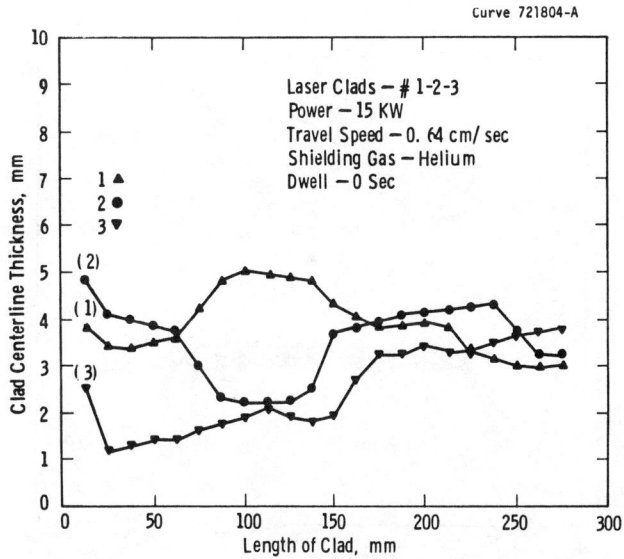

Figure 6--Thickness versus length of clad profiles, sintered S-156 alloy, helium gas shielding evaluations. 5.05 KW/cm².

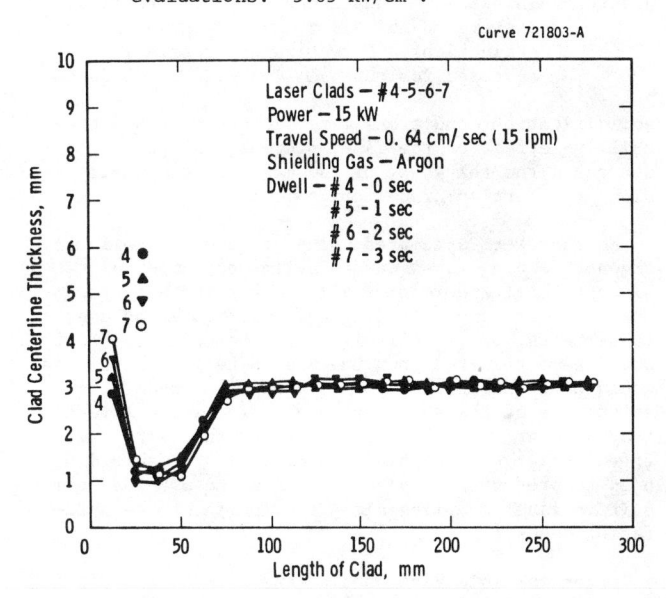

Figure 7--Thickness versus length of clad, sintered S-156 alloy, argon gas shielding evaluations. 5.05 KW/cm².

The fusion envelope would conveniently separate from the sintered bed following solidification, thus enabling a simplified measurement of the thickness profile and uniformity of the specimen from beginning to end.

A series of 12 runs were conducted to determine the effect of gas shielding species with regard to bead contour and uniformity of the hardfacing alloy deposit. Operating parameters for these runs are summarized in Table 1. Subsequent to each run, the thickness of the hardfacing deposit was determined as a function of the length of clad. Results of these thickness measurements are plotted in Figures 6, 7, and 8.

As evidenced by Figure 6, results of the trials using helium gas shielding were erratic and non-reproducible. Considerable sparking occurred through the laser window during each of these runs, and the resulting welds exhibited excessive oxidation and significant thickness variation. Off-axis gas flow rates were increased from 30 to 40 CFH for Run #2, and 40 to 50 CFH for Run #3, with no apparent improvement in coupling uniformity. Additionally, the higher off-axis jet flow rates caused excessive roughening (ripple-patterns) of the solidified weld surface.

Results for the trails using argon shielding gas were significantly improved. The hardfacing alloy deposits were uniform in thickness to within ±3% along their length (with the exception of a thin cusp which occurred within the first 5 cm of each run) and were reproducible in thickness to within ±2% from run-to-run (Figure 7). In an attempt to eliminate the thin cusp, the laser beam was permitted to dwell for increasing periods of time after ramping up to full power prior to actuating the milling table drive mechanism. However, as shown by the longitudinal thickness profiles in Figure 7, this dwell period merely thickened the starting nugget and the thin cusp uniformly repeated from run-to-run. For Runs 8-10 (Table 1; Figure 8) the travel speed was reduced from 0.64 cm/sec (15 ipm) to 0.51 cm/sec (12 ipm) to increase the energy input per unit length of weld from 18.9 KJ/cm to 23.6 KJ/cm. This increased the nominal thickness of the hardfacing deposit from 3.0 mm (Runs 4-7, Figure 7) to 3.4 mm (Runs 8-10, Figure 8), but the thin cusp continued to form within the first 5 cm of each run. The thin cusp was finally eliminated in Runs 11 and 12 by changing the direction of translation of the boat specimen from one co-current (right-to-left; Figure 5) to one counter-current (left-to-right; Figure 5) to the direction of shielding gas flow. These specimens were uniform in thickness along their entire length, and approximately 0.2 mm thicker than those generated in Runs 8-10. Apparently the off-axis shielding gas impingement on the molten puddle when translated co-current to the direction of gas flow exhibited an initial chilling effect which caused the formation of the thin cusp until steady-state

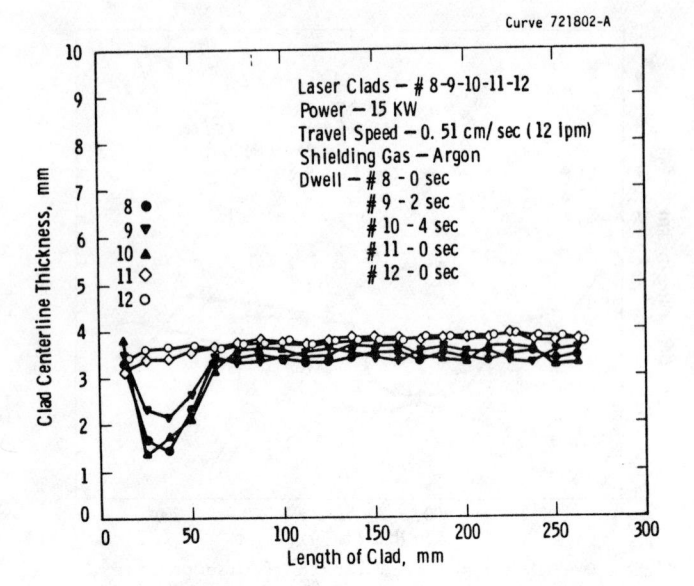

Figure 8--Thickness versus length of clad profiles, sintered S-156 alloy, argon gas shielding evaluations. Direction of translation reversed for runs #11 and #12. 5.05 KW/cm².

solidification conditions were achieved. Also, the chilling effect of the gas impingement slightly decreased the thickness of the hardfacing deposit along the portion of uniform thickness.

No runs were attempted using mixtures of helium and argon shielding gases due to a limited number of sintered boat specimens available for these tests. However, with the shielding hood configuration used in these tests, it is unlikely that a beneficial effect would have resulted from the use of helium as this gas would have vented through the open laser beam window at the top of the shielding hood. The argon shielding gas, however, with a density greater than air, layered effectively over the boat specimens in the shielding hood to prevent excessive oxidation and provided for uniform laser coupling along the length of the hardfacing deposit.

Argon gas shielding should generally be effective for most CW-CO_2 laser metalworking applications which require power densities in the range of 10^3-10^7 W/cm², i.e., transformation hardening, surface alloying, cladding, hardfacing, welding and cutting. However, applications which require power densities in excess of 10^8 W/cm² (deep penetration welding, laser glazing, drilling and shock-hardening) should employ helium gas shielding, since argon becomes subject to ionization which generates an opaque plasma that absorbs the incident laser irradiance and prevents the laser energy from reaching the workpiece.[3]

3.3 Laser Beam/Stellite 156 Coupling Efficiency Evaluations

A series of parametric studies were conducted to determine whether coupling of the laser beam was effected by the physical nature of the preplaced S-156 hardfacing alloy, i.e., loose powder, sintered preform, or fused preform. A series of 21 runs were conducted, including various combinations of material condition, laser power, beam raster configuration, power density and travel speed. The parametric data are summarized in Table 2.

Following laser processing, the specimens were sectioned and metallographically examined to determine the thickness and cross-sectional area of the resulting fusion envelope. Area measurements were determined from polar-planimeter measurements of photographic enlargements of the fusion envelope cross-sections, and these values were subsequently divided by the square of the magnification to obtain the true cross-sectional area. Unit weights of the fusion envelope segments could not be determined by direct measurement, as a thin band of partially fused material would adhere to the periphery of each segment. These weights were determined by multiplying the measured fusion zone cross-sectional area of each segment by the density (ρ = 8.32 gm/cm³). Results of the metallographic measurements and unit weight calculations are summarized in Table 2. Figures 9, 10, and 11 illustrate cross-sectional views of the fusion envelopes resulting from run numbers S-1, S-6 and S-13, respectively.

Only three runs were recorded for the loose powder boat specimens, as the slower travel speeds were the only operating parameters which would produce fusion envelopes of uniform cross-section. At speeds greater than 0.64 cm/sec (15 ipm) surface tension forces opposing the volumetric contraction due to consolidation of the low density loose powder bed would produce "nugget-chains" which would repetitively neck down to resemble a row of welded nuggets.

To evaluate laser coupling characteristics of fused preforms of S-156, machined strips approximately 3 mm in thickness were tack-welded to backing bars of 316 SS for translation under the laser beam. The intense reflection of the laser beam from the bright machined surface of the Stellite strip caused the plexiglass panel comprising the ceiling of the work-station enclosure to burst into flame, and run FS-1 was aborted. Run FS-2 was run to completion with the reflected energy diverted to a water cooled calorimeter, but run FS-3, which was to be a duplication of run FS-1, once again caused a fire in the plexiglass enclosure and this run was also aborted. At this point, parametric studies of fused S-156 preforms were discontinued. Specimen FS-2 and a metallographic section of the fused S-156 strip are illustrated in Figure 12.

By examination of the data in Table 2, it is apparent that the fusion zone thickness and cross-sectional area values decrease with increasing travel speed for both the sintered S-156 specimens and the loose powder specimens. The thickness and area versus travel speed data for the sintered S-156 specimens were mathematically evaluated using computer least squares analyses, and both the thickness and area were found to vary inversely as hyperbolic functions of travel speed. The data correlated to curves of the form

$$y = \frac{1}{AX + B} \quad (1)$$

where A and B are constants unique to each condition of laser power level and beam raster configuration. The pertinent equations and least-squares curve are illustrated in Figure 13.

The sintered S-156 thickness and cross-sectional area data for the 10 KW and 15 KW runs were normalized for comparison by plotting this data versus the laser energy available per unit length of weld. This unit energy term, $\bar{H}$ (J/cm), was calculated by dividing the laser power, P (watts), delivered to the workpiece by the travel speed, S (cm/sec), and these values are

TABLE 2

LASER PARAMETRIC STUDIES

STELLITE 156 ALLOY

Run No.	Material Condition	Laser Power (KW)	Beam Raster Dimensions (cm)	Power* Density (W/cm^2)	Travel Speed (cm/sec)	Fusion Zone Thickness (cm)	Fusion Zone Area (cm^2)	Unit Weight (g/cm)	Energy Per Unit* Length of Weld (J/cm)
S-1	Stellite 156	15	0.89 x 2.67	5.05 x 10^3	0.38	0.437	0.923	7.72	31579
S-2	(Sintered	15	"	"	0.42	0.404	0.854	7.11	28369
S-3	7.32 g/cm^3)	15	"	"	0.51	0.361	0.723	5.69	23622
S-4		15	"	"	0.63	0.318	0.615	5.12	18898
S-5		15	"	"	0.76	0.277	0.532	4.43	15748
S-6		15	"	"	0.89	0.231	0.479	3.99	13498
S-7		15	"	"	1.06	0.162	0.400	3.33	11342
S-8		15	"	"	1.27	0.137	0.338	2.81	9449
S-9		15	"	"	1.48	0.121	0.290	2.41	8097
S-10		15	"	"	1.69	0.108	0.258	2.15	7088
S-11		10	0.89 x 1.65	5.45 x 10^3	0.42	0.549	0.762	6.34	18912
S-12		10	"	"	0.63	0.439	0.438	3.65	12598
S-13		10	"	"	0.85	0.328	0.379	3.08	9445
S-14		10	"	"	1.27	0.216	0.253	2.05	6299
S-15	↓	10	"	"	1.69	0.158	0.186	1.51	4725
LP-1	Stellite 156	15	0.89 x 2.67	5.05 x 10^3	0.21	0.846	1.880	15.64	56604
LP-2	(Loose Powder	15	"	"	0.42	0.733	1.280	10.66	28369
LP-3	4.45 g/cm^3)	15	"	"	0.64	0.660	0.942	7.84	18898
FS-1	Stellite 156	15	0.89 x 2.67	5.05 x 10^3	0.21	Run aborted due to fire in work station.			
FS-2	(Fused Preform	15	"	"	0.25	0.150	0.389	3.24	47244
FS-3	8.32 g/cm^3)	15	"	"	0.21	Run aborted due to fire in work station.			

*Corrected for mirror power losses.

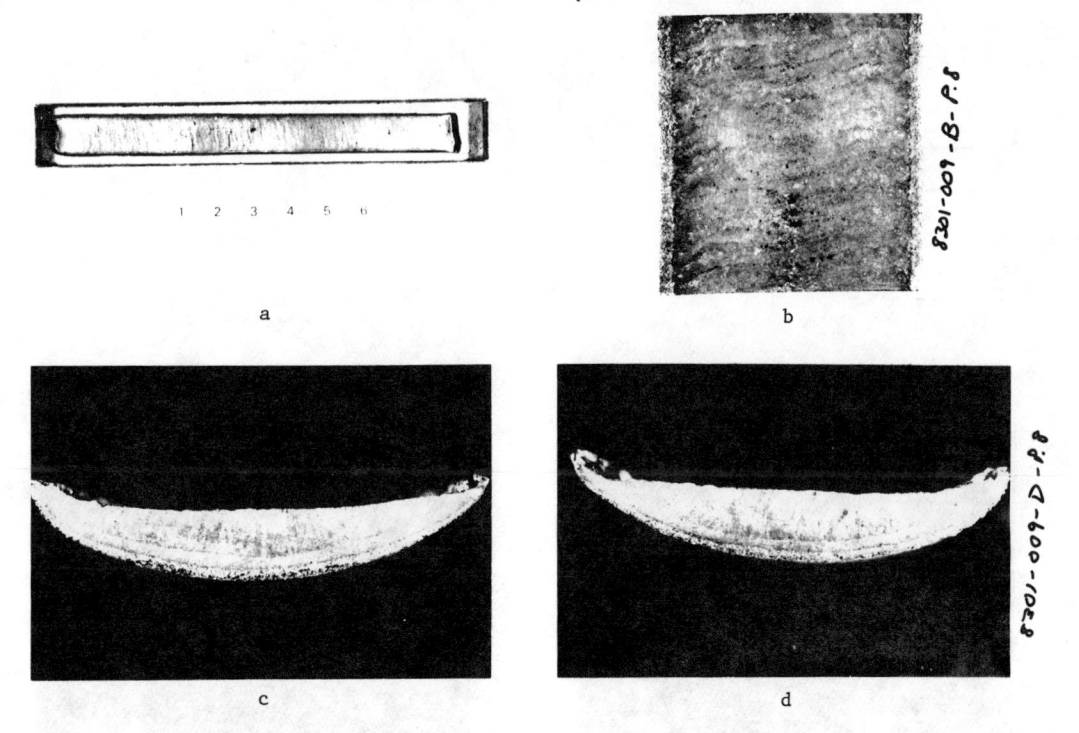

Figure 9--Fusion envelope for laser parametric run S-1, Table 2. (a) Specimen S-1; (b) Top view of segment sectioned from center of specimen; (c) and (d) Metallographic cross-sectional views of specimen S-1, 3.5X. 5.05 KW/cm^2.

listed in Table 2. The data were similarly evaluated using computer least squares analyses and found to vary as hyperbolic functions of $\overline{H}$ (J/cm) expressed as equations of the form

$$y = \frac{X}{AX + B} \quad (2)$$

Results of these calculations are illustrated in Figure 14. Notably, the cross-sectional area of fusion envelope generated by 10 KW of laser output (Curve A_{10}, Figure 14) exceeds the cross-sectional area generated by 15 KW of laser power (Curve A_{15}, Figure 14) over comparable ranges of unit energy. This result implies that the net energy efficiency to produce a fusion envelope (distinct from melting efficiency) increases with increasing power density,

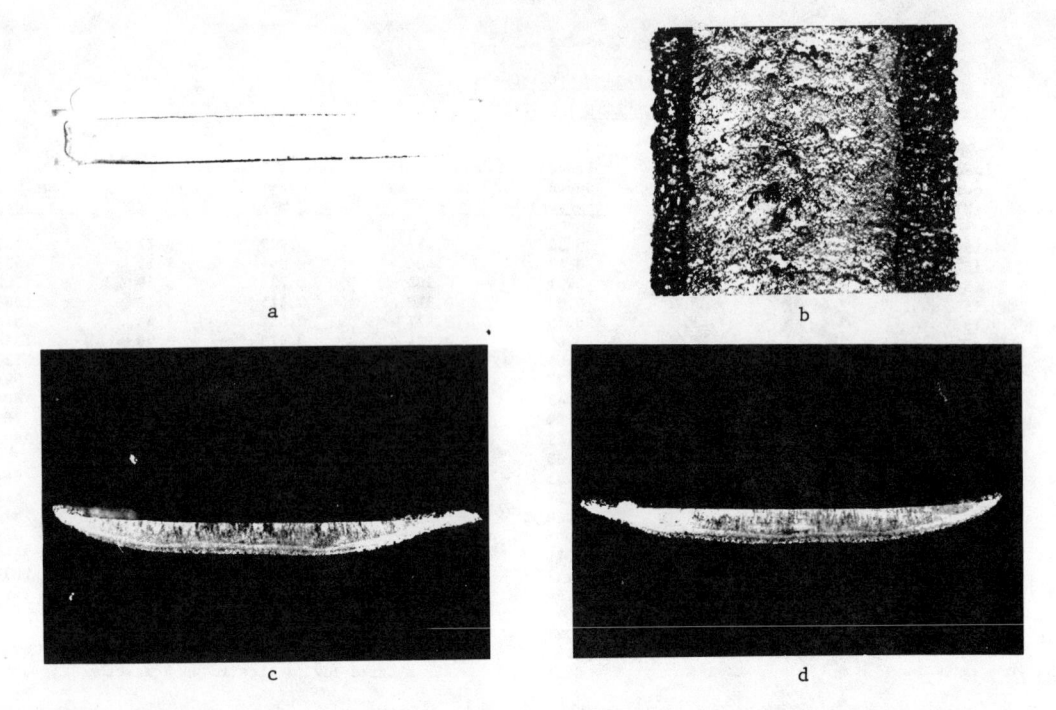

Figure 10--Fusion envelope for laser parametric run S-6, Table 2. (a) Specimen S-6; (b) Top view of segment sectioned from center of specimen; (c) and (d) Metallographic cross-sectional views of specimen S-6, 3.5X. 5.05 KW/cm^2.

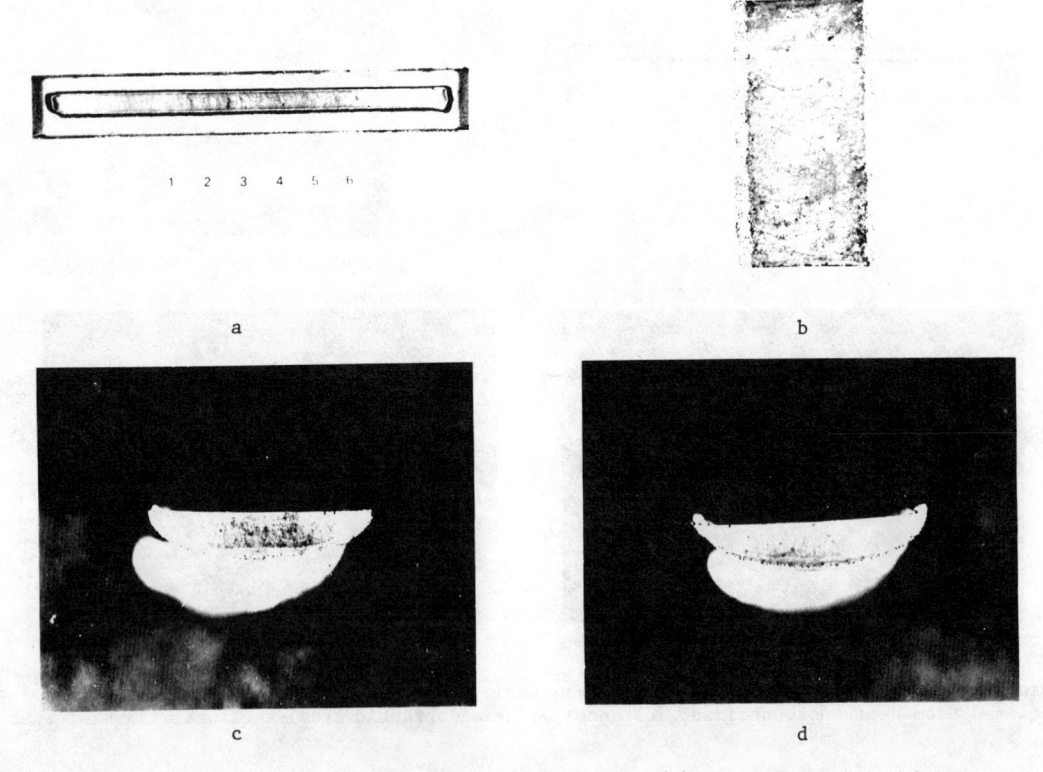

Figure 11--Fusion envelope for laser parametric run S-13, Table 2. (a) Specimen S-13; (b) Top view of segment sectioned from center of specimen; (c) and (d) Metallographic cross-sectional views of specimen S-13, 3.5X. 5.45 KW/cm^2.

i.e., 5.45 KW/cm^2 for Curve A_{10} versus 5.05 KW/cm^2 for Curve A_{15}, as will be shown in the subsequent discussion.

In order to determine the heat transfer efficiencies resulting from coupling of the laser beam to the loose powder, sintered preforms and fused preforms of S-156, a welding diagram of Log H versus Log S developed by Harth and Leslie[4] was used in conjunction with the welding heat transfer model of Christensen and Davies[5]. The Log H versus Log S diagram is based upon the fundamental welding equation

$$H = \frac{R_A Q}{S} \qquad (3)$$

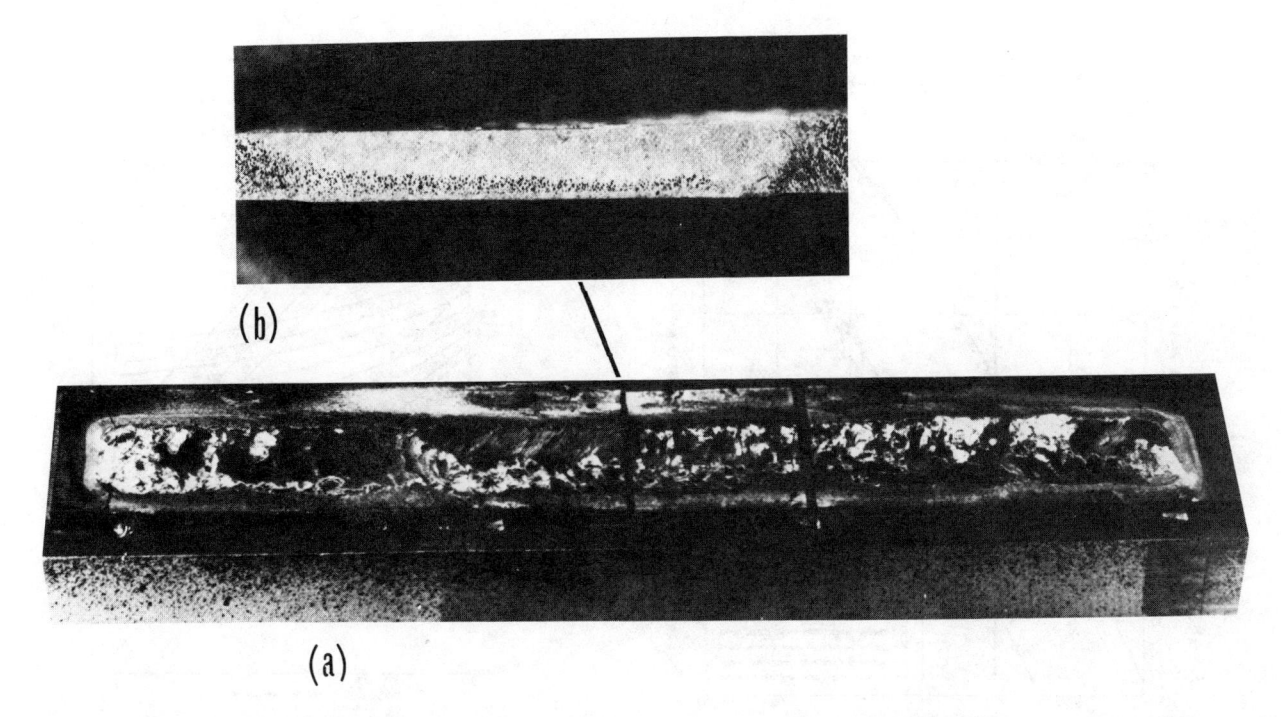

Figure 12--Laser parametric run FS-2, Table 2. (a) Fused strip attached to 316SS backing bar. (b) Metallography cross-sectional view of laser treated strip, 4.5X. 5.05 KW/cm².

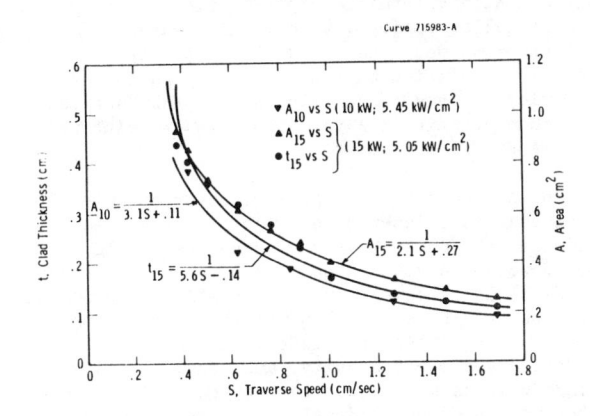

Figure 13--Thickness and cross-sectional area values of sintered S-156 fusion envelope versus travel speed.

where H is the energy absorbed per unit length of traverse, R_A is the absorption coefficient, Q is the power available in the laser beam and S is the travel speed. In logarithmic form

$$\text{Log } H = \text{Log } R_A Q - \text{Log } S \qquad (4)$$

which is the equation of a straight line on a Log H versus Log S plot. On such a plot, lines of constant $R_A Q$ appear as straight diagonal lines, and the parameters H, R_A, Q and S define a unique set of laser operating parameters which correspond to a given area of fusion envelope.

The welding heat transfer model of Christensen and Davies is based upon the original heat transfer model of Rosenthal[6] for relating welding parameters to the thermal history for bead-on-plate welds and relates to the point source theory of Carslaw and Jaeger[7] for conduction of heat in solids. In this model, Christensen and Davies developed mathematical relationships relating welding variables to such items as thermal history at a given location within a weld,

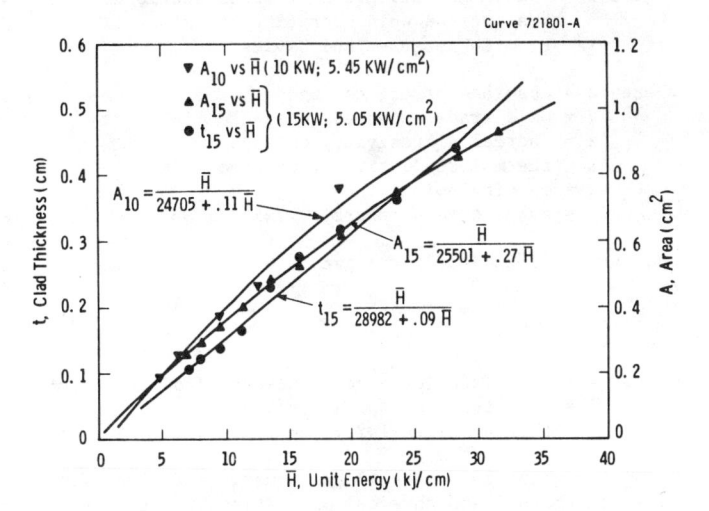

Figure 14--Thickness and cross-sectional area values of sintered S-156 fusion envelopes versus available unit energy. Curves determined by computer least squares analyses.

the cross-sectional area of molten metal in a weld, and the depth of weld penetration. These mathematical relationships, which correlated with experimental calorimetric measurements for bead-on-plate welding studies, were graphically represented as diagrams having nondimensional coordinates. The nondimensional diagram relating weld operating parameters to the fusion zone cross-sectional area is illustrated in Figure 15. In this diagram the nondimensional operating parameter, n, is defined as

$$n = \frac{qV}{4\pi\alpha K(T_c - T_o)} \qquad (5)$$

101

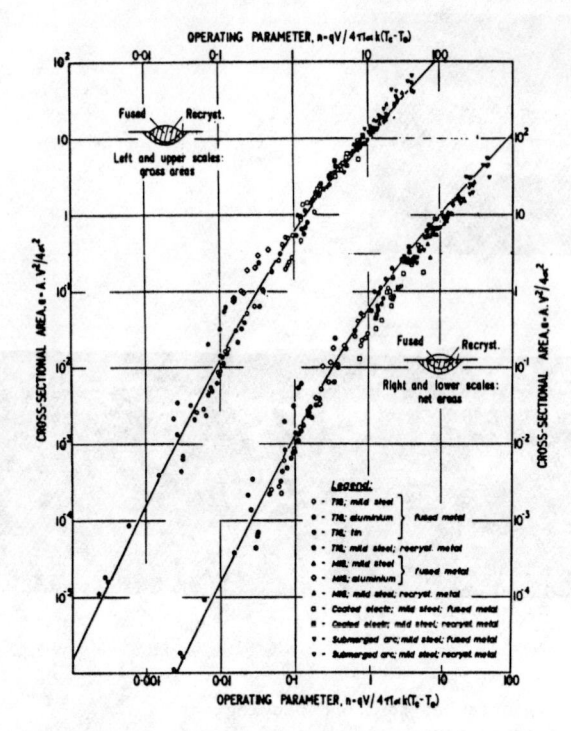

Figure 15—Nondimensional graph of cross-sectional area versus weld operating parameters. From Christensen and Davies[5].

where q = absorbed power, cal/sec
V = weld traverse speed, cm/sec
α = thermal diffusivity, cm²/sec
K = thermal conductivity, cal/sec cm°K
T_c = melting point, °K
T_o = ambient temperature, ~ 298°K

and the nondimensional area parameter, a, is defined as

$$a = \frac{AV^2}{4\alpha^2} \quad (6)$$

where A = actual fusion zone cross-sectional area, cm²
V = weld traverse speed, cm/sec
α = thermal diffusivity, cm²/sec

Using the graph in Figure 15 and Equations (5) and (6) plus the thermal and physical properties of Stellite 156 alloy (summarized in Table 3), it was possible to calculate the actual energy absorbed per unit length of fusion envelope generated for each of the parametric runs of S-156 alloy, and this data is listed in Table 4. Also, lines of constant cross-sectional area were calculated for S-156 alloy (using Figure 15 and Equations (5) and (6)) for travel speeds ranging from 0.1 cm/sec to 1.0 cm/sec. Results of these calculations plus the calculated data points corresponding to the laser parametric studies are illustrated on the Log H versus Log S diagram in Figure 16. By superimposing lines of best data fit corresponding to Eq. (4), i.e., lines of constant $R_A Q$, onto Figure 16, it is possible to determine the laser coupling efficiency, R_A, from Eq. (3) for the loose powder, sintered preforms, and fused preforms of S-156 alloy. The results of the calculations shown in Figure 16 illustrate significant differences in laser coupling efficiency. The parametric runs conducted at 15 KW of laser power and a power density of 5.05 KW/cm² demonstrate that loose powder absorbs 80% of the incident laser power, while the sintered preforms absorb 57% of the incident

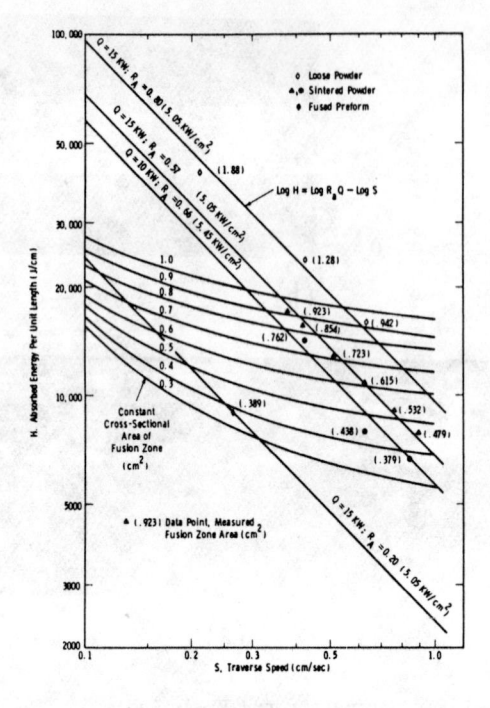

Figure 16—Log H versus Log S diagram comparing experimentally determined cross-sectional area data for S-156 alloy fusion envelopes with curves of constant cross-sectional area calculated from theoretical relationships developed by Christensen and Davies. Lines of constant $R_A Q$ approximate laser absorption efficiencies for loose powder, sintered preforms and fused preforms.

laser power and the fused preform absorbed only 20% of the available laser power. For the sintered S-156 specimens, the parametric runs conducted at 10 KW of laser power at a power density of 5.45 KW/cm² absorbed 66% of the laser power compared to 57% absorption at 15 KW and 5.05 KW/cm². This result confirms that laser coupling efficiency is strongly dependent upon incident laser power density, and suggests that optimum results may be achieved by operating at even higher values of power density and travel speed.

Laser absorption efficiencies for each of the individual parametric runs were calculated by dividing the energy absorbed per unit length of weld by the energy available per unit length of weld, and this data is listed in Table 4. The individual absorption efficiencies lie within ±3% of the average absorption efficiency values determined in Figure 16.

The cross-sectional area and thickness values for the sintered S-156 specimens were compared with calculated values of specific energy, E (KJ/cm²), and computer least-squares analyses demonstrate that the fusion zone area and thickness data also vary as hyperbolic functions of the specific energy expressed as equations similar to Eq. (2). Results of these calculations are illustrated in Figure 17.

4.0 LASER MELTING EFFICIENCY CALCULATIONS

Melting efficiency values were calculated for each of the laser parametric runs to determine what proportion of the absorbed laser energy is utilized for melting material in the fusion zone. The melting efficiency is defined as the ratio of energy required

TABLE 3

THERMAL AND PHYSICAL PROPERTIES
STELLITE 156 ALLOY

Melting Range:	1250°C-1300°C
Melting Point:	T_M = 1275°C (mid-point of melting range)
Density:	ρ = 8.32 gm/cm^3 (determined experimentally)
Average Thermal Conductivity:	$\bar{k}$ = 0.065 cal/sec·cm°K (298°K to Melting Point)
Average Heat Capacity:	$\bar{C}_p$ = 0.15 cal/gm°K (298°K to Melting Point)
Average Thermal Diffusivity:	$\bar{\alpha}$ = 0.052 cm^2/sec (298°K to Melting Point)
Heat of Fusion:	H_f = 74.2 cal/gm (calculated from elemental composition)
Enthalpy of Liquid:	H = 261.7 cal/gm

Stellite 156 Powder Alloy Composition:

C	Cr	Si	W	Fe	Ni	Co	Cu	Mn	Mo	O	N
1.65	28.32	1.36	4.28	0.61	0.33	Bal.	<.10	<.10	<.10	0.028	0.065

Powder Mesh Size: (-60 + 325)
Apparent Density: 4.44 gm/cm^3
Supplier: Metallurgical International, Inc.
Heat Number: A-76891

TABLE 4

LASER PARAMETRIC STUDIES
ENERGY ABSORPTION EFFICIENCY
VALUES VS. OPERATING PARAMETERS
STELLITE 156 ALLOY

Run No.	Laser Power (KW)	Power* Density (W/cm^2)	Travel Speed (cm/sec)	Interaction Time (sec)	Specific* Energy (J/cm^2)	Fusion Zone Thickness (cm)	Fusion Zone Area (cm^2)	Available Energy Per Unit Length of Weld* (J/cm)	(cal/cm)	Energy Absorbed Per Unit Length (cal/cm)	Absorbtion Efficiency (%)
S-1	15	5.05 x 10^3	0.38	2.34	11817	0.437	0.923	31579	7547	4063	53.8
S-2	15	"	0.42	2.10	10605	0.404	0.854	28369	6780	3816	56.2
S-3	15	"	0.51	1.75	8838	0.361	0.723	23622	5646	3107	55.0
S-4	15	"	0.64	1.40	7070	0.318	0.615	18898	4517	2629	58.2
S-5	15	"	0.76	1.17	5909	0.277	0.532	15748	3764	2199	58.4
S-6	15	"	0.89	1.00	5050	0.231	0.479	13498	3226	1912	59.3
S-7	15	"	1.05	0.84	4242	0.162	0.400	11342	2710	1525	56.3
S-8	15	"	1.27	0.70	3535	0.137	0.338	9449	2258	1271	56.3
S-9	15	"	1.48	0.60	3030	0.121	0.290	8097	1935	1089	56.2
S-10	15	"	1.69	0.53	2677	0.108	0.258	7088	1694	953	56.3
S-11	10	5.45 x 10^3	0.42	2.10	11445	0.549	0.762	18912	4520	3118	69.0
S-12	10	"	0.63	1.40	7630	0.439	0.438	12598	3011	1864	61.9
S-13	10	"	0.85	1.05	5723	0.328	0.379	9445	2257	1553	68.8
S-14	10	"	1.27	0.70	3815	0.216	0.253	6299	1505	966	64.2
S-15	10	"	1.69	0.53	2888	0.159	0.186	4725	1129	720	63.8
LP-1	15	5.05 x 10^3	0.21	4.19	21159	0.846	1.880	56604	13528	10686	79.4
LP-2	15	"	0.42	2.10	10605	0.733	1.280	28369	6780	5505	81.2
LP-3	15	"	0.64	1.40	7070	0.660	0.942	18898	4516	3635	80.5
FS-1**	15	5.05 x 10^3	0.21	4.19	21159			56604	13528		
FS-2	15	"	0.25	3.50	17695	0.150	0.389	47244	11291	2229	19.7
FS-3**	15	"	0.21	4.19	21159			56604	13528		

*Corrected for mirror absorption power losses.
**Runs aborted due to fire in work station.

to melt metal in fusion zone divided by the absorbed laser energy. The energy required to melt the metal in the fusion zone was calculated for each specimen by multiplying the unit weight (gm/cm) by the enthalpy of the liquid at the melting point, (cal/gm). The enthalpy of the liquid was determined to equal 261.7 cal/gm for the S-156 alloy. Melting efficiency values are summarized in Table 5.

Examination of the data in Table 5 indicates that the melting efficiency increases with increasing travel speed for both the loose powder and sintered specimens. Furthermore, the melting efficiencies for the loose powder and sintered specimens are comparable for comparable travel speeds, and appear to approach an upper limit of approximately 60% for each material condition. This would imply that melting efficiency is primarily dependent upon the ratio of travel speed to thermal diffusivity, i.e.,

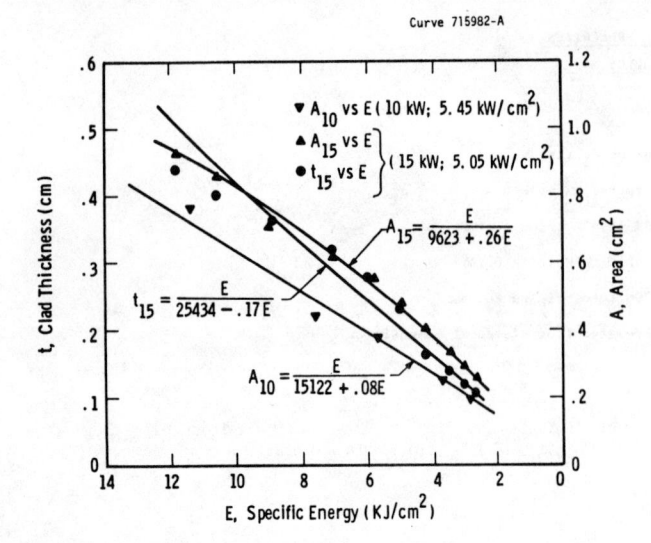

Figure 17--Thickness and cross-sectional area values of sintered S-156 fusion envelopes versus specific energy. Curves determined by computer least squares analyses.

the ratio of advance rate to the rate of energy loss by conduction, and that high values of S/α yield high melting efficiency.

Another way to compare the results of these parametric studies is in terms of total melting efficiency, i.e., that percentage of the total available energy which is utilized to form the fusion zone. The total melting efficiency is simply the product of the absorption efficiency multiplied by the melting efficiency values calculated in Table 5. Calculated values of total melting efficiency are listed in Table 6. A comparison of this data demonstrates the superiority of loose powder for applications involving laser hardfacing. At a total melting efficiency of 45%, loose powder exceeds the total melting efficiency of a comparable sintered preform (S-4) by a factor of 0.5, and exceeds the total melting efficiency characteristics of the fused preform by a factor of 6.

5.0 DISCUSSION

The results of these parametric studies have demonstrated that laser hardfacing is a highly controllable, reproducible process, providing that adequate measures are taken to shield the workpiece and the molten pool from atmospheric oxidation. The absorption of energy from the laser beam involves a complex interaction of photons and electrons at the surface of the metal such that the absorbed energy is instantly converted to heat at the point at which the quantum of optical energy was absorbed. The absorptive characteristics of the molten surface can radically change due to the formation of drossy films (slag) such that a significant portion of the optical energy is reflected away. Discontinuous, or patchy surface film formation during laser processing can cause variations in laser coupling efficiency during the course of a welding traverse, causing nonuniformity and variation of the thickness contour of the fusion envelope. This type of behavior was demonstrated during the helium gas shielding evaluations of S-156 alloy, Figure 6. Proper inert gas shielding is particularly important when operating at moderately low laser power densities (10^3 - 10^4 W/cm^2) such as those employed for surface hardfacing and cladding; at higher power densities approaching 10^6 W/cm^2 and above, there is sufficient thermal convection, vaporization of the substrate, and keyholing to minimize surface film effects.

TABLE 5

LASER PARAMETRIC STUDIES

MELTING EFFICIENCY VALUES VS. OPERATING PARAMETERS

STELLITE 156 ALLOY

Run No.	Laser Power (KW)	Power* Density (W/cm²)	Travel Speed (cm/sec)	Fusion Zone Thickness (cm)	Fusion Zone Area (cm²)	Unit Weight (g/cm)	Energy To Melt (cal/cm)	Energy Absorbed Per Unit Length (cal/cm)	Melting Efficiency (%)
S-1	15	5.05 x 10³	0.38	0.437	0.923	7.72	2020	4063	49.7
S-2	15	"	0.42	0.404	0.854	7.11	1861	3816	48.8
S-3	15	"	0.51	0.361	0.723	5.69	1489	3107	47.9
S-4	15	"	0.64	0.318	0.615	5.12	1339	2629	50.9
S-5	15	"	0.76	0.277	0.532	4.43	1158	2199	52.7
S-6	15	"	0.89	0.231	0.479	3.99	1043	1912	54.6
S-7	15	"	1.06	0.162	0.400	3.33	871	1525	57.1
S-8	15	"	1.27	0.137	0.338	2.81	735	1271	57.8
S-9	15	"	1.48	0.121	0.290	2.41	631	1089	57.9
S-10	15	"	1.69	0.108	0.258	2.15	563	953	59.1
S-11	10	5.45 x 10³	0.42	0.549	0.762	6.34	1660	3118	53.2
S-12	10	"	0.64	0.439	0.438	3.65	955	1864	51.2
S-13	10	"	0.85	0.328	0.379	3.08	806	1553	51.9
S-14	10	"	1.27	0.216	0.253	2.05	537	966	55.6
S-15	10	"	1.69	0.159	0.188	1.51	404	720	56.2
LP-1	15	5.05 x 10³	0.21	0.846	1.880	15.64	4093	10686	38.3
LP-2	15	"	0.42	0.733	1.280	10.66	2790	5505	50.7
LP-3	15	"	0.64	0.660	0.942	7.84	2052	3635	56.5
FS-1**	15	5.05 x 10³	0.21						
FS-2	15	"	0.25	0.150	0.389	3.24	847	2229	38.0
FS-3**	15	"	0.21						

*Corrected for mirror absorption power losses.

**Runs aborted due to fire in work station.

TABLE 6

LASER PARAMETRIC STUDIES

TOTAL MELTING EFFICIENCY VALUES VS. OPERATING PARAMETERS

STELLITE 156 ALLOY

Run No.	Laser Power (KW)	Power* Density (W/cm^2)	Travel Speed (cm/sec)	Fusion Zone Thickness (cm)	Fusion Zone Area (cm^2)	Unit Weight (g/cm)	Absorption Efficiency (%) X	Melting Efficiency (%) =	Total Melting Efficiency (%)
S-1	15	5.05 x 10^3	0.38	0.437	0.923	7.72	53.8	49.7	26.7
S-2	15	"	0.42	0.404	0.854	7.11	56.2	48.8	27.4
S-3	15	"	0.51	0.361	0.723	5.69	55.0	47.9	26.3
S-4	15	"	0.64	0.318	0.615	5.12	58.2	50.9	29.6
S-5	15	"	0.76	0.277	0.532	4.43	58.4	52.7	30.8
S-6	15	"	0.89	0.231	0.479	3.99	59.3	54.6	32.4
S-7	15	"	1.06	0.162	0.400	3.33	56.3	57.1	32.1
S-8	15	"	1.27	0.137	0.338	2.81	56.3	57.8	32.5
S-9	15	"	1.48	0.121	0.290	2.41	56.2	57.9	32.5
S-10	15	"	1.69	0.108	0.258	2.15	56.3	59.1	33.2
S-11	10	5.45 x 10^3	0.42	0.549	0.762	6.34	69.0	53.2	36.7
S-12	10	"	0.64	0.439	0.438	3.65	61.9	51.2	31.7
S-13	10	"	0.85	0.328	0.379	3.08	68.8	51.9	35.7
S-14	10	"	1.27	0.216	0.253	2.05	64.2	55.6	35.7
S-15	10	"	1.69	0.159	0.188	1.51	63.8	56.2	35.9
LP-1	15	5.05 x 10^3	0.21	0.846	1.880	15.64	79.4	38.3	30.4
LP-2	15	"	0.42	0.733	1.280	10.66	81.2	50.7	41.4
LP-3	15	"	0.64	0.660	0.942	7.84	80.5	56.5	45.4
FS-1**	15	5.05 x 10^3	0.21						
FS-2	15	"	0.25	0.150	0.389	3.24	19.7	38.0	7.5
FS-3**	15	"	0.21						

*Corrected for mirror absorption power losses.

**Runs aborted due to fire in work station.

The results of the absorptivity calculations (Figure 16, Table 4) demonstrate that laser coupling efficiency varies with the physical nature of the hardfacing deposit. Loose powder, with an absorption coefficient of 0.8, is a significantly better light trap than sintered powder ($R_A = 0.57$) and a factor of 4 superior to the absorptivity of the fused preform. The observation that absorption efficiency increases with increasing powder density (sintered specimens, Table 4, Figure 16) has been confirmed by a number of studies[8,9]. Figure 18 illustrates results from a study by Gregson, et al.,[10] which show that reflectivity decreases (absorptivity increases) with increasing laser power density for a number of dissimilar alloys. The specimens evaluated in this study were solid bars with polished surfaces. The sharp transitions in reflectivity which occur at power densities in excess of 10^6 W/cm^2 for alloys AISI 1045, O-1, and 3% C cast iron correspond to the onset of surface melting.

The heat transfer efficiencies (absorption efficiencies) determined from these laser parametric studies are, in general, comparable to the arc heat transfer efficiencies reported[5] for submerged-arc welding ($\sim$ 90%), shielded metal-arc welding ($\sim$ 75%), and gas-metal-arc welding ($\sim$ 66%). The laser absorption efficiencies are decidedly superior to heat-transfer efficiencies for gas-tungsten-arc welding ($\sim$ 35%), and slightly superior to transferred-plasma-arc welding ($\sim$ 50%). However, the major advantage of laser processing lies in the high total melting efficiency, i.e., 35-45% of the total available energy is utilized for melting, compared to the conventional arc-welding processes. Typically, the corresponding total melting efficiencies for the arc-welding processes range from 5-25%, with the exception of submerged-arc-welding, which yields a total melting efficiency comparable to that of the laser, i.e., 40-50%.[11] Moreover, the higher travel speeds

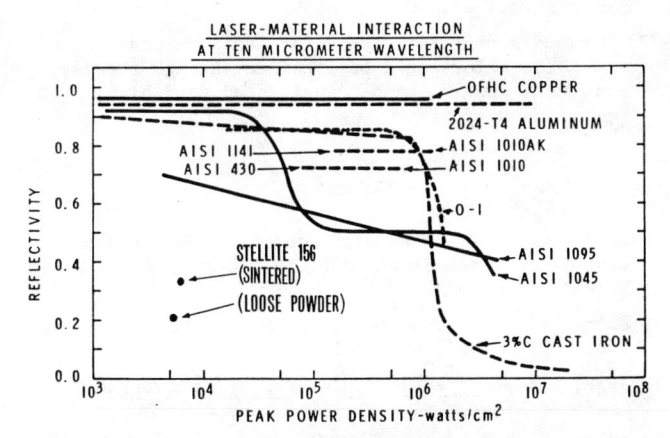

Figure 18--Laser reflectivity measurements versus peak power density for a number of dissimilar alloys. After Gregson, et al.[10]

attainable with the laser contribute to less heating of the substrate by conduction (high values of $S/\bar{\alpha}$), and a proportionally greater amount of sensible heat is lost through radiative heat loss from the surface of the workpiece.

6.0 CONCLUSIONS

A series of parametric studies have been conducted to characterize laser beam/clad coupling efficiencies for use in hard-facing applications. Laser heat transfer efficiencies were determined for cobalt-base Stellite 156 specimens prepared in the form of loose powder, sintered preforms and fused preforms. Process variables investigated included laser power density, beam interaction time and shielding gas composition. From the results of these studies it may be concluded:

(1) The laser hardfacing process is energy efficient, empirically predictable, and reproducible providing that adequate inert gas shielding practices are employed.

(2) Argon gas shielding may effectively be employed for hardfacing applications at laser power densities in the range of $10^3 - 10^4$ W/cm^2.

(3) The fusion zone cross-sectional area and thickness both decrease in proportion to 1/S (hyperbolic speed dependence) for specimens of S-156 alloy. Fusion zone area and thickness values can similarly be expressed as hyperbolic functions of $\bar{H}$, the laser energy available per unit length of weld, and E, the specific energy density, where $\bar{H}, E \sim f(1/S)$.

(4) At constant laser power density (5.05 KW/cm^2), the laser absorption efficiency varies with the physical nature of the S-156 hardfacing alloy, i.e., loose powder - 80%; sintered preform - 57%, and fused preform - 20% (decreases with increasing density).

(5) Laser absorption efficiency increases with increasing power density (for sintered preforms of S-156 alloy).

(6) Laser melting efficiency increases with increasing travel speed (for loose powder and sintered preforms of S-156 alloy). This is attributed to less heating of the substrate by conduction in proportion to the ratio $S/\bar{\alpha}$. Limiting values of melting efficiency approach 60% for both specimen configurations.

(7) Maximum energy efficiency for laser hardfacing of S-156 alloy would be attained through the use of loose powder, high travel speed, and high power density.

REFERENCES

1. E. V. Locke, D. Gnanamuthu, and R. A. Hella, Avco-Everett Research Report 398, Avco Corp., Everett, MA, March 1974.

2. E. M. Breinan, G. H. Kear, and C. M. Banas, Physics Today, Nov. 1976, pp. 44-50.

3. J. F. Ready, Effects of High-Power Laser Radiation, Academic Press, New York, 1971, pp. 213-225.

4. G. H. Harth and W. C. Leslie, Welding Research Supplement, April 1975, pp. 124s-128s.

5. N. Christensen, V. deL. Davies, and K. Gjermundsen, British Welding Journal, February 1965, pp. 54-75.

6. D. Rosenthal, Welding Journal, 1941, Vol. 20, Research Supplement, pp. 220s-225s.

7. H. S. Carslaw and J. C. Jaeger, Conduction of Heat in Solids, Clarendon Press, Oxford, 1959.

8. N. G. Basov, et al., Sov. Phys.-Tech. Phys., 13, 640 (1968).

9. J. F. Ready, op.cit., pp. 114-118.

10. V. G. Gregson, Jr., F. A. Koltuniak, and C. W. Knakal, General Motors Manufacturing Development Report No. MD 76-001, General Motors Technical Center, Warren, MI, February 1976.

11. J. M. Barry, Z. Paley, and C. M. Adams, Jr., Welding Journal, 1963, Vol. 42, Research Supplement, pp. 97s-104s.

LASER SURFACE TREATMENT BY RAPID SOLIDIFICATION

A. Tiziani, L. Giordano and E. Ramous
Institute of Industrial Chemistry
University of Padova

1. INTRODUCTION

The use of high power lasers in surface treatments (hardening,coating,alloying) has become conventional. Recently,attention has been devoted to the study of surface melting of alloys:the laser surface melting, followed by rapid self-quenching of the melted layer can produce very refined microstructures with unusual properties, directly on the workpiece surface.

Owing to the very high heating and quenching rates, trasnformations obtained may be very different from traditional ones, giving new and metastable microstructures which can be subsequentely transformed by heat treatment.

This process was first examined in non-ferrous alloys and superalloys (1).

Strutt and others (2,3,4) investigated both the structure of the melted layers and the subsequent structure modifications by heat treatment of tool steels (5).

The present study has been carried out to define the applications of laser melting treatment designed to obtain hardened surface layers,avoiding the drawbacks which traditional methods of surface hardening presents.

We have studied the treatment of some high-carbon alloy steels and of pearlitic cast-iron.

2. EXPERIMENTAL

2.1. Laser Equipment

An AVCO 15 KW continuous laser has been used for this work,with optics that supply on the plane of work a focused beam with a circular spot of 10mm, in which is concentrated 81% of the energy. Two systems can be used in order to obtain a uniform beam,rectangular and of variable dimensions:

a) scanning the beam with two mirrors oscillating around two perpendicular axes with variable aplitudes and frequencies respectively of 125 and 600 Hz (the scanning speed is so high that the energy profile may be considered uniform);

b) focusing the beam with a special mirror called "beam integrator".

Then a table for movement at variable speed is provided for in the equipment, so that the piece can be moved under the laser beam which remains fixed.

The parameters, that consequently will be considered in order to define the treatment to which the pieces have been subjected, are:

A (mm^2)   area of the laser beam or spot;
P (KW/cm^2) Laser power at the point of work within the scanning area of the beam;
τ (s)   interaction time defined as the time ne - cessary for the spot to cover a space corresponding to its dimension in the direction of displacement.

The incident power was measured by a calorimeter(and to obtain some information on the distribution of energy and on the shape of the beam) a box of plexiglass has been used as a reference.

2.2 Reflectiveness

It is well-known that the reflectiveness of metals in the solid state is very high and varies with the wavelength of the incident ligth (fig.1).

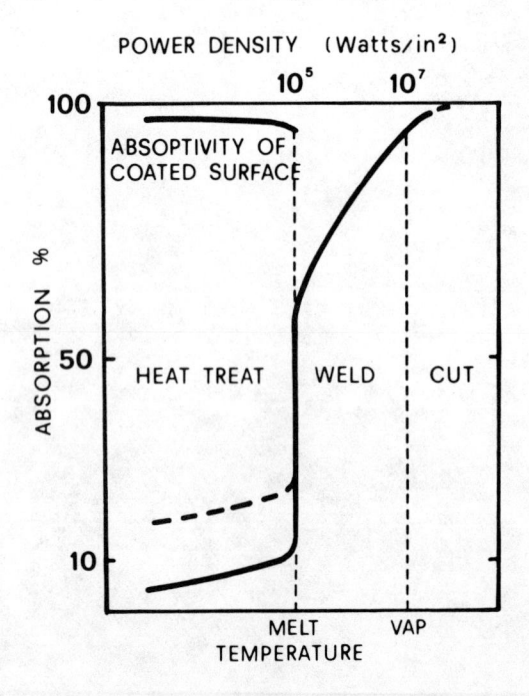

Fig.1-Absorption energy in metal surface.

At 7 μm CO_2 laser wavelength, it has been assessed that the reflectiveness of the steel surface is around 85% of the value of the incident energy but diminishes when the metal passes to the molten state. Should the surface be covered with absorbent materials the value of reflectiveness falls down to 15÷20%.

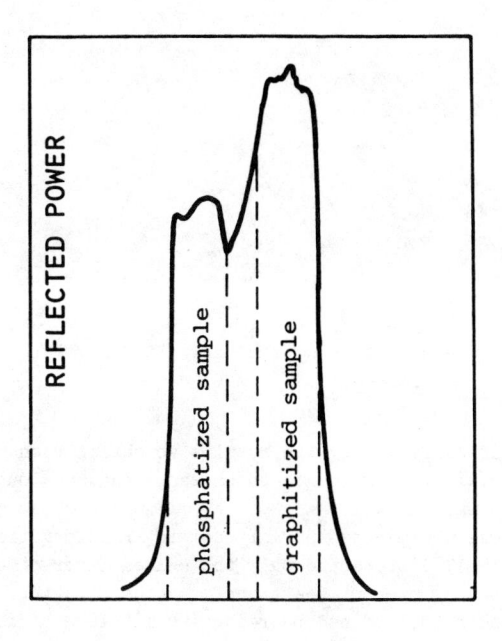

Fig.2 - Reflected power (unfiltered radiation).

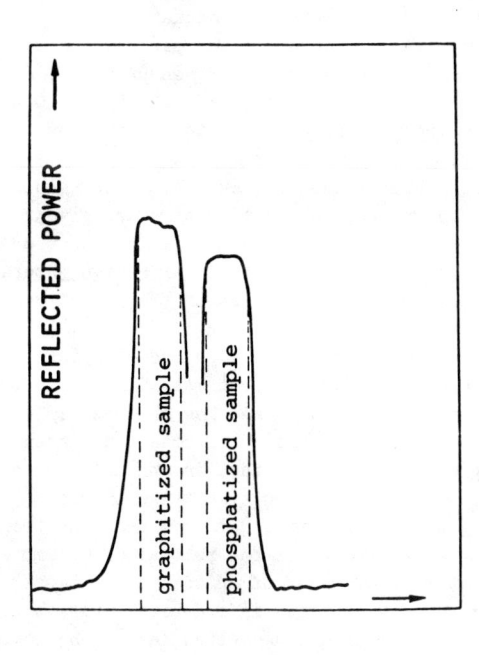

Fig.3 - Reflected power (measured wavelength: 10,6 μm).

It is necessary therefore to cover the surface with a material of high absorbency as long as the material is in a solid state. The coatings usually adopted are graphite slag and phosphatizing. Information on the influence of the latter on thermic treatments with surface remelting is not to be had in written form. A preliminary study on the above cited two coatings has, therefore, been rendered necessary (6). Some of these results given in fig.2 and fig.3 show that graphite coated samples reflect more than phosphatized ones. A direct consequence of such a phenomenon is a deeper zone of fusion for the latter.

2.3 Gas Covering

As well as coating we have tried to individualize the influence of a gas covering on the remelting treatment. In particular the influence of gases such as CO_2, H_2, N_2, air on the depth of the melted zone has been thoroughly studied. The results (7) have given evidence that laser parameters being equal the thickness of the molten part obtainable in air is definitely the highest at least in the cases we considered.

3. CAST-IRON TREATMENT

3.1. Materials

The tests have been carried out on samples of grey iron with flake graphite, the composition of which is given in table 1.

Table 1

C	Si	Mn	Cr	Ni	S	P	Sn
3.10	2.16	0.50	0.19	0.08	0.041	0.028	0.11

The structure turns out to be type A graphite and of dimensions contained between 3 and 5 of the ASTM scale in a pearlitic matrix.

3.2 Treatment conditions

The laser parameters, power densities and interaction time, haven been chosen on the basis of data derived from mathematical model of the thermal transient induced by laser beam on the surface layers. Obviously more than one pair of values (power density/interaction time) could give the same molten depth concerned (1 - 2 mm).
After some preliminary tests, we chose, to work with a power density of about 4,5 KW/cm^2 and consequent interaction time between 1 and 8 s. to obtain suitable surface melted layers.

3.3 Results and discussion

In the fig.4 the microhardness profiles along a transverse section are reported.

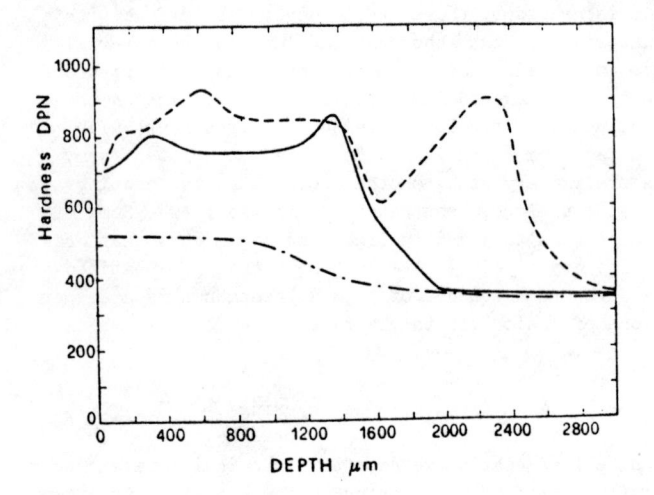

Fig.4 - Microhardness profile.

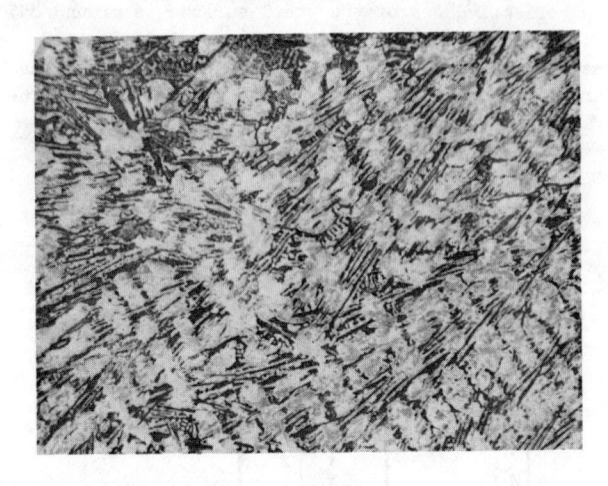

Fig.5 - structure of the melted layer. (O.M.- 40 μ)

In the samples of lower thickness, the hardness improvement of the surface melted layer is very reduced. Only in the samples of sufficient thickness, over about 10 mm, the laser melting, and subsequent self-quenching induces a noticeable surface hardening. But with the highest samples thickness, over about 20 mm, the maximum hardness decreases and the hardness profile appears more discontinuous.

These different patterns correspond to very different microstructures, in samples of various thickness.

In effect, the treatment by laser involving surface melting, treatment parameters (power density and interaction time), determine mainly the depth of the melted layer. But the final microsctructure (and hardness profile) is determined by the self- quenching rate obviuously depending on the workpiece thickness.

In the thinner samples the quenching rate is not sufficient to reach the critical rate for martensite transformation.

Then the structure of the melted layer, is only slightly modified and contains small cementite and fine pearlite quantities.

Indeed the structure obtained in larger workpieces is typical of white cast-iron with cementite, dendrites of transformed austenite and in the melted layer, as in fig.5.

But in these samples also the unmelted transition layer between the melted zone and the base material appears transformed into martensite with variable quantities of retained austenite.

This latter zone was that favourable for seeding of microcracks going towards the surface of the workpiece. Obviously, the martensite structure appears in the transition zone, under the melted layer, only where the critical quanching rate for hardening was overcarried.

The formation of these macroscopical faults can be therefore avoided by chosing the treatment conditions in order to limit the layer that cool with velocity superior to the critical one, only within the melted zone.

This is confirmed by the results we obtained on workpieces with intermediate thickness, ranging about 10 and 20 mm.

In these workpieces the martensitic transformation had practically interested only the melted layer, the transition zone presented a transformation structure but with bainite, refined pearlite and little martensite.

In these samples, with intermediate thickness, both the structure and the hardness profile of the superficial layers were satisfying, and microcracks had not been noticed.

In pieces of high thickness, the microcracks formations can be avoided by reducing adequately the cooling velocity by pre-heating of the pieces.

In the conditions we examined, in practise, cast-iron workpieces with thickness, ranging between 10 and 20mm, can be directly hardened in the surface by laser melting.

For this type of samples, in fig.6 the values of penetration vs. the interaction time are given.

With low interaction times in the formation of microcracks was ascertained, whilst with high interaction times a decrease of hardness in the melted zone was noted (fig.7).

The structure obtained for low and high interaction times are analogous to those met with in the preceding case respective to high and low thickness. In the samples where the quantity of retained austenite is negligible it was noted that in the melted zone, increasing the interaction time, also increases lamellar spacing of the pearlite, deriving from the transformation of austenite. After the removal of 0,1 mm of surface material by means of grinding, the latter were subjected to wear tests in order to verify the tribological behaviour of the melted layer subject to low specific load values but a high number of alternate cicles of stesses typical of the slide-bed couple (8). As a material for comparison cast-iron G32 was chosen. The behaviour of laser melted cast-iron has been compared with that of samples of induction hardened steel and sulfurized steel. In our test condition wear of laser remelted cast-iron was reduced from 2 or 4 times respect other materials examined.

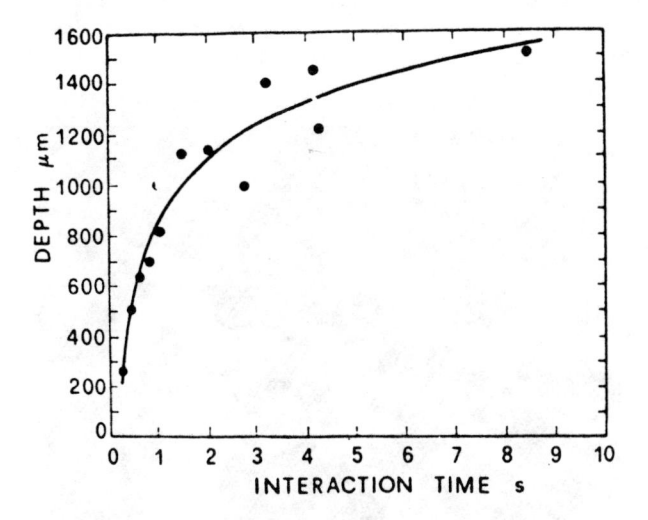

Fig.6 - Interaction time-depth curve.

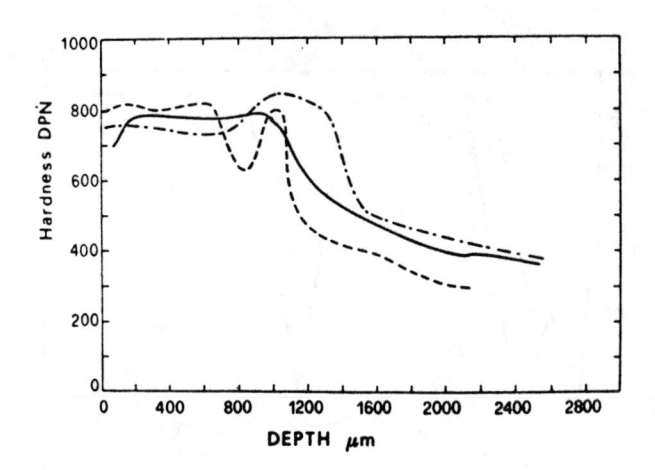

Fig.7 - Microhardness profile.

4. HIGH CARBON ALLOY STEEL TREATMENT

The surface melting by laser was examined on the same high carbon alloy steels: 52100, AISI A2 and D3. The aim of the work was to study the possibility of obtaining metastable surface layers with mainly austenitic structure. This being metastable and carbon oversaturated, would be transformed with subsequent heat treatment to structures of improved hardness and toughness. Reported here are the results concerning the AISI A2 steel, as typical example, composition of which is shown in table 2.

Table 2 - Average composition of AISI A2 steel.

C	Cr	Mn	Si	Mo	V
1,0	5,0	0,7	0,3	1,0	0,2

4.1 Treatment Conditions.

The treatment conditions investigated were a specific power of 15 KW/mm^2 and interaction time between 0.1 and 0.8 s., the spot area was 8 x 10 mm.
Layers only melted by laser and after subsequent heat treatment have been examined by light microscope, and SEM. Metastable oversaturated austenite was also examined by X-ray diffraction and Mössbauer spectroscopy.

4.2 Results

Surface melting.
Micrographic observations of all the samples show a molten surface layer characterized by a remarkable variety of structures: needle-shaped, dendritic, cellular. The outer part of the molten layer of sample treated with interaction time of 0,2 s., for instance, was observed to have present ferrite needles oriented more or less perpendicularly to the surface and a zone directly below and adjacent formed by a network of δ-ferrite and carbides surrounded by austenitic islands (Fig.8 and 9).

Fig.8 - Outer zone of melted layer of the sample treated with 0,2 s. (S.E.M. - 30 μ)

The prevalent structure of specimens treated at interaction time of 0,6 and 0,3 s. is undoubtedly the dendritic one, as can be seen from the optic micrographs reported in fig.10 and 11.
On the other hand specimen at 0,6 s. besides presenting a more considerable melted region, shows a very homogeneous cellular structure formed by austenitic grains surrounded by δ-ferrite and carbides (fig.12).
The thickness of the melted region, correlated to the Steen and Coll. parameter (1/V^{0,5}) curve, displays for

this steel a linear movement already met with in other steels (9).

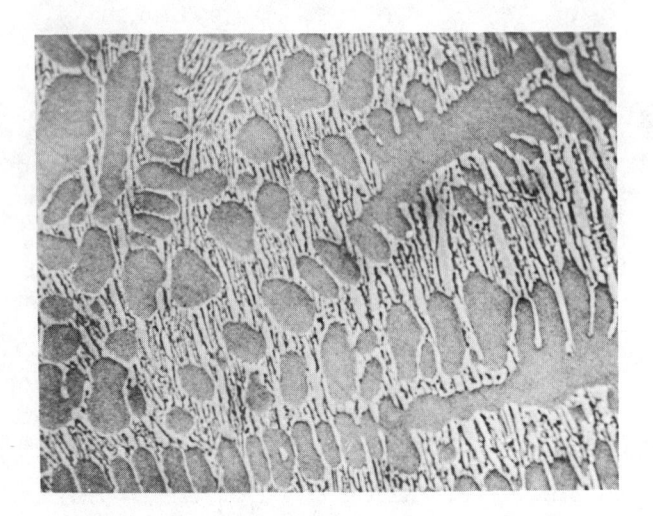

Fig.9 - Inner zone of melted layer of the sample treated with 0,2 s.(S.E.M.-20 μ).

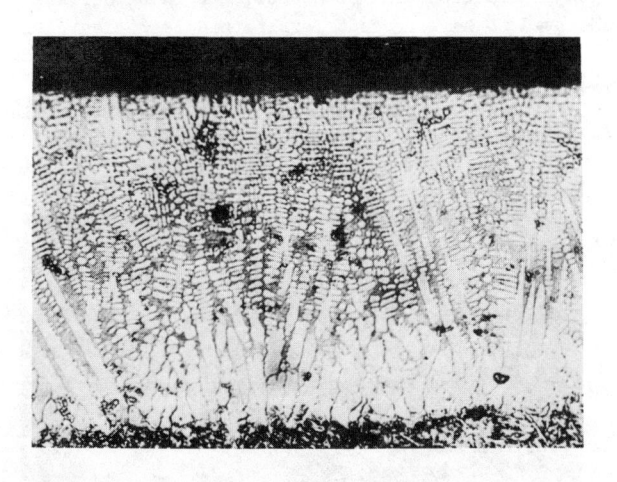

Fig.10 - Melted zone of specimen treated at 0,6 s.(O.M.-200 μ).

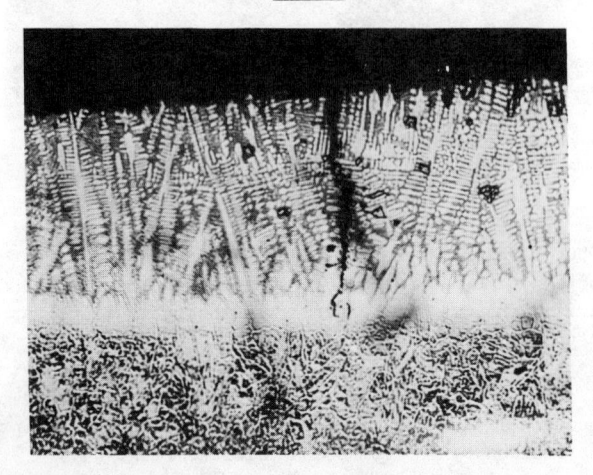

Fig.11 - Melted zone of specimen treated at 0,3 s.(O.M.-200 μ).

Analysing the melted region of specimens treated at 0,3 and 0,2 s.respectively we can differatiate three zones: the first one composed of austenitic dendrites and austenite plus carbides, the second one formed only by austenite and the third one consisting of the heat treated zone. As noted above the first zone presents both transverse and longitudinal variation that justify the irregular movement of the micro-hardness profile (fig.13).

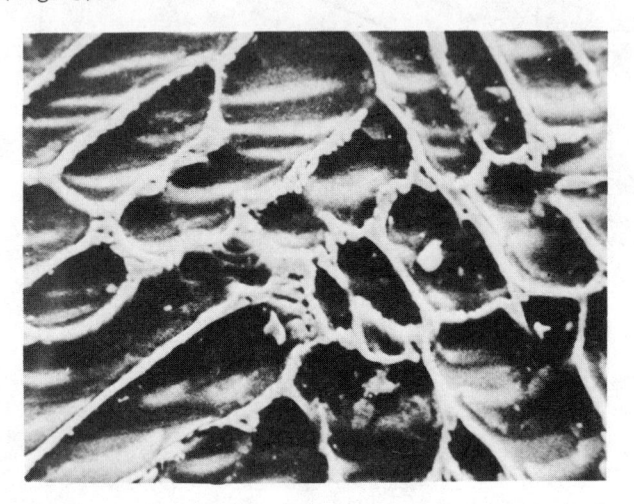

Fig.12 - Melted zone of the specimen treated at 0,6 s. (S.E.M.-20 μ).

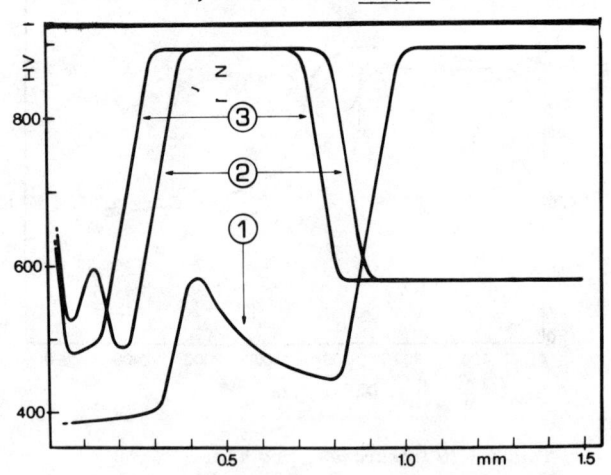

Fig.13 - Microhardness profile of only laser melted specimens.

N.B. : numbers 1,2 and 3 refer to interaction times of 0,6 - 0,3 and 0,2 respectively.

Only specimen at 0,6s., with a very homogeneous structure, displays a regular microhardness profile. The X-ray diffraction analysis of specimens at 0,3 and 0,2 s. established that the dendritic structure is composed of carbides and little ferritehigher,however, in the needle-shaped zones. The predominance of austenite was confirmed; and the Mössbauer spectroscopic analysis brought out as well the carbon supersaturation of that phase.

550° Treatment

The destabilizing treatment at this temperature gave rise to different results on various specimens.

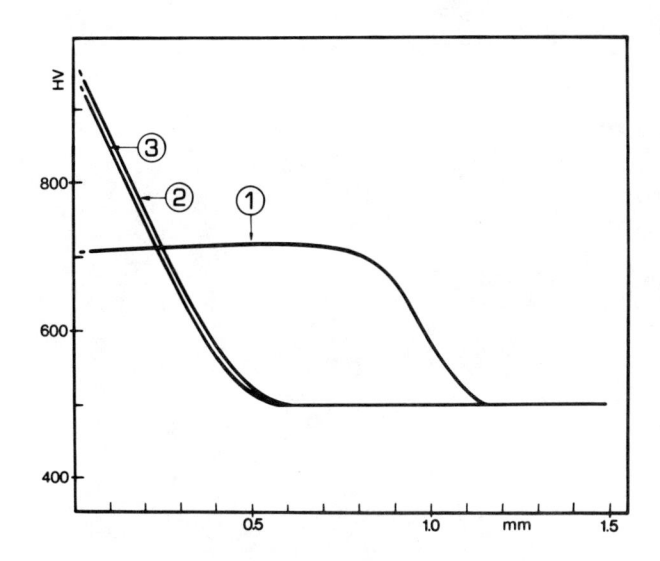

Fig.14 - microhardness profile of specimens treated at 550° C.

In fact the microhardness profile (fig.14) of specimens 2 and 3 show e decreasing trend moving from the outside to the inside of the melted structure, with surface-microhardness values that are higher than 900 HV.
On the other hand, for specimen 1 the transformation of the cellular austenite confirms its higher homogeneity, with microhardness values higher than 700 HV. affirmed. In specimens 2 and 3 the micrographic observation (fig.15) shows how there is a further carbide precipitation at the same time as austenite is transformed into ferrite. On the other hand a lamellar structure is the result of the transformation of the mainly austenitic structure of sample 1 (fig.16).

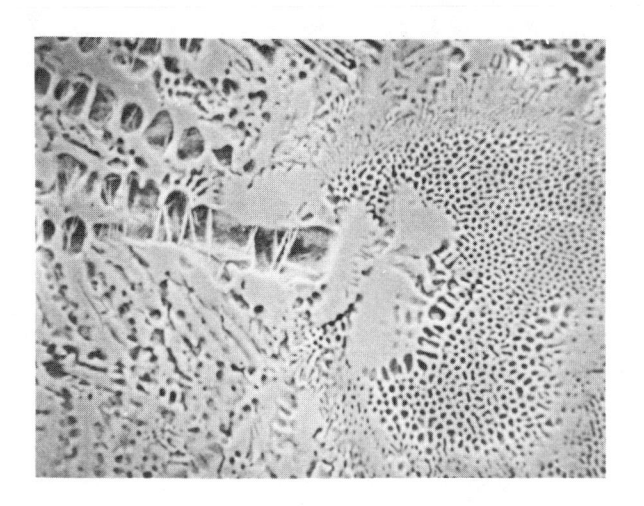

Fig.15 - specimen 2 treated at 555°C (S.E.M. 20 μ).

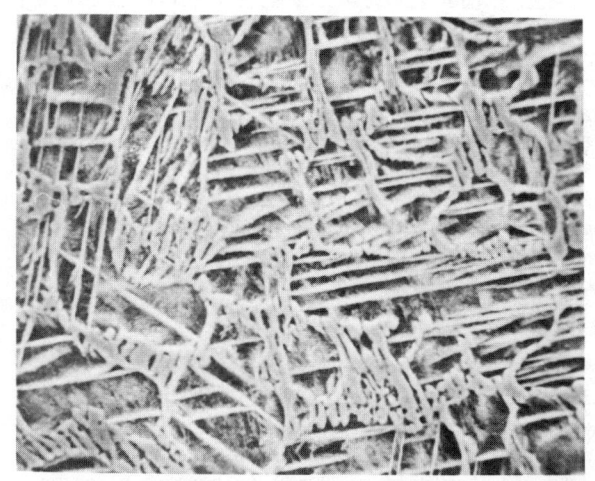

Fig.16 - Specimen 1 treated at 550°C (S.E.M. 20 μ).

Tentative treatment below 500°C proved not sufficient to obtain significative transformation of the metastable austenite. Treatments of many hours produced only partial transformation of the cellular austenite to bainite.

5. DISCUSSION

From our tests we can observe that it is possible to obtain, by laser melting and subsequent heat treatment surface layer with increased hardness on high carbon alloy steels, maintaining a high toughness in the bulk of the workpiece.
The best structure for the laser melted layer to seek for an useful application of such process is the cellular one. The reasons for such a choice can be summarized in few comments: the possibility of removing a surface layer during the finishing process, the homogeneity of thestructure of the hardness, the absence of cracks. In order to obtain such results one needs to operate with higher interaction times and lower cooling speeds. These latter parameters obviously depend also on the thickness of the pieces, and it is obvious that the cellular structure is the more difficult treatment.
An explanation of the multiple structures that are present in the melted layer is obviously difficult, since the metallurgy of rapid cooling has not yet produced in this field mathematical interpretative models that relate chemical and thermic processes. The interpretation of structures transformed by means of subsequent thermic treatment is on the other hand more linear and traditional, even if these structures have not been released from their previous history and in particular from the distribution for example of carbon and chrome. It is however, evident that the transformation that is observed is connected to the classical mechanism:

$$\gamma \rightarrow \alpha + \text{carbides}$$

The same dependence of such a process on time and temperature finds an easy interpretative scheme in the analysis of the isothermic transformation of austenite at cooling. The same isothermal transformation diagrams of the AISI steel present very long times for the transformation of austenite at 350°C. It is therefore evident that destabilization treatment at low temperature, necessary for obtaining greater hardness, would offer better results for steels of the same type but with percentages of chrome below 4%. In fact with these compositions the times for destabilization treatment are compatible with the actual necessities of industry.

6. CONCLUSIONS

Laser surface fusion followed by heat treatment for the destabilization of austenite are ideal processes for obtaining surface hardening of steels rich in C and Cr. It is necessary though to have recourse to laser parameters that ensure the formation of homogeneous cellular structures. This is possible only with high times of interaction and obviously with high specific power outputs and with relatively low cooling speeds. The subsequent thermic treatments must be carried out in such a way to make the transformation occour of austenite into inferior (if possible) bainite. Hardness definitely superior to 900 HV can be achieved with high C steels containing a percentage of Cr below 4%. For the AISI A 2 steels transformation is possible in industrially significant times only of the pearlitic type which ensures a hardness superior to 700 HV.

This process presents the following advantages in comparison with traditional treatments:
- a greater tenacity
- a greater structural stability
- absolute absence of residual austenite.

REFERENCES

1. E.M.Breiman, B.H.Kear, C.M.Bones and L.E.Greenwald; Superalloy Metallurgy and Manufacture. 1975.

2. P.R.Strutt, M.Tuli, H.Nowotny and B.H.Kear; Mat. Sci.Eng. 1978.

3. M.Tuli, P.R.Strutt and B.H.Kear; Confe rence of Rapid Solidification Processing, 1977.

4. Young-Wonkim, P.R.Strutt and H.Nowotny; Laser melting and heat treatment of M2 tool steel: A microstructural characterization. Metallurgical Translation A.Vol.10A, 1979.

5. W.M.Steen and C.Courtney; Metals Technology, 1979.

6. Report n° 201/82 - University of Padua.

7. Report n° 202/82 - University of Padua.

8. L.Locati - "Macchine e tribologia" - Ed.ERIS.

9. same as n°5.

LASER SURFACE TREATMENT OF CHROMIUM ELECTROPLATE ON MEDIUM CARBON STEEL

Christodoulou, G. & Steen, W. M.

Department of Metallurgy and Materials Science
Imperial College of Science and Technology
London SW7 2BP

ABSTRACT

An investigation is reported of the laser surface treatment of four different thicknesses of chromium electroplate on a medium carbon steel.

Three basic conditions were noted that of total melt through, partial melt through and no melting. The microstructures, hardness, degree of adhesion, and quality of the surface for these three conditions are reported.

The apparent reduction in cracks in the plated layer and improved adhesion are noted together with the operating conditons required to produce these beneficial effects for different initial plate thicknesses.

INTRODUCTION

Hard chromium plating is well known for its success in corrosion protection and wear resistance at high temperatures. However it is prone to spalling and corrosion through the vertical cracks present in this type of casting due to its brittleness (1).

Attempts have been made to improve its durability by (a) heat treatment (2), (b) laser treatment (1,3).

The laser treatment of Montgomery (3) was performed with a low power density beam, travelling slowly in an imperfect shroud of helium. This treatment did not melt the chromium layer completely, only the intermetallic layer. The treated layer showed severe cracking and failure in a rubbing wheel wear test. This has been attributed to nitrogen embrittlement of the chromium. Such nitrogen embrittlement is well known. Snavely and Faust (4) found that the nitrogen reaction would take place if the temperature was sufficiently high, even though the concentration of nitrogen was low. Wain et al (5,6) showed the strong embrittling effect of nitrogen and Cairns and Grant (7) found that the effect is much greater for water quenched samples than for slowly cooled samples.

It appears that no work has been reported on the laser heat treatment of chromium layers over the whole range of heating cycles possible. Reported here is such a survey for chromium thicknesses in the range 20 - 240µm deposited on a medium carbon steel. The resultant shape of melt zone, movement of chromium, microstructures and hardnesses are discussed.

EXPERIMENTAL PROCEDURE

The experimental arrangement using a control laser, a 2kW CW CO_2 laser and a controlled atmosphere chamber is shown in Figure 1.

The chromium plated samples were mild steel of composition :

C	Ni	Cr	Mo	Mn	Si	Cu	V	Al	S	P
0.36	2.7	0.72	0.52	0.61	0.12	0.22	0.01	0.04	0.02	0.015

The steel was in the full heat treated and tempered state prior to plating.

The plating thicknesses used were :

(a) 20-30µm; (b) 30-40µm; (c) 100-120µm;
(d) 220-240µm.

To reduce the reflectivity of the plating, some samples were painted with manganese dioxide, since it was necessary to take care to avoid carbon or phosphorous pick up. However, manganese dioxide will give off some oxygen on heating which might be disadvantageous. Thus the bulk of the samples had their reflectivity reduced by reverse etching in a bath of chromic and sulphuric acid to give a dull grey surface.

For this analysis the traverse speed was varied from 2 - 2000 mm/s and the beam diameter on the work surface was set at either 0.2mm or 1.0mm. The beam diameter was measured by acrylic prints and a rotating rod sensor (8). The laser power was held constant at 1500 ± 20W. The laser processed samples were sectioned and the size of affected zones, their hardness, and microstructure examined both optically and with electron microscopes.

RESULTS AND DISCUSSION

Shape and size of affected regions

Three types of melt zones are observed after laser treatment :

(i) complete melting of chromium layer and its dilution with the substrate steel, Figure 2a;

(ii) melting of only part of the chromium layer and some of the interface zone, Figure 2b;

(iii) no melting of the chromium, but transformation hardening of the substrate steel, Figure 2c.

The melting beneath the unmelted chromium layer as in Figure 2b and the overhang of the chromium layer onto the melt zone as in Figure 2a are the result of the higher melting point of chromium ($\sim 1903°C$) compared to mild steel ($\sim 1375°C$).

The range of microstructure observed with various speeds for a focussed 1.5kW beam (0.2mm diam) are shown in Figure 3, where complete melting is seen in micrographs a-d, partial melting in micrograph e, and no melting in micrograph f.

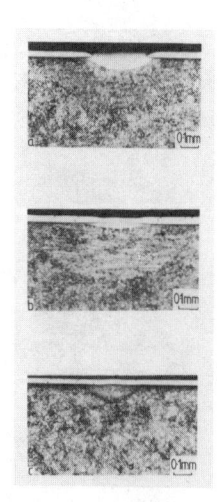

Figure 2. Laser treated zones, showing the three cross-sectional structures : (a) complete melting (b) part melting and (c) no melting of the chromium plating but transformation hardening of the substrate steel. [1.5kW, 0.2mm B.D, 30μm plating thickness, (a) 2.0mm/s; (b) 6.0mm/s; (c) .60mm/s]

Surface upset is small compared to similar runs on unplated steel.

The effect of increasing the beam diameter to 1.0mm is shown in Figure 4, where it is seen that better control of the partial melting process is obtained with good surface finish.

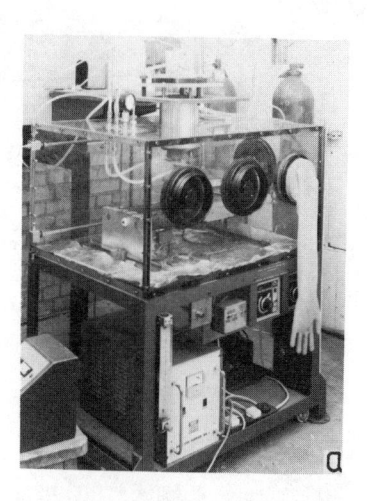

Figure 1(a). The controlled atmosphere chamber
(b). A side view of photograph (a) showing the x-y table supporting the rotating disc.

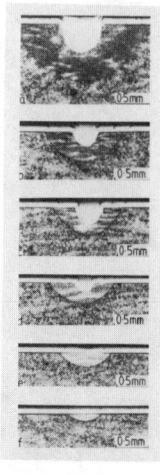

Figure 3. Profiles of laser treated zones of chromium plated steel (1.5kW, 0.2mm B.D., 100-120μm plating thickness, (a) 2.5mm/s, (b) 4.0mm/s, (c) 6.0mm/s, (d) 11.0mm/s, (e) 18.0mm/s, (f) 38.0mm/s).

The power density (P/DV, J/mm²) required to achieve the three types of melt zone is a function of electroplate thickness, as shown in Figure 5.

Depth of laser affected zones

The plots of total depth of laser affected zone (i.e. HAZ + LMZ) and only LMZ vs transverse speed are

117

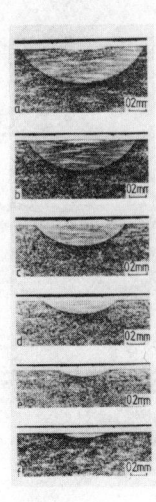

Figure 4. Profiles of laser treated zones of chromium plated steel (1.5kW, 1.0mm B.D., 100-120μm plating thickness, (a) 4.0mm/s, (b) 6.0mm/s, (c) 14.0mm/s; (d) 26.0mm/s, (e) 40mm/s; (f) 64.0mm/s).

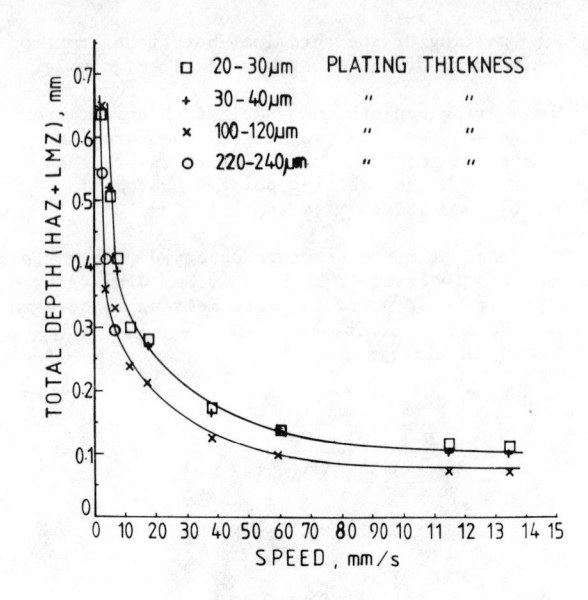

Figure 6. Total depth (i.e. HAZ + LMZ) of chromium plated steel as a function of speed and plating thickness. (1.5kW and 0.2mm B.D).

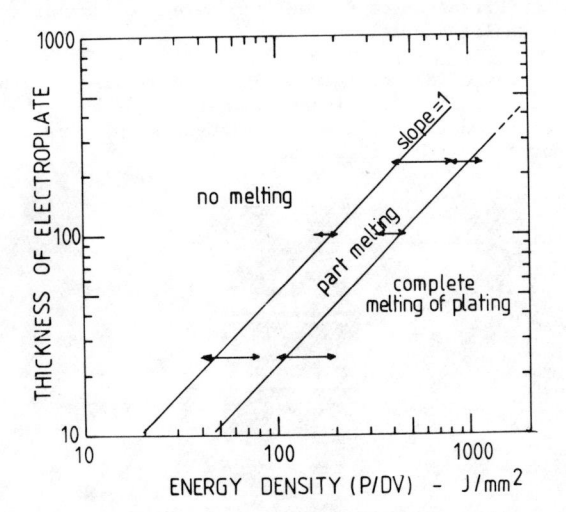

Figure 5. Range of energy densities observed for the production of different surface melting effects as a function of chromium plating thickness.

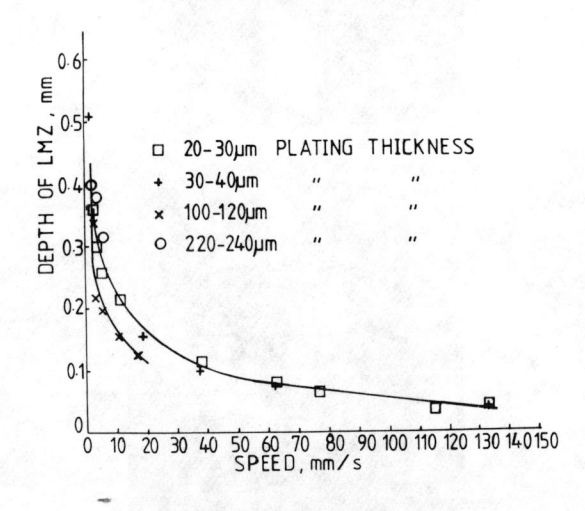

Figure 7. Depth of laser melted zone (LMZ) as a function of speed and plating thickness for chromium plated steel (1.5kW and 0.2mm B.D).

shown in Figure 6 and Figure 7 respectively for four different thicknesses of electroplate and constant beam diameter of 0.2mm. Similar plots with 1.0mm beam diameter and thicknesses of electroplate 30μm and 100μm were also obtained (9), the latter is shown in Figure 8.

$$\text{depth} \propto \left(\frac{1}{\text{speed}}\right)^n$$

or log/log plot as in Figure 9 shows the good fit to such a relationship.

The actual equation being:

LMZ + HAZ - 0.2mm beam diameter

for 20 - 30μm plate : $\log_{10}(\text{depth}) = 0\ 0.097 - \frac{1}{2}\log_{10}(\text{speed})$

for 100-240μm plate : $\log_{10}(\text{depth}) = -0.06 - \frac{1}{2}\log_{10}(\text{speed})$

LMZ only - 0.2mm beam diameter

for 20-30μm plate : $\log_{10}(\text{depth}) = -0.167 - \frac{1}{2}\log_{10}(\text{speed})$

for 100-240μm plate : $\log_{10}(\text{depth}) = -0.284 - \frac{1}{2}\log_{10}(\text{speed})$.

LMZ + HAZ - 1.0mm beam diameter

for 100μm plates : $\log_{10}(\text{depth}) = 0.10 - 0.55 \log_{10}(\text{speed})$

for 20-30μm plates : $\log_{10}(\text{depth}) = 0.11 - \frac{1}{2} \log_{10}(\text{speed})$

This implies : $\text{depth} = \dfrac{A}{V^{\frac{1}{2}}}$

where A is a constant whose value varies with the plating thickness.

Theoretically one can predict (10) that the penetration depth of a particular isotherm, x, is proportional to the Fourier diffusion length of $\sqrt{\dfrac{4\alpha D}{V}}$ and that the value of the isotherm is determined by the available energy and heating time. Thus for a given value of

$$\sigma^* = \dfrac{\sqrt{4\alpha t}}{x^2} = \dfrac{\sqrt{4\alpha D}}{x^2 V}$$

there is a fixed value of $T^* = \dfrac{T \pi k D}{P(1-r_f)}$

therefore $\dfrac{\sqrt{4\alpha D}}{x^2 V} = B \dfrac{T \pi k D}{P(1-r_f)}$

Therefore the depth of penetration of a particular isotherm, T, (m.pt or eutectoid) could be given by a relationship of the form :

$$\text{depth of T isotherm} \propto \dfrac{P(1-r_f)}{\pi k D T} \sqrt{\dfrac{4\alpha D}{V}}$$

$$\alpha \;\; \dfrac{P}{\sqrt{DV}} \;\; \dfrac{(1-r_f)}{\pi T B} \;\; \dfrac{1}{\sqrt{k\rho C_p}}$$

Thus $A = \dfrac{(1-r_f)}{\pi \sqrt{k\rho C_p} \; TB}$

The variation of A with plating thickness can thus be seen as possibly due to the variation in $\sqrt{k\rho C_p}$, since the value of the surface reflectivity, r_f, is independent of the plating thickness as in B and π. Table I shows the general properties of the chromium electroplate and the gun steel substrate.

TABLE 1. PHYSICAL PROPERTIES OF ELECTROPLATE AND SUBSTRATE

Property material	Latent ht vap. kJ/Kg	Latent ht fusion kJ/Kg	Sp. gr.	m.pt. °C	b.pt. °C	Specific ht C_p kJ/Kg°K	Density ρ Kg/M³	Thermal conductivity k W/M°K	Coef. of thermal exp. α
Chromium	615	338	7.1	1903	2480	0.419	7100	91.3	6.2 x 10⁻⁶
300M steel	-	15.2	7.6	1375	3000	0.448	7600	57.7	6.5 x 10⁻⁶

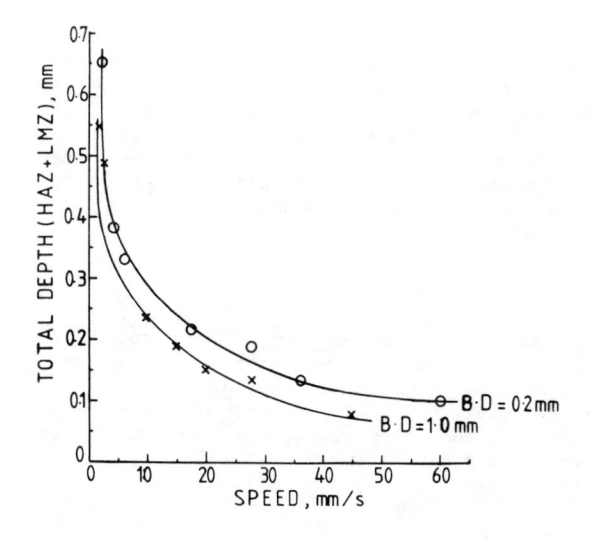

Figure 8. Total depth (i.e. HAZ + LMZ) of chromium plated steel as a function of speed and beam diameter (1.5kW and plating thickness 100-120μm).

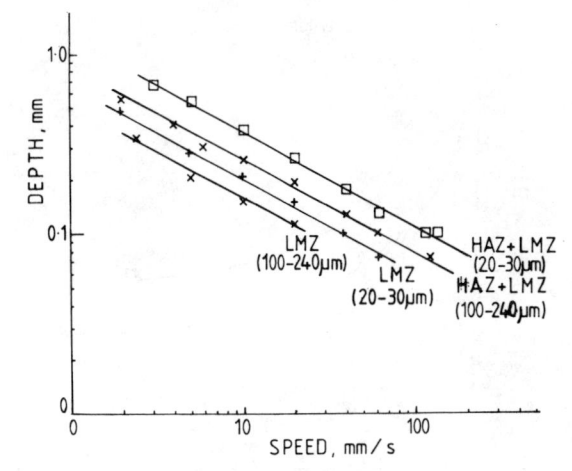

Figure 9. Depth of laser affected zone as a function of speed and plating thickness for chromium plated steel (1.5kW and 0.2mm B.D).

Thus the ratio of $\sqrt{K_1 \rho_1 C_{p_1}}_{Cr} : \sqrt{k_2 \rho_2 C_{p_2}}_{steel}$

is 1.20 : 1.0

In fact the expected variation in depth of LMZ + HAZ due to the material properties would be 1.20 times more deep in steel than in chromium. The observed difference in depth for 30μm plate and 200μm plate (Figure 5) is 1.18:1 at 135 mm/s and 1.29:1 at 20 mm/s. Considering the expected accuracy of the order of magnitude calculation this agreement is good. It is necessary to also remember that the physical properties should vary with penetration depth, (i.e. dilution) and that there is a variation in melting point between chromium and steel which will be having an influence on the LMZ.

The difference of nearly 600°C between the melting point of chromium and that of steel means that it is possible to melt the substrate without melting the clad layer. Indeed such might be the case in Figure 2b. Whereas Figure 3, a, b, c and d show the unmelted edges of the plating protruding into the mixed melt.

The dependence of the depth on $1/\sqrt{\text{beam diameter}}$ is observed at the higher speed regions for the beam diameters and plating thicknesses used. At slower speeds, less than 10 mm/s another mechanism appears to operate. This may be due to the high convection in the melt zone giving an effectively high thermal conductivity, k, for small beam diameters which would negate the effects of the smaller beam diameter. Generally the depth may be dependent on

$$\left(\frac{1}{\sqrt{kDV}}\right)$$

Width of laser affected zone. The total width of the LMZ + HAZ and LMZ only as functions of the traverse speed are shown in Figures 10 and 11. respectively for a beam diameter of 0.2 mm. The widths of the LMZ + HAZ, W_{LH} for a 1.0 mm diameter beam as a function of traverse speed, V, for the 20-30μm plating thickness is shown in Figure 12.

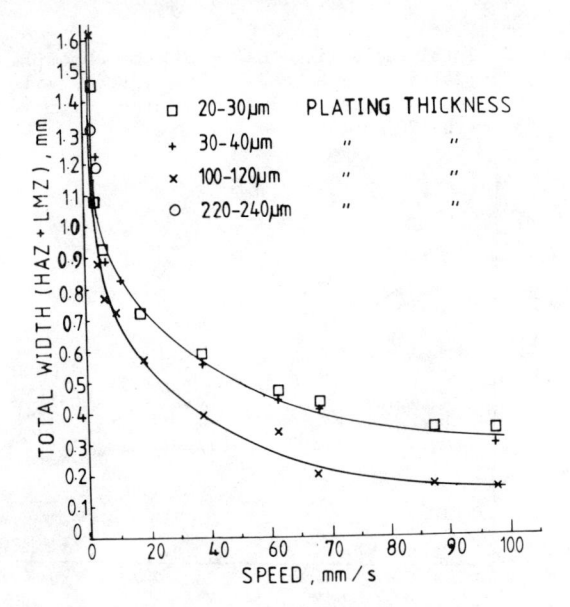

Figure 10. Total width (i.e. HAZ + LMZ) of chromium plated steel as a function of speed and plating thickness (1.5kW and 0.2mm B.D).

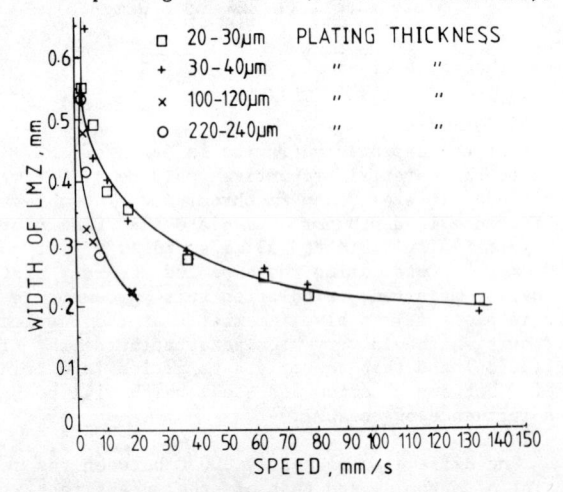

Figure 11. Width of laser melted zone (LMZ) as a function of speed and plating thickness for chromium plated steel (1.5kW and 0.2mm B.D).

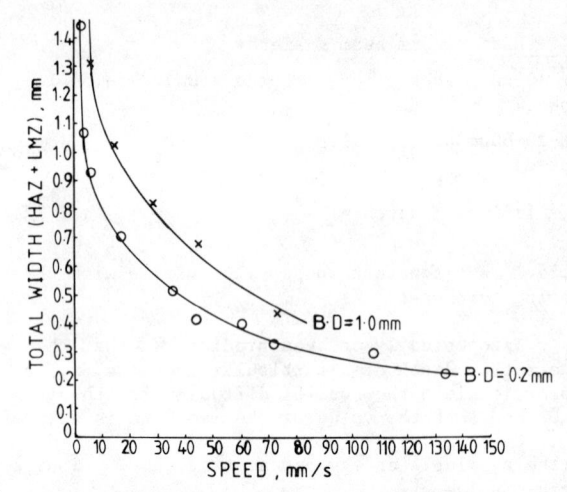

Figure 12. Total width (i.e. HAZ + LMZ) of chromium plated steel as a function of speed and beam diameter (1.5kW and plating thickness 20-30μm).

Once more a hyperbolic relationship is observed, as well as a greater spread for the samples on thinner plating.

The relationships observed are:

for the focussed beam, 0.2mm and then 20-30μm plate : width for LMZ + HAZ, denoted as W_{LH} :

$$\log_{10}(W_{LH}) = 0.477 - \tfrac{1}{2} \log_{10}(V).$$

Width for LMZ only, denoted as W_L:

$$\log_{10}(W_L) = -0.149 - \tfrac{1}{4} \log_{10}(V)$$

for a focussed beam, 0.2mm, and thick plate 100-120μm.

LMZ + HAZ : $\log_{10}(W_{LH}) = 0.38 - \tfrac{1}{2}\log_{10}(V)$

LMZ only

$$\log_{10}(W_L) = 0.2 - \tfrac{1}{3} \log_{10}(V)$$

for a focussed beam, 0.2mm, thick plate, 220-240μm

LMZ + HAZ

$$\log_{10}(W_{LH}) = 0.25 - \tfrac{1}{2} \log_{10}(V)$$

for a defocussed beam, 1.0mm, thin plate 20-30μm

$$\log_{10}(W_L) = 0.146 - \tfrac{1}{4} \log_{10}(V)$$

Within the range and order of accuracy of this work, it is possible to conclude that the width of LMZ + HAZ, W_{LH}, is related to traverse speed, V by an equation of the form :

$$W_{LH} = \frac{A'}{V^{\tfrac{1}{2}}}$$

whereas the width of LMZ only W_L has a relationship of the form :

$$W_L = \frac{B}{V^{\tfrac{1}{4}}}$$

It is also seen from the sparse data on two beam diameters that increasing the beam diameter by a factor of (1.0/0.2) = 5, increases the width W_L of the LMZ by a factor of around 2, i.e. width $\alpha \sqrt{D}$.

Thus width LMZ $W_L = \dfrac{B''D^{\frac{1}{2}}}{V^{\frac{1}{4}}}$

whereas $W_{LH} = \dfrac{A'}{V^{\frac{1}{2}}}$

increases with beam diameter as seen in Figures 10 and 12.

For a focussed beam $A' = \dfrac{4.46}{d^{0.3}}$

where d = plating thickness.

It is important to note that correlation with the beam diameter and width is not expected to be simple. Since keyholing increases width with decrease of beam diameter, and very large beams give no HAZ, in between thermal conduction effects increase width with increase in diameter (11). Thus the widths W_L and W_{LH} are seen to be dependent on traverse speed, beam diameter, plating thickness and presumably laser power, and coating reflectivity.

Structural Examination and Hardness

Mapping the distribution of chromium in the fully laser melted zones shows that chromium from the electroplate is diluted with steel to form an iron-chromium alloy, see Figures 13 and 14. Higher concentration of chromium is observed in stereoscans (a) and (b) than in (e) and (f) of Figure 13, even though of similar melt depth, because in the first stereoscans the sample has larger thickness of chromium plating (100-120μm) than in the latter (20-30μm). The above figures also show the very rapid diffusion of chromium which can only be accounted for by a highly turbulent pool. Figure 13(c) shows the small lenticular pool beneath the partly melted chromium electroplate. The chromium appears to be moving by diffusion and so one concludes the lenticular pool is tranquil, see Figure 13(d). Micrograph (a) of Figure 14 shows overlapping of laser traces with processing parameters chosen to produce complete melting of the 37μm thickness chromium plating, whilst stereoscan (b) in the same figure shows mapping of the chromium distribution along the laser melted layer.

Figure 14.
(a) The iron-chromium alloy layer produced by overlapping laser tracks on chromium plated steel.
(b) The chromium concentration in the melted layer of the micrograph (a). (1.5kW, 20mm/s, 1.0mm B.D. and plating thickness 37μm).

It is observed, that by overlapping laser traces on a chromium plated steel, a dilution of chromium with melted steel takes place to form a uniform layer of iron-chromium alloy. By geometrical consideration this alloy would be expected to be more dilute than that observed in a single track.

The chromium electroplate which was partly melted or did not melt at all during laser treatment was rendered considerably softer. Before treatment it had a hardness of about 1000 VHN, whilst after treatment the hardness of electroplate dropped to the value of about 280 VHN, see Figure 15.

Softening of chromium electroplates on heat treatment was also reported by Montgomery (1). Snavely and Faust (4) have attributed the rapid decrease in hardness on heating the chromium electroplate over 500°C, to recrystallization and grain growth in the plate. New grains formed by recrystallization and grain growth have also been observed, see Figure 15 (a), (b) and (c). The recrystallized grains are shown to exhibit characteristics usually found in a rapidly solidified structure, such as columnar growth in the direction of thermal gradient.

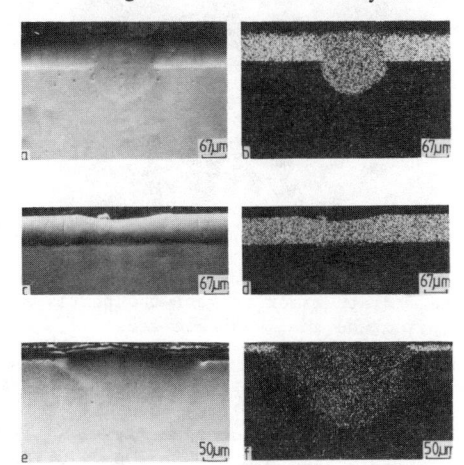

Figure 13. Stereoscans (a), (c) and (e) show the profiles of laser melted zones of chromium plated steel. Stereoscans (b), (d) and (f) show the chromium concentration in the melted zones (a), (c) and (e). (1.5kW, 0.2mm B.D. and (a) 11mm/s, (c) 18mm/s and (e) 11mm/s).

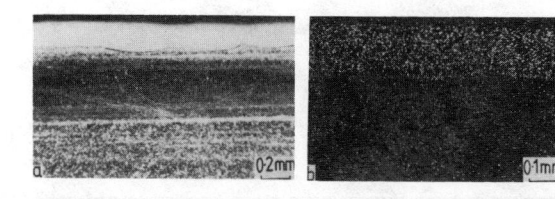

Figure 15(a).

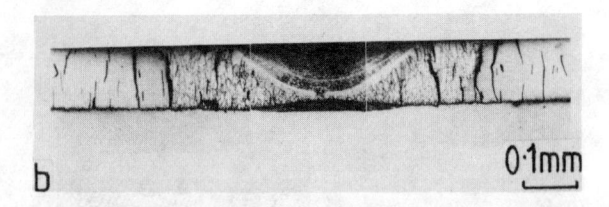

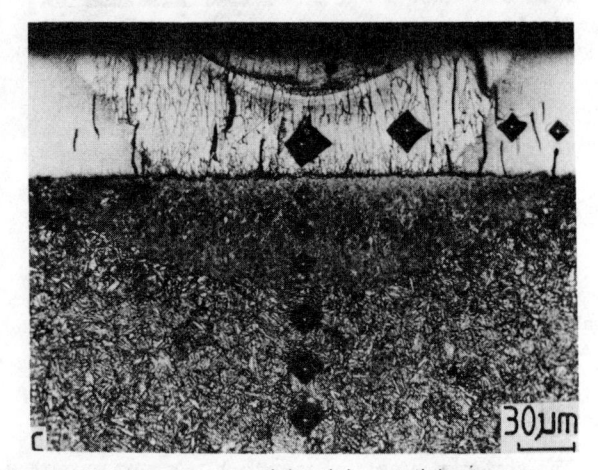

Figure 15. Micrographs (a), (b) and (c) show the laser treated chromium electroplate with new grains formed, which are much softer than the untreated electroplate [(a) and (b), 1.5kW, 18mm/s, 0.2mm B.D. and plating thickness 100-120μm, (c) same as (a) and (b) but at 40mm/s].

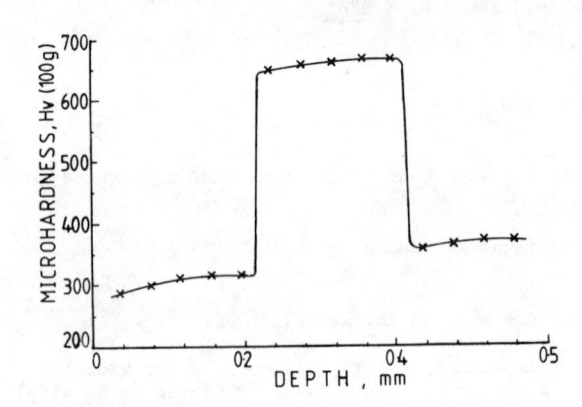

Figure 16. Hardness of laser treated chromium plated steel as a function of depth (1.5kW, 20mm/s and 37μm plating thickness).

Overlapping laser traces, with the process parameters selected to melt completely the chromium plating, produced a uniform layer of iron-chromium alloy; its hardness variation with depth measured from the laser melted surface, is shown in Figure 16. The new iron-chromium alloy formed is relatively soft, with an average hardness of about 320 VHN (100g load) having underneath it a transformation hardneed steel (untempered martensite) of hardness about 370 VHN (100g load). The hardness of the new iron-chromium alloy layer depends on the concentration of chromium in the layer which is a function of the plating thickness and the laser processing parameters.

An intermediate layer is formed between the partly melted chromium electroplate and the underneath steel. Stereoscans (c) and (d), Figure 13 show the partly melted electroplate with the small lenticular pool beneath it and the distribution of chromium in the pool; previously it was suggested that there was little flow within this region. This point is emphasised by the two layers seen in this region, Figure 17 (a) and (b), with (b) being an enlargement of (a) to show the microcrystalline structure of the rapidly solidified steel underneath the lenticular pool. An intermediate layer of intermediate hardness, about 420 VHN (100g load) was produced in the region of the lenticular pool, which is an alloy of chromium and iron in agreement with Montgomery's work (3). Micrograph (b) of Figure 18 is an enlargement of (a) in the same figure, showing the intermediate layer to have a microcrystalline structure at the chromium side of the layer, which is likely to be recrystallized chromium grains.

Figure 17.
(a) Intermediate layer between the partly melted chromium and the substrate steel
(b) Enlargement of micrograph (a) to show the lenticular shape of the intermediate layer and the microcrystalline structure of the rapidly solidified substrate steel (1.5kW, 0.2mm B.D., 18mm/s and 100-120μm plating thickness).

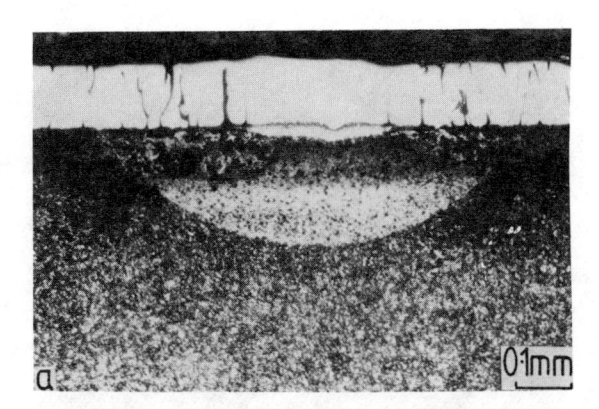

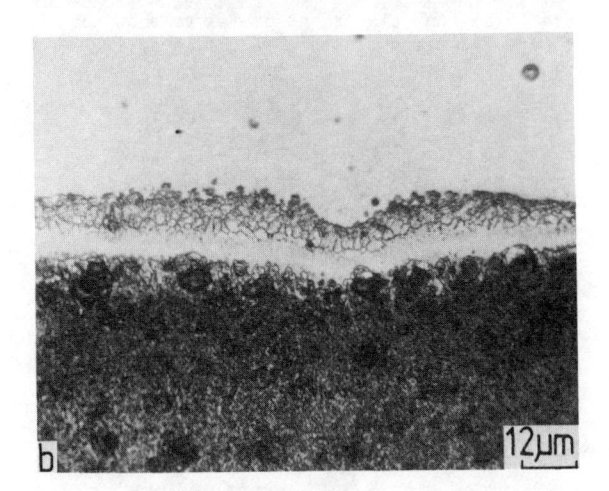

Figure 18. (b) Enlargement of micrograph (a), to show a microcrystalline structure on the chromium side of the intermediate layer.

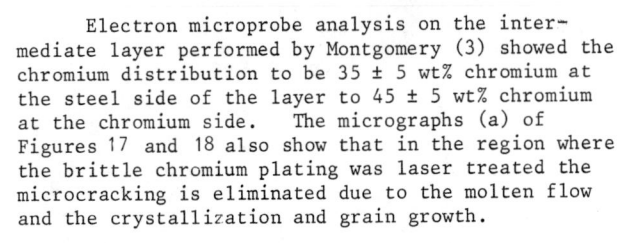

Figure 19.(a) Iron-chromium alloy layer, produced by overlapping laser trace, on chromium plated steel (1.5kW, 20mm/s, 1.0mm B.D. and 30μm plating thickness).

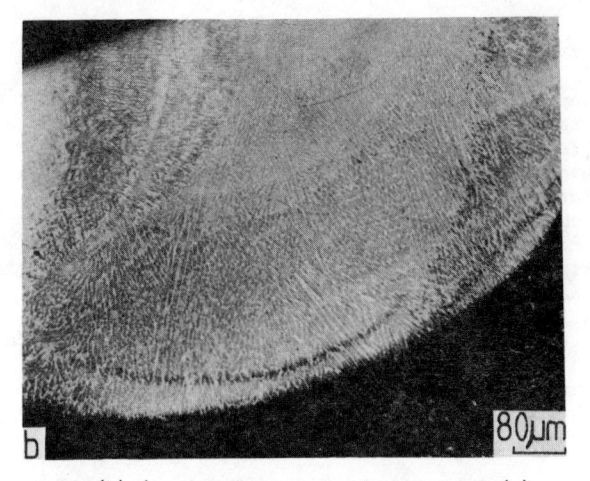

Figure 19.(b) is an enlargement of part A of (a), to show the microcrystalline structure of the rapidly solidified iron-chromium alloy layer

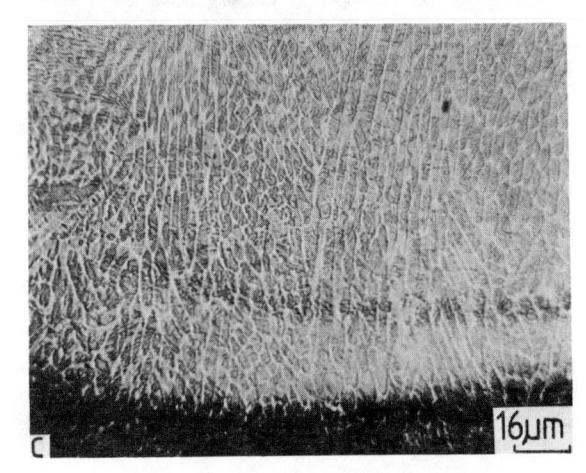

Figure 19(c) is an enlargement of (b) to show epitaxial growth from previously existing grain boundaries.

Electron microprobe analysis on the intermediate layer performed by Montgomery (3) showed the chromium distribution to be 35 ± 5 wt% chromium at the steel side of the layer to 45 ± 5 wt% chromium at the chromium side. The micrographs (a) of Figures 17 and 18 also show that in the region where the brittle chromium plating was laser treated the microcracking is eliminated due to the molten flow and the crystallization and grain growth.

Overlapping laser traces with processing parameters chosen to melt completely the 30μm chromium plating is shown in Figure 19(a). Enlargements of part A of micrograph (a) in Figure 19, are shown in micrographs (b) (c) and (d) in the same figure, illustrating the microcrystalline structure of the rapidly solidified iron-chromium alloy. Micrographs (c) and (d), Figure 19 show the growth to be epitaxial extending from pre-existing boundaries. Similar observations were reported by Weinman et al (12) when studied the laser treatment of chromium plated low carbon steels. The small dendrite sizes (∿2μm) observed in micrographs (b) and (c) of Figure 19 are evidences that the cooling rates are extremely rapid.

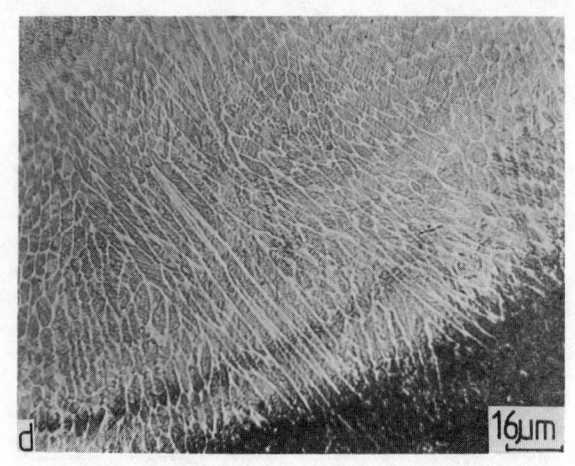

Figure 19(d). An enlargement of (b) to show epitaxial growth from previous existing grain boundaries.

Figure 20(b). An enlargement of (a) to show the microstructure of the layer.

The columnar growth in Figure 19(b) and (c) extends in the direction of the thermal gradient. An intermediate layer was produced between the rapidly solidified iron-chromium alloy layer and the transformation hardened steel as shown in the micrographs (a) and (b) of Figure 20. Its hardness is about 420 VHN, and showed no tendency to crack at hardness impressions loads, see Figure 20(c), suggesting that it is not excessively brittle as the untreated electroplate. The micrograph (d) in Figure 20 is an enlargement of micrograph (a) in the same figure to show the rapidly solidified microstructure of the iron-chromium layer produced by overlapping laser traces to melt completely the 37μm thickness chromium plating. The newly formed grains are seen to be extending in the direction of thermal gradient of the last laser pass.

Figure 20(c). An enlargement of (a) to show that hardness impressions reveal an intermediate layer produced between iron-chromium alloy layer and steel.

Figure 20(a). Iron-chromium alloy layer produced by overlapping laser traces on chromium plated steel (1.5kW, 20mm/s, 1.0mm B.D. and 37mm plating thickness).

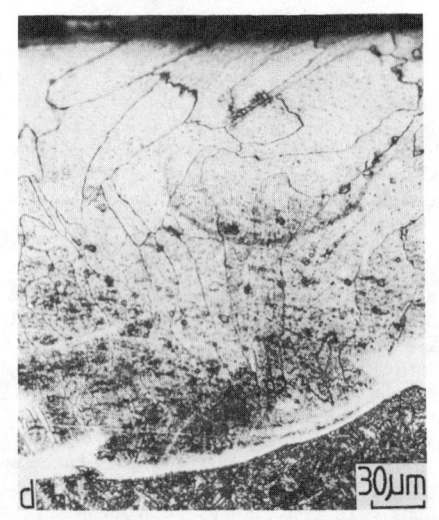

Figure 20(d). An enlargement of (a) to show extended grains in the direction of thermal gradient.

CONCLUSIONS

1. **Shapes and size of affected regions**

    (a) The operating laser parameters to obtain the three basic structures, i.e. (i) complete melting (ii) part melting and (iii) no melting of chromium electroplate for four different plating thicknesses and two beam diameters are shown in Figure 5.

    (b) The depth of laser affected zone, i.e. HAZ + LMZ and LMZ only, was found to be linearly related to the parameter $\frac{P}{\sqrt{DV}}$ and dependent on the variation in thermal properties.

    (c) Correlations between width of laser affected zone and traverse speed, beam diameter and plating thickness have been established for the range of variables used.

2. **Structural examination and hardness**

    (a) Hardness values in the melted zones vary with the level of chromium dilution which itself depends on the plating thickness and the other processing parameters.

    (b) Laser surface treatment softens the originally hard chromium plating forming new grains by the process of recyrstallization and grain growth. Microcracking usually present in the brittle chromium plating is eliminated by laser treatment due to the molten flow and recrystallization.

    (c) By controlling the laser processing parameters, it is possible to obtain partial melting of the electroplate and melting of the underneath steel to form a lenticular shape intermediate layer. The intermediate layer is an iron-chromium alloy of hardness about 420 VHN and has a fine grain size microstructure. Underneath the layer on the side of the steel there is a region of microcrystalline structure which is produced by rapidly solidified steel. On the other side of the intermediate layer lies the recrystallized chromium plating.

    (d) Overlapping laser traces with the processing parameters chosen to melt completely the electroplate, a uniform layer of an iron-chromium alloy is formed having a smooth surface finish. Between the relatively soft iron-chromium layer and the transformation hardened steel, an intermediate layer, having similar properties as the one seen above, is produced.

ACKNOWLEDGEMENTS

This work has been carried out with the support of procurement executive of the Ministry of Defence.

We also acknowledge the interest and financial support from the US Army European Office.

REFERENCES

1. R.S. Montgomery, Muzzle wear of cannon, Wear, 33 (1975) 359-368.
2. A. Brenner, P. Burkhead and C. Jennings, Physical properties of electrodeposited chromium, J.Res.Nat.Bur.Stand., 40 (1948) 31-59.
3. R.S. Montgomery, Laser treatment of chromium plated steel, Wear, 56 (1979) 155-166.
4. C.A. Snavely and C.L. Faust, Studies on the structure of hard chromium plate, Trans. Electrochem.Soc. 97 (1950), 99-108.
5. H.L. Wain, F. Henderson and S.T.M. Johnstone, A study of the room-temperature ductility of chromium, J.Int.Met., 83 (1954-55) 133-142.
6. H.L. Wain, F. Henderson, S.T. Johnstone and N. Lovat, Further observations on the ductility of chromium. J. Inst.Met. 86 (1957-58) 281-288.
7. R.E. Cairns and N.J. Grand, The effects of carbon, nitrogen, osygen and sulphur on the ductility-brittle fracture temperature of chromium. Trans. A.I.M.E., 230 (1964) 1150-1159.
8. G.C. Lim and W.M. Steen, Optics and Laser Technology, June 1982, p.149.
9. G. Christodoulou and W.M. Steen, February 1982. MOD (P.E.) unpublished report.
10. W.M. Steen and C. Courtney, Metal Technology, December 1979, p.456.
11. G. Christodoulou and W.M. Steen, Proc. Laser '81 Opto-Electronik 5th International Congress, p.119, Munich.
12. L.S. Weinman, J.N. DeVault and P. Moore, in Applications of Lasers in Materials Processing. American Society for Metals, Cleveland 1979, Metzbower, E.A., Editor, 245-257.

LASER SURFACE ALLOYING LOW CARBON STEEL

T. Chande and J. Mazumder

Department of Mechanical and Industrial Engineering
1206 West Green Street
University of Illinois at Urbana-Champaign
Urbana, IL 61801

Abstract

Laser surface alloying, the process of altering surface elemental compositions, has been of interest for some time now. This paper presents the results of an initial systematic study of LSA using a statistical experimental design technique to determine the relationship between process variables and the dimensions, solute content and microstructure of the LAZ. Interactions between process variables are discussed and reproducibility of the process assessed. This information is used to develop contour plots for solute content of LAZ. Dimensionless plots are developed that provide a link between microstructural refinement, solute content and processing conditions. This information can be used to obtain improved metallurgical control in LSA. The mechanism of mass transport in the pool is discussed.

1. Introduction

Laser surface alloying (LSA) is the process of altering surface elemental compositions by adding small quantities of alloying elements to a pool of molten metal produced on a substrate by controlled laser irradiation. Convection in the melt pool and subsequent solidification establish the modified surface compositions. A wide range of solute concentrations can be easily obtained by changing processing conditions, including extended solid solutions. Herein lies the vast potential of LSA for creating surfaces for diverse applications.

Laser surface alloys are used in applications that are composition and structure sensitive, such as those relating to corrosion or wear. Thus, laser alloyed microstructures are of considerable metallurgical and engineering interest. Development of laser alloyed structures is influenced by the nature and relative amounts of the alloyed species, their distribution and the cooling rates that occur during laser processing. The precise combination of these factors is determined by processing conditions. A knowledge of the effects of the many process variables is thus desirable to gain a fundamental understanding of laser alloyed structures. This understanding will help us take advantage of the benefits available from LSA.

The biggest advantage of LSA is that it simultaneously protects the surface against chemical, mechanical and thermal forces. Firstly, the alloyed layer physically separates the substrate from the environment. Secondly, the laser alloyed zone has a refined microstructure that provides protection against chemical attack as well as thermal forces.

Further, the refined microstructure and the strong adhesion between LAZ and substrate due to the metallurgical bond between them provides excellent resistance to mechanical forces. LSA also enjoys other advantages over conventional coating processes. After LSA, due to the small heat affected zone, bulk characteristics of the substrate are retained. Very small amounts of alloying element are actually used in LSA so that scarce alloy materials are conserved. In addition, the high cooling rates that exist during LSA could help generate novel microstructures.

While LSA has been of interest for some time now, a timely survey of the state of the art [1] found no systematic study of LSA to establish the effects of the many process variables on the dimensions of the laser alloyed zone (LAZ), its solute content, microstructure and the resultant mechanical and metallurgical properties. While the feasibility of producing highly alloyed, fairly uniform LAZ using high power CW CO_2 lasers on ferrous substrates seems well established [2-4], only limited process related information is available. Salient features of LSA have been identified [2,4] and some data of varying traverse speed on chromium content and dimensions of LAZ are available [3]. An acceptable processing region of power-speeds for cast-irons has been reported [5] (See Table 1). Efforts to explain surface topography have also been made [6,7].

Thus, insufficient information is available on the influence of the magnitude of the principal process variables and on interactions between them. Fluid flow in the melt pool could play an important role in bulk mass transport in LSA, but its nature and behavior have not been clarified. Boundary conditions governing local diffusion and its rate have not been fully explored. There exists no rational basis upon which to select a set of operating conditions and the question of repeatability has not been addressed. Very little information is available by way of correlating LAZ microstructure and the resultant mechanical and metallurgical properties.

This paper presents the results of an initial systematic study of LSA using a statistical experimental design technique to determine the relationship between process variables and the dimensions, solute content and microstructure of the LAZ. Interactions between process variables are discussed and reproducibility of the process assessed. This information is used to develop contour plots for solute content of LAZ. Dimensionless plots are developed that provide a link between microstructural refinement, solute content and processing conditions. This information can be used to obtain improved metallurgical control in LSA. The mechanism of mass transport in the pool is discussed. Additional details are available elsewhere [10-12].

2. Experimental Procedure

In this study, the power, beam diameter and traverse speed were studied using a factorial experimental design [8,9]. Each factor was studied at two levels, the power at moderately high levels, the beam diameter at focus and as defocused, the speed in the low range and the thickness at the modest level (see Table II). The raw data was analyzed as outlined in Box [9].

Bars of AISI 1020 steel (50 mm wide, 22 mm thick and 300 mm long) were used as substrates. These bars were ground and electroplated with 1 micron layer of nickel, a copper flash being used to facilitate the electroplating. Before laser irradiation, all samples were held at 773 K for 90 minutes for degassing.

During laser processing, a closed loop monitor was used for continuous sampling of output power using a fast infrared dector and adjustment of electrical input to the laser to maintain a constant ±3 percent output power. The beam diameter was determined by burning lucite. A layer of helium gas flowing transverse to the direction of traverse parallel to the surface was used to shield the bead.

Samples for metallographic examination were collected 10 mm from the edge of the bar, sections examined being transverse to the laser traverse direction. Width and depth of the LAZ and coating thickness were measured from low magnification photographs. Nickel content was measured by electron probe microanalysis technique. Resolution of the beam was about 1 to 2 microns. Two traces, perpendicular to one another, from bottom to top and across the bead were made.

3. Results of Factorial Experiment
3.1 Width of LAZ

An increase in total beam power increased the width of LAZ, while increasing the speed decreased the width. The main effect of doubling the speed from 25 to 50 mm/s was 14 percent greater in absolute magnitude than the main effect of an increase in total beam power from 4 to 6 kW. With an increase in total power, more energy will be dumped on the surface, this will steepen the temperature gradients, strengthening surface-tension driven fluid-flow [10]. Heat transfer along the substrate by convection and conduction will increase, thus increasing the width. On the other hand, for a given incident beam power, an increase in traverse speed will restrict the duration of laser material interaction. The substrate will absorb less of the incident power, leading to a decrease in the width. In the upper limit, as speed is monotonically increased, the laser-material interaction will have too little time to produce any appreciable heating effects, and the substrate and laser are than said to decouple. This is to be seen in the data of Weinman, DeVault and Moore [3].

3.2 Depth of LAZ

An increase in the power from 4 to 6 kW increased the depth, on an average, by 0.66 mm, irrespective of the levels of diameter and speed. An increase in total power will supply more energy at the surface that will enhance heat flow into the substrate increasing the depth of LAZ.

An increase in both, the speed and diameter caused the depth to decrease. An increase in the speed restricts the laser-material interaction time resulting in a lowered energy transfer to the substrate. The same effect is produced by an increase in beam diameter due to the decrease in incident power density. Both lead to a decrease in depth. Speed and beam diameter were found to interact, i.e. the change in depth produced due to a variation in one depended on the level of the other.

A closer look at the DS (diameter x speed) interaction (see Fig. 1), reveals that while increasing the speed at either setting of the beam diameter produces a lower average depth, the relative decrease in the average depth is greater when the focused beam is in use (-1.14 mm) than when the defocused beam is used (-0.5 mm) as the speed is doubled from 25 to 50 mm/s. This points to a transition in the sensitivity of the process to changes in the interaction time over ranges of power density. This may be related to the overall rate of energy deposition at the surface and to the nature of the laser-material interactions, such as "keyholing". The onset of "keyholing" depends on a critical absorbed power, and leads to improved heat transfer from laser beam to substrate. Thus the temperature gradients could change to a greater extent with an increase in speed at the high power density level that at the lower level of power density.

3.3 Ni Content of LAZ

An increase in the power level decreases the solute content, while an increase in diameter and speed increases it. This is also related to the melt volume. An increase in the power increases both width and depth of the LAZ. As the amount of alloying element (determined by thickness of the coating and the irradiated area) must now be dispersed through a greater melt volume, a lower average solute content results. An increase in the diameter significantly increases the width, while an increase in the speed decreases the depth and the width. Both result in lower melt volumes and greater average solute contents.

An increase in the diameter causes a greater increase in nickel content (1.48 percent) at a lower power level than at the higher one (0.37 percent) (see Fig. 2). This is related to the melt volume, a reduced melt volume resulting from a decrease in power density at the lower power level than at the higher one.

At higher power levels, energy is still available at a sufficient rate to melt relatively large volumes with lower interaction times at higher speeds than at the lower power level. While an increase in the speed does decrease the melt volume and increase the solute content at low and high diameters, the effect on solute content (See Fig. 3) is greater at the higher diameter (1.4 percent) than at the lower one (0.57 percent) consistent with the picture of reduced melt volumes and sensitivity transitions mentioned above.

3.4 Fluctuations About the Average Composition

An analysis of the composition printed out every 10 seconds during an EPMA trace helps to quantify the mean fluctuations about the average composition of the LAZ. On analysis of this data, the fluctuations about the mean composition were found to increase as the diameter and speed increased (See Fig. 4). Thus,

Table 1 Summary of Reported Operating Conditions

	Gnanamuthu [2]	Weinman [3]	Belomondo [5]
Substrate material	AISI 1018 steel	AISI 1018 steel	Mild and stainless steel; cast irons super alloys
Alloy element	Cr;Cr+C;Cr+C+Mn Cr+C+Mn+Al	Cr;Cr+Ni	Mo+Cr+Cr-carbide+Ni+Si
Method of Application	Slurry/Spray organic binder	Sputter deposit electrodeposit	Paste/slurry, organic/inorganic binders
Coating thickness (mm)	0.025-0.75	0.002-0.018	Not known
Laser power (kW)	3.4, 5.0, 12.5	7.5 (max)	8.0 - 12.0
Power density W/cm^2	2796 - 4470	10^7	12500 - 18750
Processing speed (mm/s)	1.69 - 21.17	100 - 3750	5 - 15
Beam configuration	Square/rectangular top-hat stationary/ oscillating	Focussed/Gaussian stationary	Square/top-hat oscillating
Interaction time	0.3 - 0.2	$10^{-4} - 10^{-6}$	0.5
Shielding gas	None, He, He+Ar	He	CO_2, Ar, He, N_2
Nozzle configuration	Not known	Nozzle with porous metal insert	25 - 35 mm from fusion at 35 - 45 degrees
Alloy content Percent	0.9-43Cr, 0.5Al 1.4-4.4C;0.5-1.3Mn	1-80 Cr	Not known

Table 2 Design Matrix

			Level	
Variable	Unit	Symbol	−	+
Beam Power	kW	P	4	6
Beam Diameter	mm	D	0.8	1.28
Beam Speed	mm/s	S	25	50

	Levels		
Treatment No.	P	D	S
1	−	−	−
2	+	−	−
3	−	+	−
4	+	+	−
5	−	−	+
6	+	−	+
7	−	+	+
8	+	+	+

Thickness of nickel coating was 1 micron.

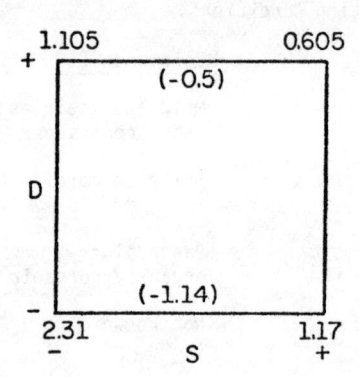

Figure 1 Analysis of D x S Interaction of Dept of LAZ (figures at vertices are recorded depths (mm) averaged over eight samples, irrespective of power level. Figures in parenthesis are changes in average depth resulting from an increase in speed at respective diameter settings)

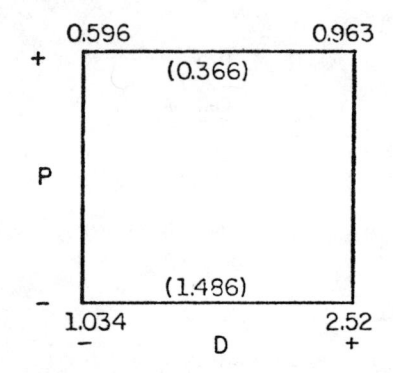

Figure 2 Analysis of P x D Interaction for Nickel Content of LAZ with Same Conventions as in Fig. 1

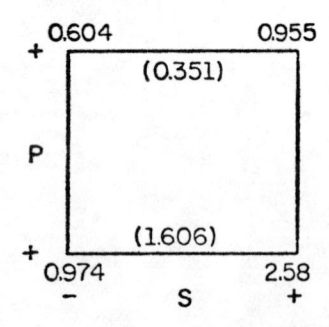

Figure 3 Analysis of the P x S Interaction for Nickel Content of LAZ with Same Conventions as in Fig. 1

the uniformity in the composition increased as the diameter and speed decreased. A decrease in speed implies that the interaction time was greater, the pool was thus molten for a longer time and diffusion on a local scale could be expected to help even out composition gradients to a greater degree, leading to lower fluctuations about the average composition. A decrease in beam diameter steepens the tempeature gradients above the melt pool, which enhances the vigor of fluid flow in the pool. This produces a finer dispersion of solute rich pockets, and local diffusion works to produce uniformly alloyed LAZ's [11].

3.5 Reproducibility of LSA

Replication during this experiment provides an independent measure of the experimental error. The reproducibility of the process may be assessed by comparing the standard error in the mean solute content of all the runs to the mean solute content itself. This figure is ± 9 percent for these set of runs. Repeatability of the process depends upon the stability of the laser and its support systems and the mode of supply of the alloy elements, with the former being the more dominant.

4. Contours for Nickel Content of LAZ

If X and Y are the scaled process variables of interest, contours for the measured response may be plotted (See Figs. 5,6) using an equation of the form

$$Y = (A + BX)/(X + DX) \qquad (1)$$

where A, B, C, and D are constants deduced from experimental data and determined by the particular combination of variables plotted. The value of the contour to be plotted, the mean measured response and the main effect of the variable being held constant add up algebraically to determine A. B and C depend on the algebraic sum of the main effects of X and Y respectively and their two factor interaction with the variable being held constant. If X and Y do not interact, with one another and the variable held constant, then D is zero, and the contours will be linear (See Fig. 6). It follows from Eq. (1) that

$$(dY/dX) = (BC - AD)/(C + DX)^2 \qquad (2)$$

the actual variation being dependent upon the values of the constants. Thus, the change in Y required to maintain a given solute content depends upon the level of the other variables through their interactions and the solute content.

5. Dimensionless Process Parameter Plots

A dimensional analysis of LSA using measurable process variables [11] and known substrate thermophysical properties yields the dimensionless laser surface alloying parameter (LASAP)

$$LASAP = P/(pud^2) \qquad (3)$$

where P is the incident power (W), p is the shielding gas pressure at the surface (N/m^2), u is the traverse speed (m/s) and d is the beam diameter (m). LASAP represents dimensionless operating conditions and may be used to summarize experimental data. A decrease in power density, an increase in speed and an increase in surface shielding gas flow, that will increase surface

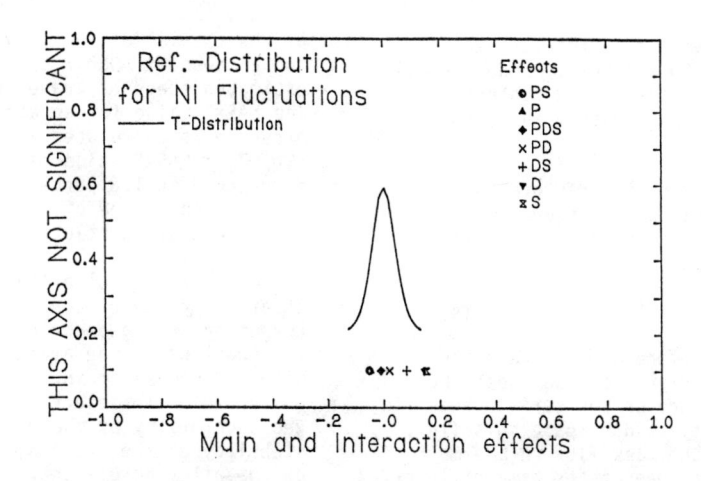

Figure 4   T-Distribution for Fluctuations about Average Composition (effects that lie to the edges of the distribution are statsitically significant, fluctuations increased significantly with an increase in speed and beam diameter)

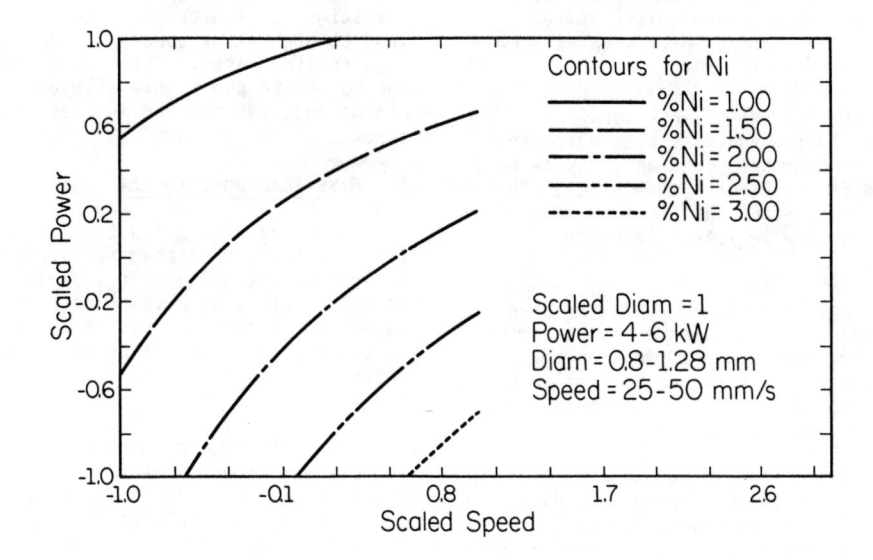

Figure 5   Contour Plots for Nickel Content of LAZ at Constant Scaled Diameter of 1

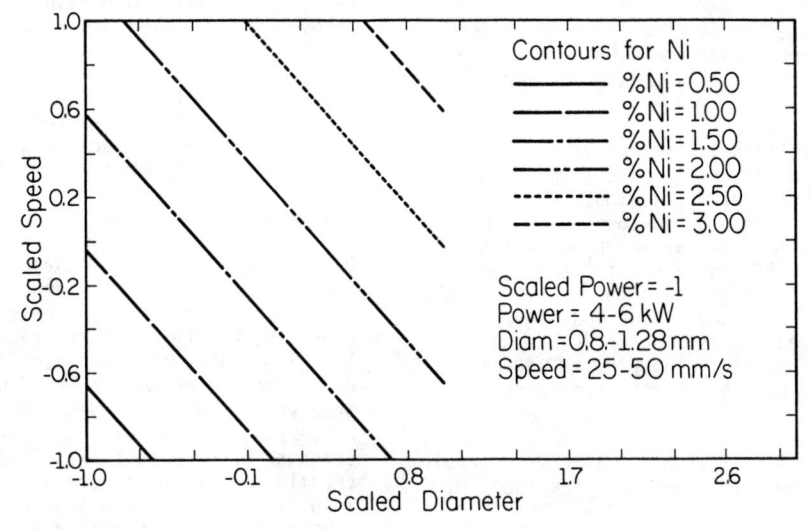

Figrue 6   Contour Plots for Nickel Content of LAZ at Constant Scaled Power of -1.

convective heat losses, will all diminish heat transfer from laser beam to substrate, reducing melt volumes and increasing average solute contents. LASAP is proportional to the energy density in the melt pool.

Another dimensionless number of interest, LAPP also yields a dimensionless set of operating conditions, but scales these by the thermophysical properties of the substrate

$$LAPP = Pd/(\alpha \rho u^2 d^2) \qquad (4)$$

where P, d and u are as before and ρ, α are the density and thermal diffusivity of the substrate. The ratio $(P/d^2\rho)/(\alpha)$ is measure of the surface heat flux, to the substrate's ability to dissipate this flux. $(1/\rho u)$ is a measure of the mass flux in the melt pool, while (d/u) is a measure of the time of interaction. This number quantifies the complex transport processes in LSA vying with the interaction time. Recognizing that $P/\alpha u$ is a natural scale for viscosity during laser processing, LAPP may be viewed as the inverse of the effective Reynold's number during laser processing. We note that a dimensionless three-dimensional equation for chemical diffusion has the inverse Reynold's number in it.

Data from the present experiments vary hyperbolically when plotted against these parameters (See Figs. 7,8). The geometric constants may be estimated and are noted on the figures themselves. The LASAP plot is made at constant gas pressure, and represents the effect of a variation in energy density.

Data from Weinman, DeVault and Moore [3] have also been plotted against these parameters, assuming an atmospheric gas pressure and a 200 micron beam diameter. Their hyperbolic variation is similar to trends in present data, pointing to the generality of these plots (See Figs. 9,10).

A possible reduction in melt volume with a decrease in LASAP may be detected using this data (see Fig. 11). The ratio of the maximum width of LAZ to its maximum depth may be used to characterize the melt volume. This ratio is seen to increase with a decrease in LASAP, the depth decreasing more rapidly than the width, resulting in reduced melt volumes.

A theoretical prediction of pool geometry from a three-dimensional temperature conduction model [13] using constant properties shows trends similar to that seen above (See Fig. 12).

Higher (W/D) ratios occur at low values of LASAP because a decrease in the energy density pool subjects it to a higher rate of heat extraction. Finer microstructures should result for lower values of LASAP. Microstructures from samples processed at LASAP values of 0.5×10^6 and 2.5×10^6 are shown in Figs. 13 and 14 to illustrate this point. The sample processed as the higher diameter and the higher speed shows the more refined microstructure at the same magnification (see Fig. 13). The LAPP, LASAP and contour plots can be used to select processing conditions.

As a numerical illustration, consider the problem of generating an alloy with 2 percent nickel on 1020 steel. From Fig. 8 for 2 percent nickel, the corresponding value of LAPP is approximately 3.5×10^{10}. Using a value of the density of 7863 kg/m^3 and of the thermal diffusivity of 0.6×10^{-5} m^2/s, choosing a power of 5000 W and beam diameter of 1 mm, the estimated speed of traverse of the beam is 55 mm/s. The LASAP value for an atmospheric surface gas pressure is calculated as about 0.9×10^6. From Fig. 7, a LASAP value of 0.8×10^6 is seen to correspond to 1.8 percent nickel. Thus, one can estimate the range of experimental variables that can give the desired solute content.

Actual control of the composition may now be obtained by use of the contour maps and recognizing the underlying geometrical nature of these plots. The fluctuations in the average composition will increase with a decrease in interaction time. This can be quite useful in selecting the operating conditions. Reproducibility of the process will be governed by the stability of the laser system and the method of supply of the alloying element. LAPP and LASAP plots can thus be used to plan LSA experiments.

The composition of laser alloys is perhaps the principal metallurgical characteristic governing their successful application. As the result of this investigation indicate, processing conditions that lead to high alloy specie concentrations also produce high cooling rates. This happy coincidence enables one to obtain the highly alloyed and refined microstructures desired in laser surface alloys in one stroke.

6. Mass Transport in LSA

Assuming Fick's law to be applicable, the thin film solution, if fitted for a typical LSA run, yields an approximate value of $D_{eff} = 10^6$ cm^2/s [10]. This unrealistically high value of D (by 10 orders of magnitude) indicates that there must be a basic change in the mechanism of solute redistribution with the onset of fluid flow which significantly accelerates the process.

Two principal components of fluid flow may be identified, front-to-back, and across the melt pool. The advancing beam melts fresh material and fluid motion tends to finely disperse the newly melted alloy-rich layer from the surface throughout the pool. Chemical diffusion works to even local composition gradients. Given a fine enough dispersion of solute-rich volume elements, the higher diffusivity in the liquid state would help to produce a uniformly alloyed LAZ. A fine dispersion results as the vigor of fluid flow, defined as the volume of liquid moved per unit time, increases. The cross flow is the component that brings about mixing affects and is generated by surface tension gradient mechanism. The magnitude of the surface tension gradients is related to the surface temperature gradients as [10]

$$\partial\sigma/\partial x = (\partial\sigma/\partial T)(\partial T/\partial x) = [-C_p(1 + \ln(T/T_0))] \partial T/\partial x$$

where C_p is the specific heat and T_0 is a reference temperature. While the temperature coefficient of surface tension is only slowly varying with temperature, the temperature gradient approaches 10^5 K/cm. Thus the magnitude of $(\partial\sigma/\partial x)$ is primarily dependent on the magnitude of the temperature gradient $(\partial T/\partial x)$. A decrease in beam diameter will raise power densities and temperature gradients, strengthening cross flow. This increases the vigor of the fluid flow and a more uniformly alloyed LAZ should result. This has been seen experimentally (see section of fluctuations in average composition).

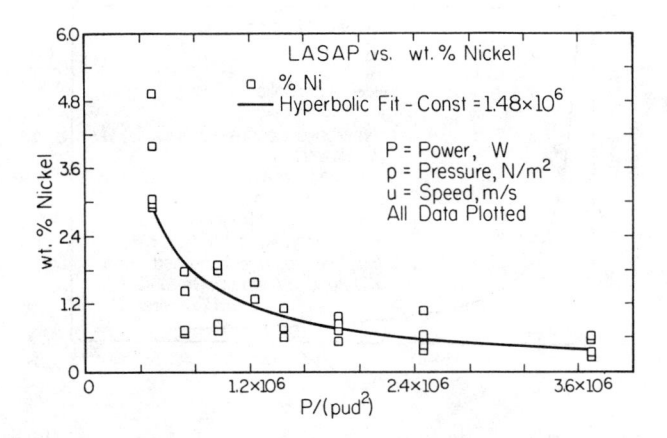

Figure 7 Nickel Content plotted against the Laser Surface Alloying Parameter, LASAP, at Constant Gas Pressure

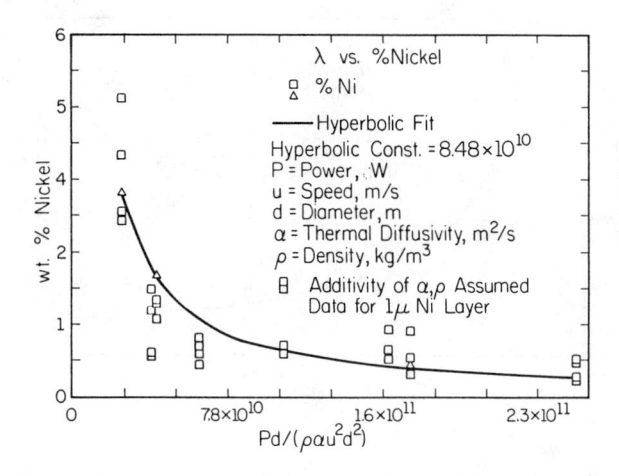

Figure 8 Nickel Content versus LAPP, the Laser Processing Parameter.

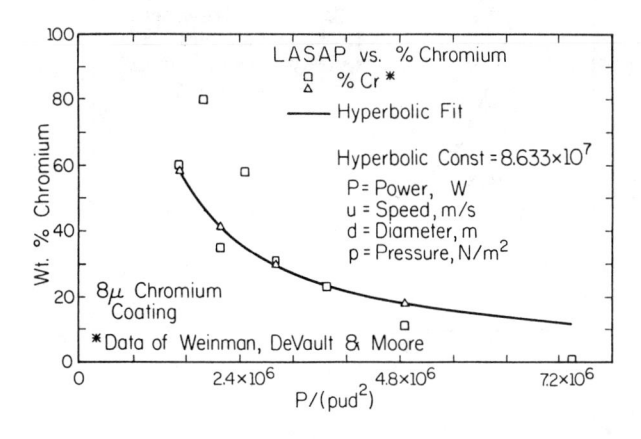

Figrue 9 Chromium Content versus LASAP from Weinman [3] at Constant Gas Pressure

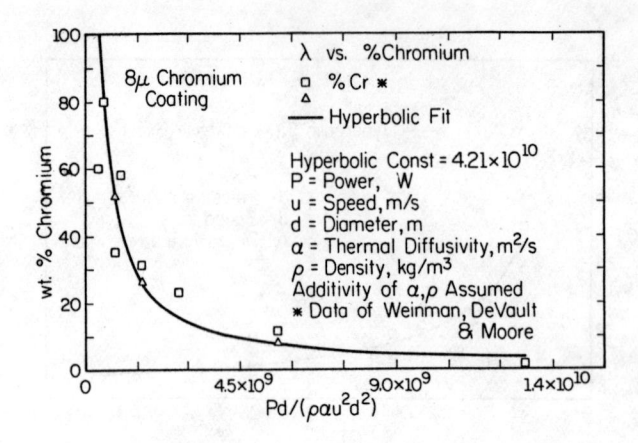

Figure 10 Chromium Contents from Weinman [3] versus LAPP.

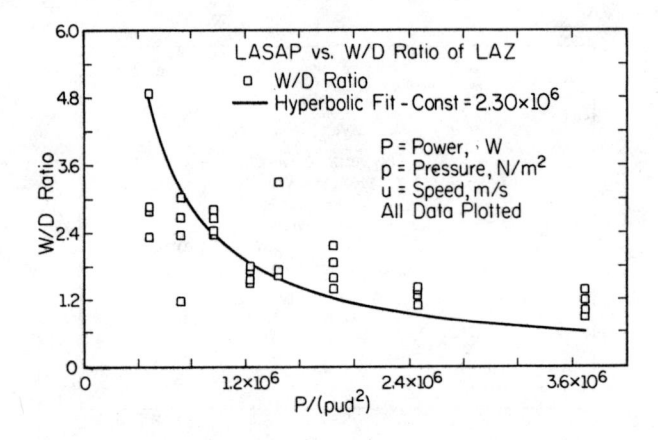

Figure 11 The Ratio Width to the Depth of the LAZ plotted versus LASAP with Constant Gas Pressure

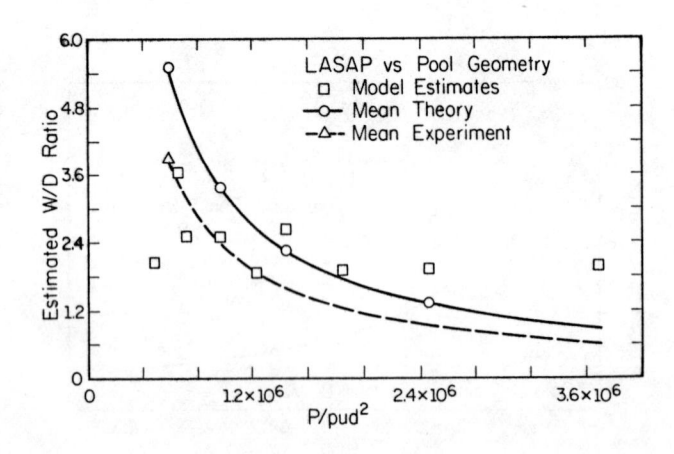

Figure 12 Theoretical Prediction of Pool Geometry versus LASAP at Constant Gas Pressure from Three-Dimensional Heat Conduction Model [13]

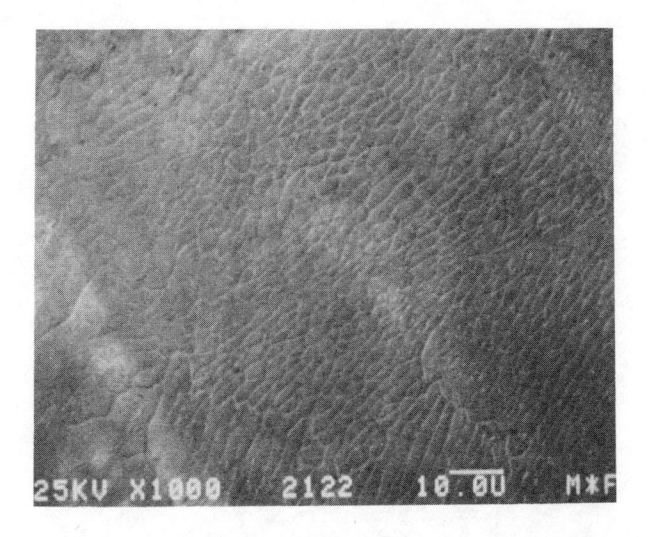

Figure 13 Cross Section of a Bead processed at 4 kW, 1.28 mm Beam Diameter, and 50 mm/s Traverse Speed at an LASAP Value of 0.5×10^6

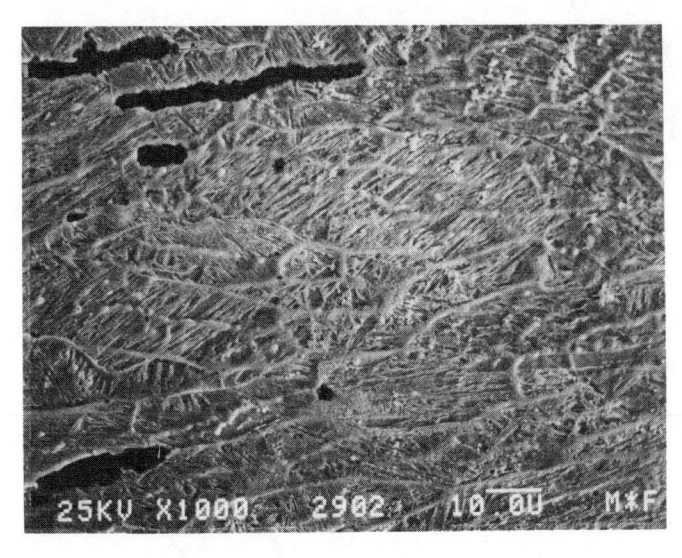

Figure 14 Cross Section of a Bead processed at 4 kW, 0.8 mm Beam Diameter and 25 mm/s Traverse Speed at an LASAP Value of 2.5×10^6

7. Conclusion

The results of this study provide specific answers to questions a prospective user of LSA may have on the effects of a variation in processing conditions, on the effects of the magnitude of process variables, on the mechanisms involved, on the reproducibility of LSA, on the selection of processing conditions and the resultant alloy composition and refinement. This information can be used for a systematic utilization of LSA.

References

1. C. Draper: Lasers in Metallurgy, K. Mukherjee and J. Mazumder, eds., p. 67, AIME, 1981.
2. D. S. Gnanamuthu: Applications of Lasers in Materials Processing, E. A. Metzbower, ed., p. 177, ASM, 1979.
3. L. S. Weinman, J. N. DeVault and P. Moore: Applications of Lasers in Materials Processing, E. A. Metzbower, ed., p. 245, ASM, 1979.
4. P. M. Moore and L. S. Weinman, Proc. Soc. of Photo.-Opt. Instrum. Eng., Vol. 198, San Diego, p. 120, 1979.
5. A. Belomondo and M. Castagna, Thin Solid Films, 1979, Vol. 64, p. 249.
6. S. M. Copley, D. Beck, O. Esquivel and M. Bass: AIP Conf. Proc. No. 50, Symposium on Laser-Solid Interactions and Laser Processing, S. D. Ferris, H. J. Leamy and J. M. Poate, eds., p. 161, 1978.
7. T. R. Anthony and H. E. Cline, J. Appl. Phys., 1977, Vol. 48, No. 9, p. 3888.
8. O. L. Davies: The Design and Analysis of Industrial Experiments, p. 253, Hafner Publishing Co., New York, 1954.
9. G. E. P. Box, W. G. Hunter and J. S. Hunter: Statistics for Experimenters, p. 306, John Wiley & Sons, New York, 1978.
10. T. Chande and J. Mazumder, Appl. Phys. Lett., 1982, Vol. 41, No. 1, Vol. 12b, p. 42.
11. T. Chande and J. Mazumder, Opt. Eng., 1983, In Press.
12. T. Chande and J. Mazumder, Met. Trans. B, 1983, (To be published).
13. T. Chande and J. Mazumder: "Lasers in Metallurgy," K. Mukherjee and J. Mazumder, ed., p. 165, AIME, Warrendale, PA, 1981.

LASER FUSING OF HARDFACING ALLOY POWDERS

S. J. Matthews
Cabot Corporation

INTRODUCTION

The use of multi-kilowatt lasers for surface cladding has been acknowledged as a possible method of hardfacing for at least seven years. A 1976 patent[1] entitled "Cladding" teaches the use of direct wire feed under a scanning laser beam. However, British researchers claim the "shadowing" effect of solid wire does not seem to couple the laser energy as effectively as powder, and, hence, powder would be the preferred consumable for surfacing. Steen, for example, has published work using Ni-Cr-Si-B powder fed continuously under a laser beam with a manual powder torch modified to use argon carrier gas rather than oxy fuel combustion.[2] Other variations have been devised for direct powder feed into a laser beam but require powder injection nozzles of special design.[3]

Preplacement of powders onto a substrate prior to laser fusing is another possibility rather than attempting to solve problems of "feeding" a powder consumable directly under or in front of an advancing laser beam. Preplacement of loose powders has been reported in laboratory studies[4] but may not be acceptable for production environments requiring some degree of workpiece handling prior to hardfacing. Thermal spray methods, such as oxyacetylene flame spray[5] and plasma spray[6] have been used to preplace powder onto a substrate for subsequent laser fusing. However, in industrial applications, preplacement of hardfacing powders using these methods may mean increased labor cost and lower powder recoveries since most thermal spray processes are not 100% efficient.

One method for preplacement of powders that appears to be attractive is the use of a bindered paste. The bindering vehicle can usually be thinned with an appropriate solvent so as to produce a spreadable paste of the desired consistency. The use of preplaced pastes is well known in the brazing industry. Interestingly, the use of a laser for brazing has already been reported in the literature.[7] The use of laser to fuse hardfacing alloy powder pastes onto industrial components has also been reported.[8][9]

OBJECTIVE AND SCOPE

The purpose of this investigation was to explore laser fusing of preplaced powder paste as a viable hardfacing method. The scope of the program was limited to metallographic examination of simple bead-on-plate type samples. The objective was to obtain first-hand laboratory experience and understanding of important laser operating parameters which influence hardfacing deposit quality of fused powder pastes.

PROCEDURE

A 1200 watt CO_2 laser was used at full power in all experiments. The beam (TEM_{01}*) was focused through a 15-inch (381mm) focal length zinc selenide lens. A variety of hardfacing alloy powders were obtained for study. The nominal chemical compositions are listed in Table 1. Each powder was blended with an organic binder and then thinned to the desired consistency using tap water. The resulting paste was spread onto a 1/8-inch (3.2mm) thick steel coupon to a thickness of about .045 inch (1.1mm). The surface of each coupon was cleaned by belt grinding prior to application of the paste. Samples were dried for 2 hours in a 300°F (150°C) oven to remove moisture prior to laser fusing. Figure 1 is a schematic illustration showing the experimental specimen arrangement. The beam oscillation amplitude control was set to achieve a fused deposit width of about 1/4-inch (6.4mm). During fusing, an argon cover gas (about 10 ℓ/min.) was passed parallel over the surface of the specimen.

A matrix of nine experiments was conducted using 3 different specimen traversing speeds (4, 8 and 12 IPM*) and 3 different beam oscillation frequencies (10, 50 and 100 Hz). Two cobalt base powders (alloy Nos. 6 and SF6) were used in this portion of the study.

Based upon the results of the oscillation and travel speed studies, additional pastes using all of the hardfacing alloy powders listed in Table 1 were prepared. Specimens were laser fused under consistent parameters of 6 IPM travel speed and an oscillation of 75 Hz.

RESULTS AND DISCUSSION

Figures 2 and 3 are a collection 10X photomacrographs showing (respectively) the

---
* 1 IPM = 0.423 mm/sec

TABLE 1

Nominal Chemical Composition of Hardfacing Alloy Powders Used for Laser Fusing Experiments

Alloy	Cr	C	Si	Mo	Fe	Ni	Co	B
STELLITE® alloy No. SF6	19.0	0.7	2.3	–	3	13.5	Bal	1.7
STELLITE® alloy No. 158	26.0	0.7	1.2	–	–	–	Bal	0.7
STELLITE® alloy No. 6	28.0	1.1	1.0	–	–	–	Bal	–
DELCROME® alloy No. 90	27.0	2.7	–	–	Bal	–	–	–
TRIBALOY® alloy No. T-700	15.5	–	3.4	32.5	–	Bal	–	–
DELORO® alloy No. 60	16.0	0.7	4.2	–	4	Bal	–	3.3
HAYSTELLITE Composite Powder No. 1	50% tungsten carbide, 50% Ni-Cr-Si-B alloy							

®STELLITE, DELCROME, TRIBALOY, DELORO, and HAYSTELLITE are registered trademarks of Cabot Corporation

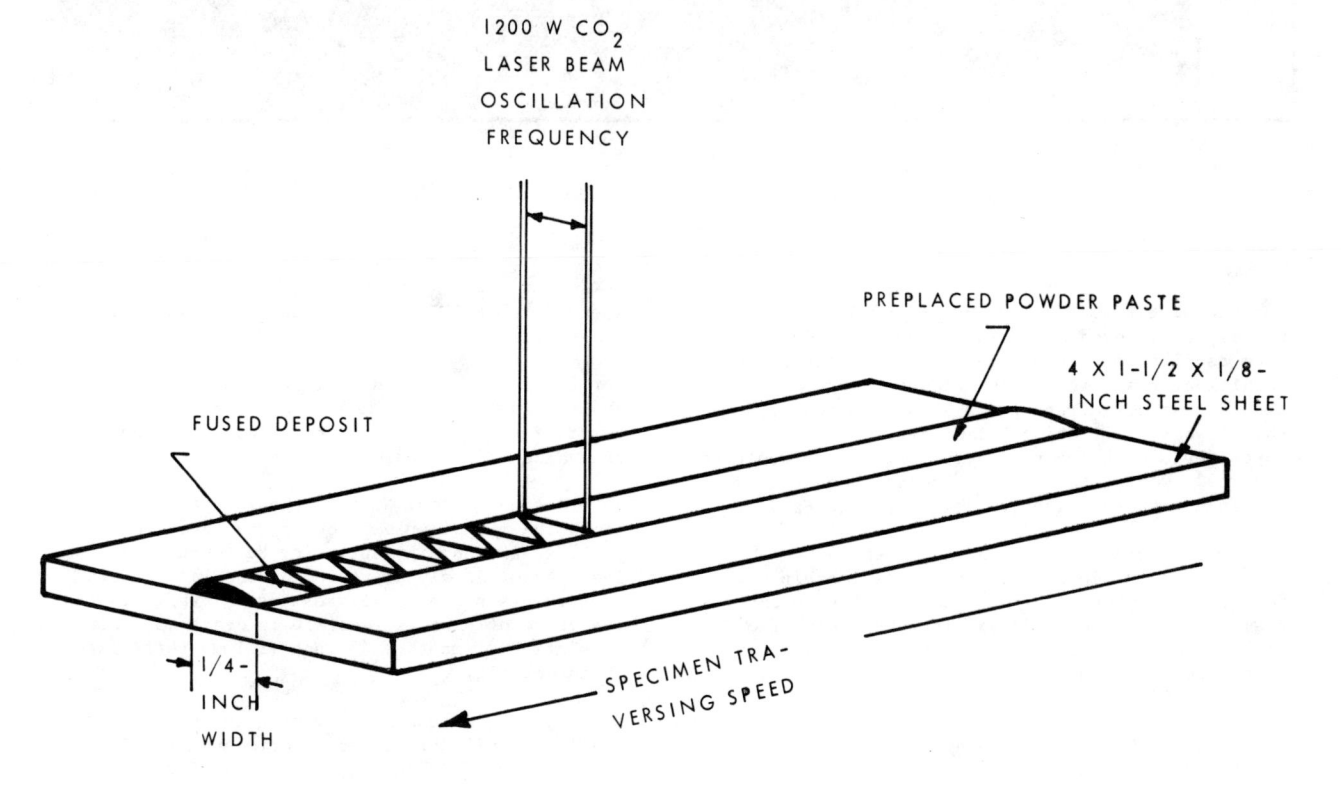

FIGURE 1. Schematic Illustration of Laser Fusing Experiments

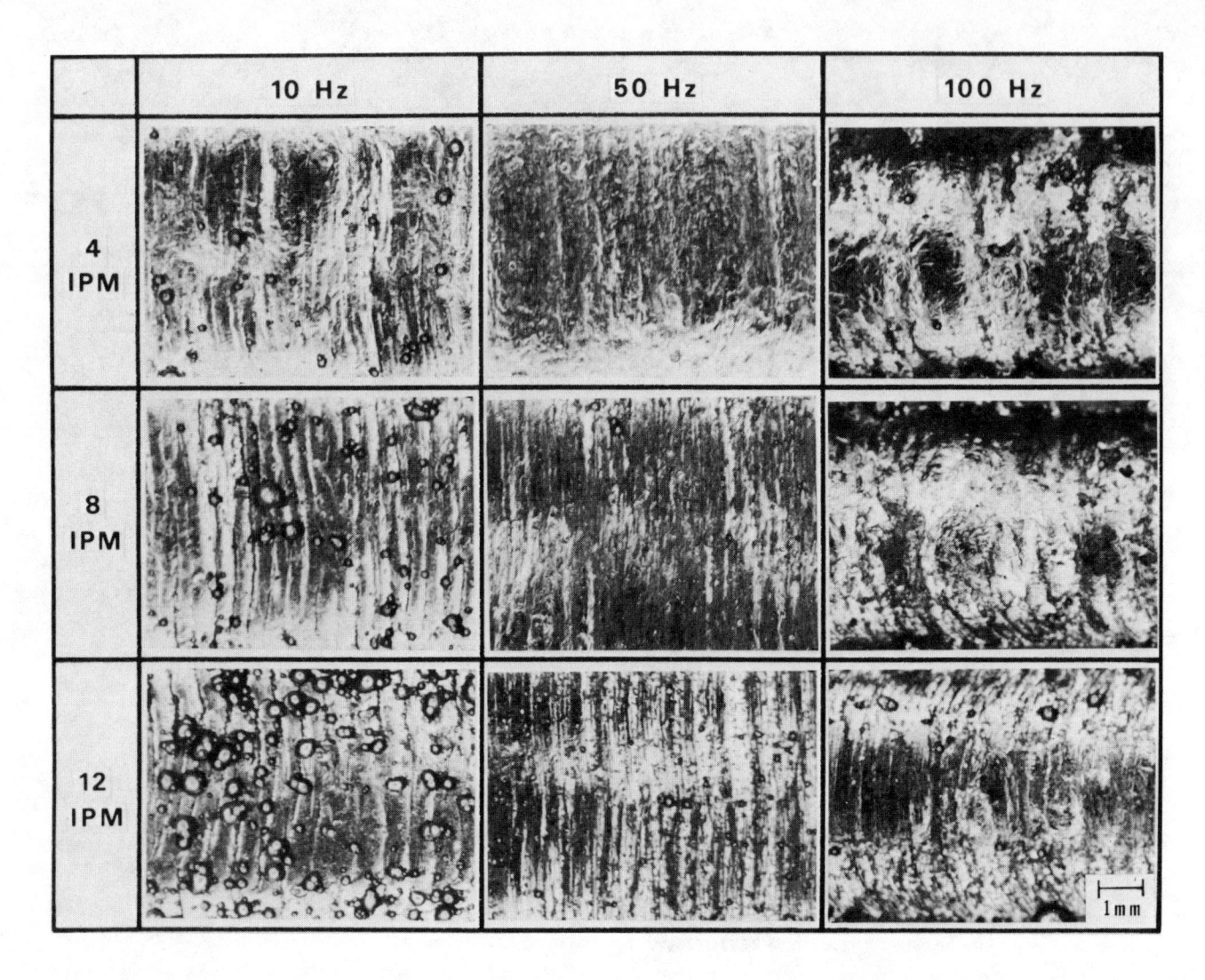

FIGURE 2. Surface of Laser Fused STELLITE alloy No. 6 Powder

surface of STELLITE alloy Nos. 6 and SF6 powder pastes laser fused at various travel speeds and beam oscillation frequencies. In both cases, the conditions resulting in the roughest surface were the fastest travel speed (12 IPM) combined with the slowest oscillation frequency (10 Hz). Experiments with the smoothest surface, as would be expected, were fused at the slowest speed (4 IPM) combined with the highest frequency (100 Hz).

In order to place these two variables in perspective, a simple numerical relationship was derived by recognizing that the oscillating laser beam advances a small linear increment down the centerline of a path each time the beam oscillates back through the middle of its linear path. If the path width is very large compared to beam diameter (d), then the number of times (N) the beam passes through a given point located along the center line can be expressed as

$$N = \frac{120\ fd}{t} \quad \text{(Equation 1)}$$

where f is oscillation frequency (Hz) and t is traversing speed (IPM).

For the laser system used in this study, d is about .020 inches (0.5mm). Substituting this and experimental values of f and t into equation 1, the values of N studied ranged from 2 to 60.

In practical terms, N corresponds to the number of times the midpoint location on any laser fused path "sees" the high intensity heat source of the laser beam. A visual review of the laser fusing experiments shown in Figures 2 and 3 suggests that the value of N be at least 30 in order to achieve a relatively smooth .045 inch (1.1mm) thick laser fused deposit.

Differences in fusing behavior were noted for the 2 alloys studied in those experiments characterized by highest value of N (t = 4 and f = 100). The alloy No. 6 deposit was not as smooth as the alloy No. SF6 deposit. This is because of

	10 Hz	50 Hz	100 Hz
4 IPM			
8 IPM			
12 IPM			

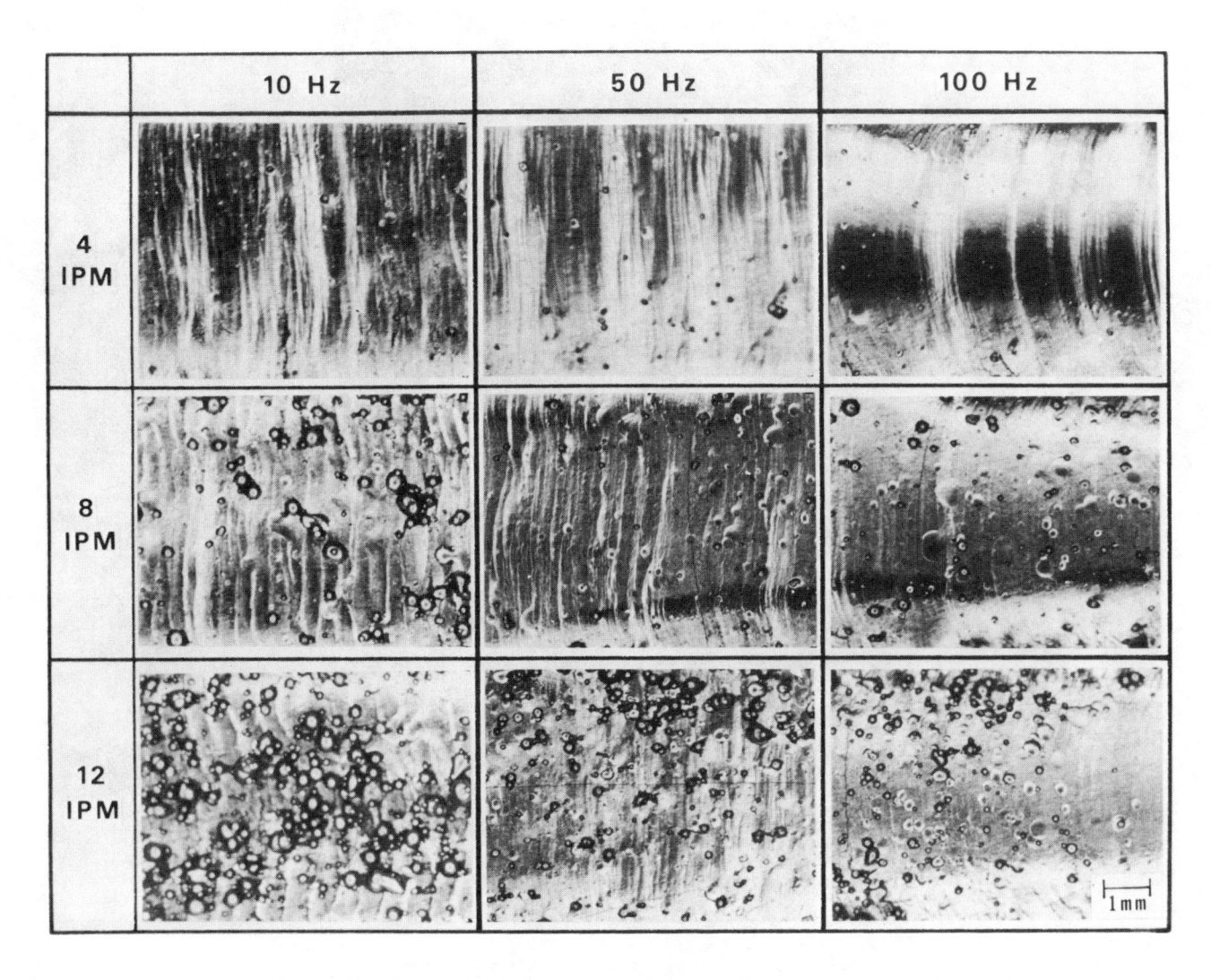

FIGURE 3. Surface of Laser Fused STELLITE alloy No. SF6 Powder

the solidus temperature differences between the two alloys (1247°C versus 1083°C). A lower solidus temperature of the powder being fused however does not seem to offer any benefit when values of N were low (< 10). For example, agglomerated beads of fused powder were observed on the surface of both alloy 6 and alloy SF6 experiments run at 12 IPM/10 Hz (N = 2).

In addition to the observed effect on surface smoothness, the variation in N from 2 to 60 also was found to influence deposit microstructure. Figure 4 shows the etched microstructures of alloy No. 6 laser fused at values of N equals 2, 15 and 60. The coarseness of the solidification microstructure (as evidenced by qualitatively comparing dendrite arm spacing) increased as N increased. The observed effect is consistent with solidification theory which predicts dendrite size to increase with increased superheat associated with the molten metal. Since N is related to the amount of energy received by a unit volume along the centerline of a fused path, high values of N mean higher heat input, hence higher superheat and coarser solidification structure.

The coarse solidification structure of alloy No. 6 shown in Figure 4c is very similar to the solidification structures achieved by the gas tungsten arc and plasma transferred arc hardfacing processes. This observation is significant since it is generally recognized that solidification structure influences wear resistance, with a coarse structure being advantageous over a fine structure in certain wear environments. Opponents of laser hardfacing could argue solidification microstructures may be "too fine" since the process is usually characterized by minimum heat input and rapid solidification. These findings suggest solidification microstructures, however, can be controlled by proper selection of travel speed and oscillation.

a) N = 2   12 IPM/10 Hz

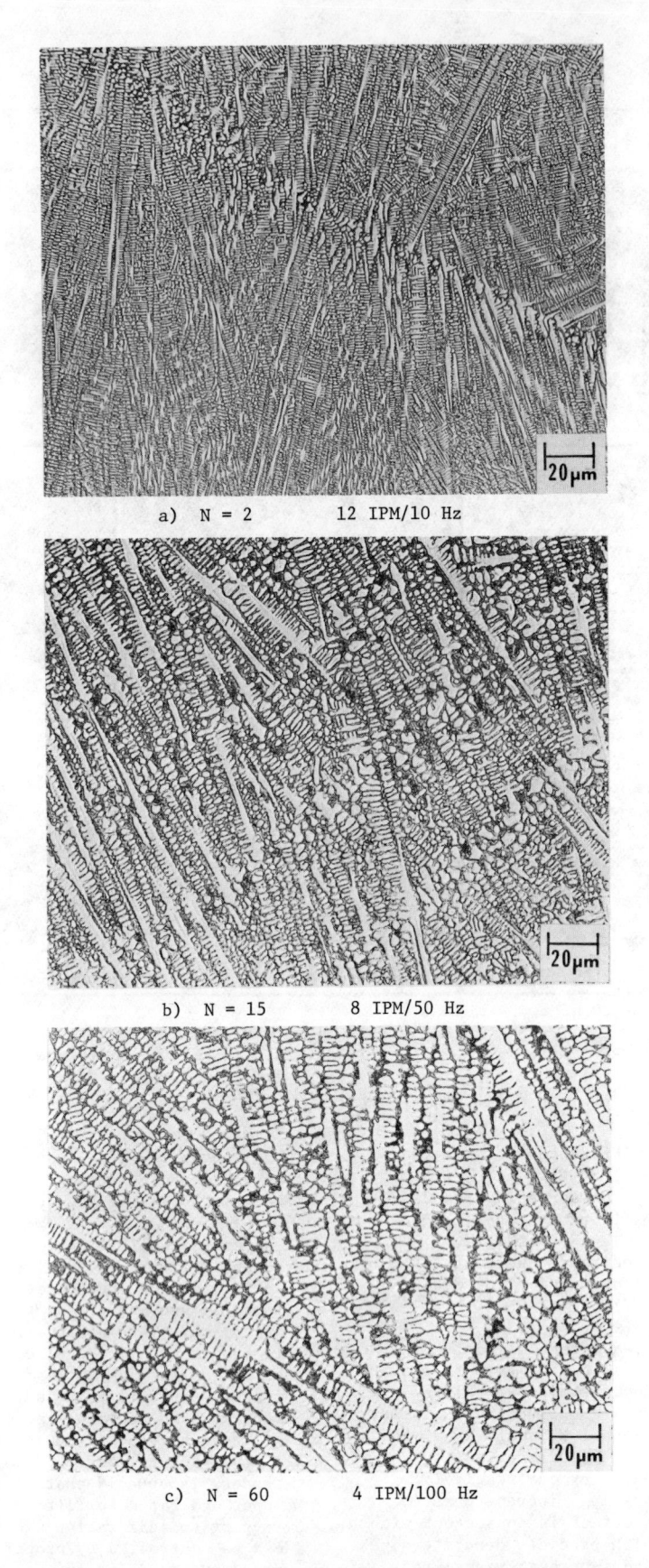

b) N = 15   8 IPM/50 Hz

c) N = 60   4 IPM/100 Hz

FIGURE 4. Microstructure of Laser Fused STELLITE alloy No. 6

Further Laser Fusing Experiments

Following the oscillation versus travel speed studies, further laser fusing experiments were conducted on the following powder pastes:

STELLITE® alloy No. 6
STELLITE® alloy No. 158
STELLITE® alloy No. SF 6
TRIBALOY® alloy T-700
DELCROME® alloy No. 90
DELORO® alloy No. 60
HAYSTELLITE® Composite Powder No. 1

All experiments were conducted using the constant conditions of 6 IPM travel speed and 75 Hz oscillation frequency. These conditions were selected based on the minimum value of N considered necessary for good fusion (i.e. N = 30). A travel speed of 6 IPM (2.5mm/sec) was selected since this is a speed typical of other automatic powder fusion processes such as plasma transferred arc.

It is of interest to note that the above powders represent a broad range of alloy types: cobalt-base, nickel-base, iron-base and, even a tungsten carbide composite alloy. Figures 5 through 11 document both the surface appearance and metallographic microstructure of each laser fused specimen. All of the samples produced good results indicating that laser fusing of hardfacing powders need not be limited to compositions characterized by the addition of elements added to lower the melting point of the alloy. The oxyacetylene spray and fuse process, for instance, cannot be used to deposit alloy Nos. 90 and T-700.

Another feature observed while examining the metallographic cross-sections at low power was the absence of any appreciable base metal melting (i.e. little or no dilution). Some evidence of base metal melting can be seen in the high magnification photomicrographs, for instance in Figure 5b. On a macroscale, however, base metal dilution was estimated to be nil.

® STELLITE, TRIBALOY, DELCROME, DELORO and HAYSTELLITE are registered trademarks of Cabot Corporation

a) Top View Photograph

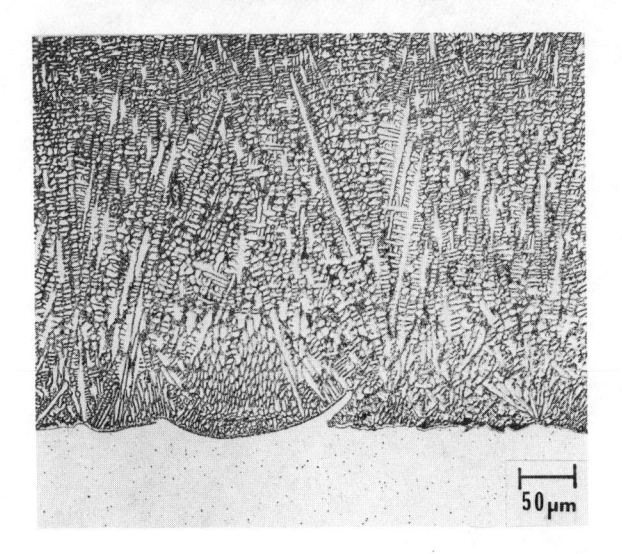

b) Metallographic Cross-Section

FIGURE 5. Laser Fused STELLITE alloy No. 6 Powder Paste 1200 Watts 6 IPM Traverse, 75 Hz Oscillation

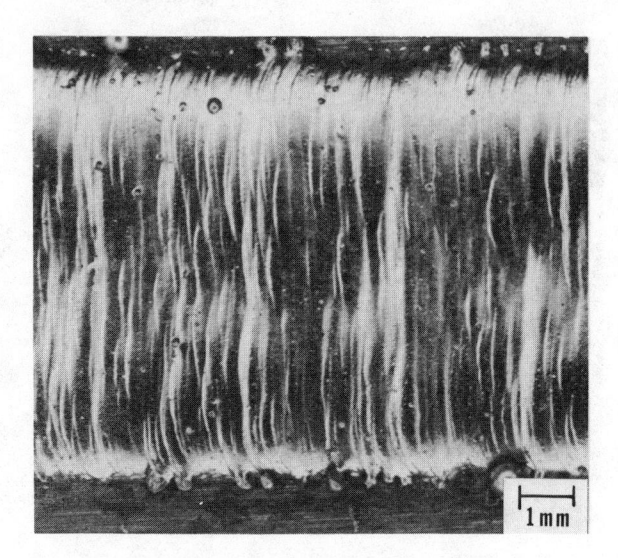

a) Top View Photograph

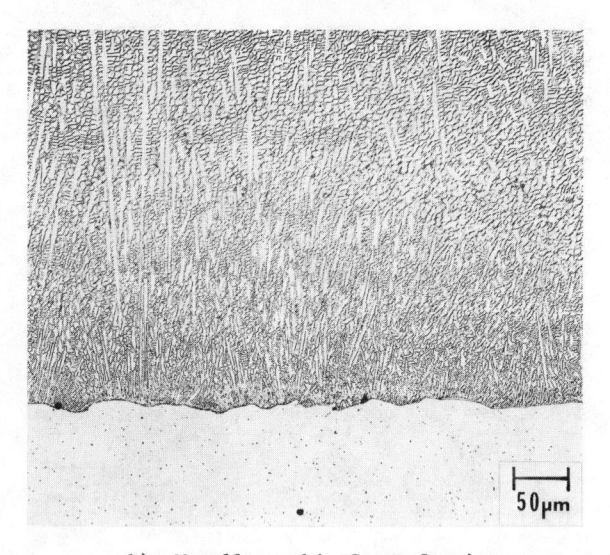

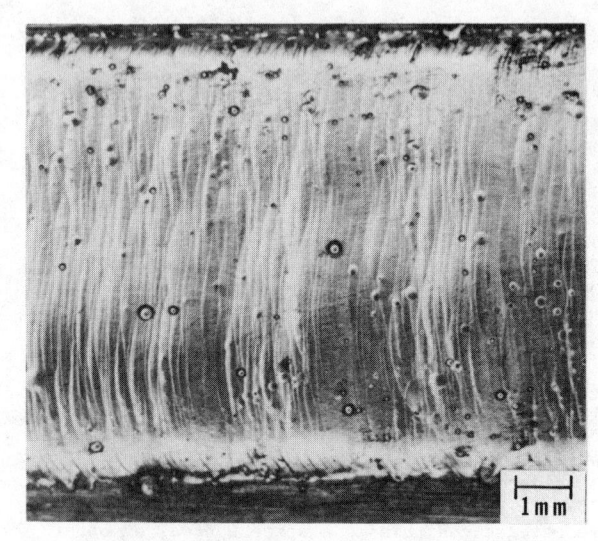

a) Top View Photograph

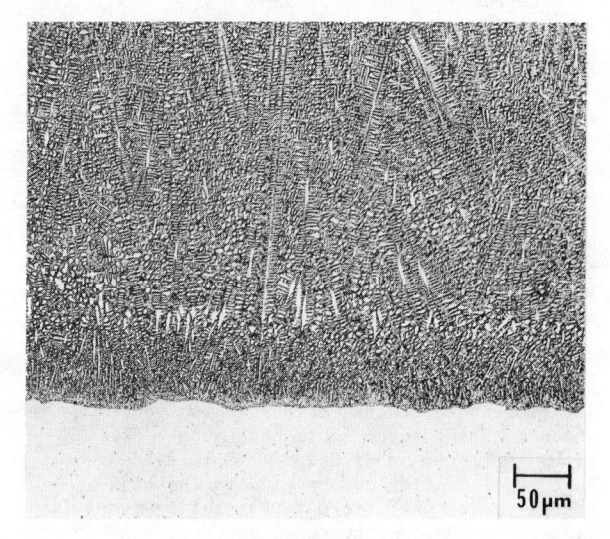

b) Metallographic Cross-Section

b) Metallographic Cross-Section

FIGURE 6. Laser Fused STELLITE alloy No. 158
Powder Paste   1200 Watts
6 IPM Traverse, 75 Hz Oscillation

FIGURE 7. Laser Fused STELLITE alloy No. SF6
Powder Paste   1200 Watts
6 IPM Traverse, 75 Hz Oscillation

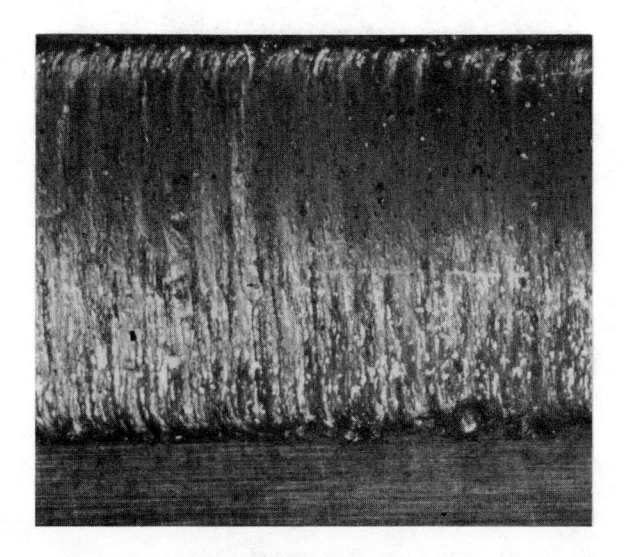

a) Top View Photograph

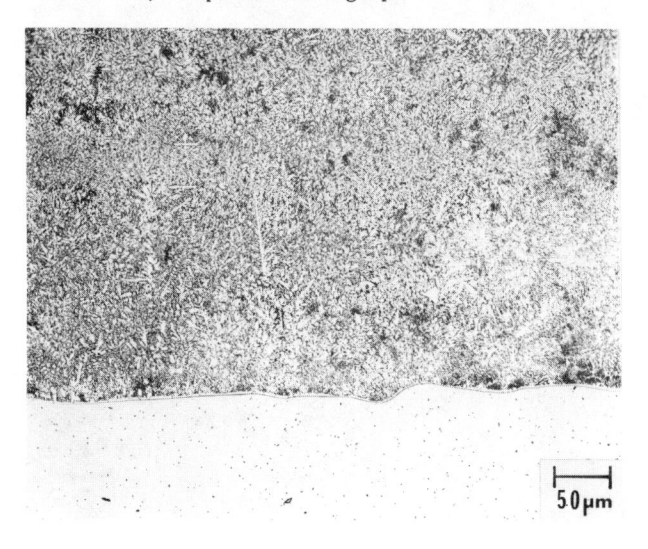

a) Top View Photograph

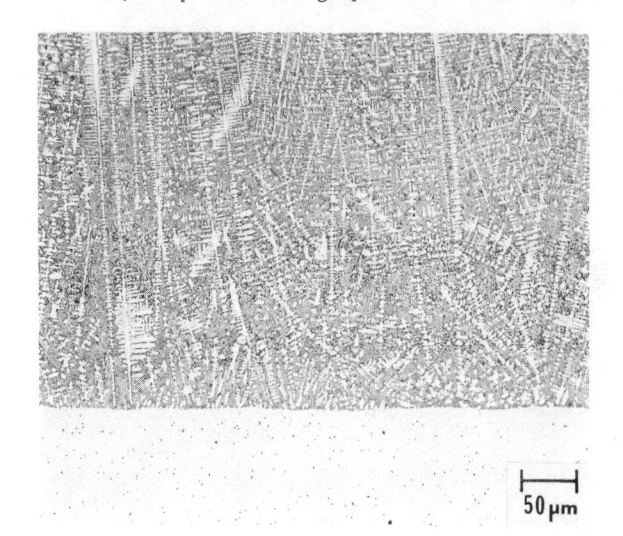

b) Metallographic Cross-Section

b) Metallographic Cross-Section

FIGURE 8. Laser Fused TRIBALOY alloy T-700
Powder Paste 1200 Watts
6 IPM Traverse, 75 Hz Oscillation

FIGURE 9. Laser Fused DELCROME alloy No. 90
Powder Paste 1200 Watts
6 IPM Traverse, 75 Hz Oscillation

145

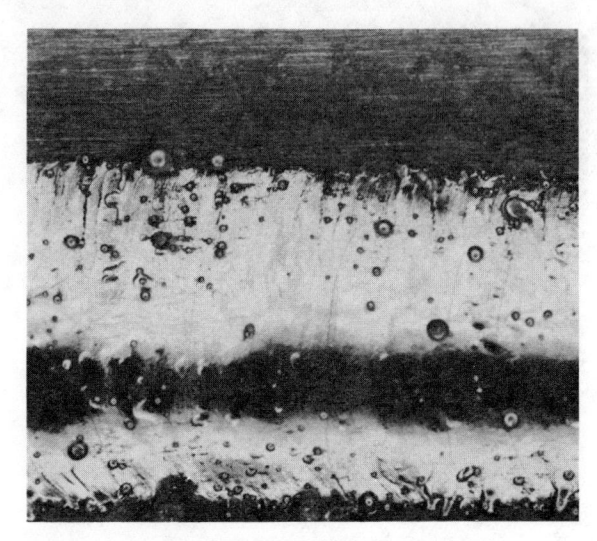

a) Top View Photograph

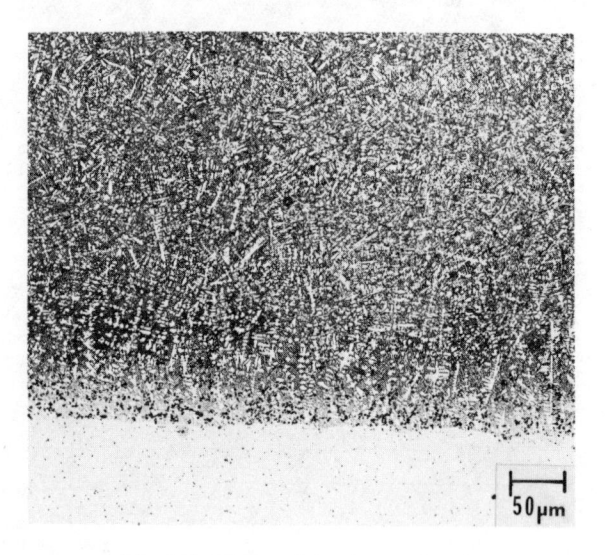

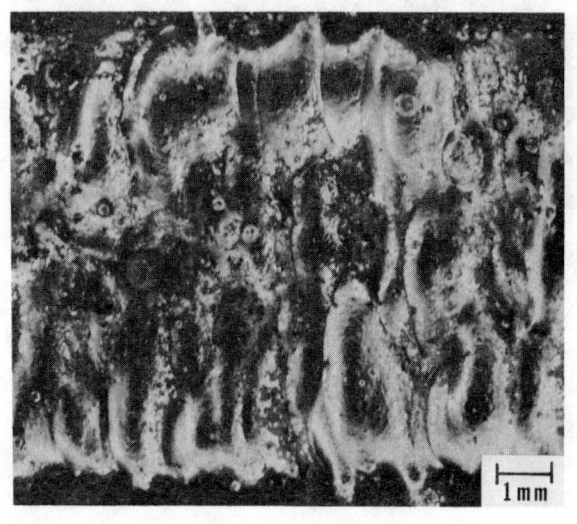

a) Top View Photograph

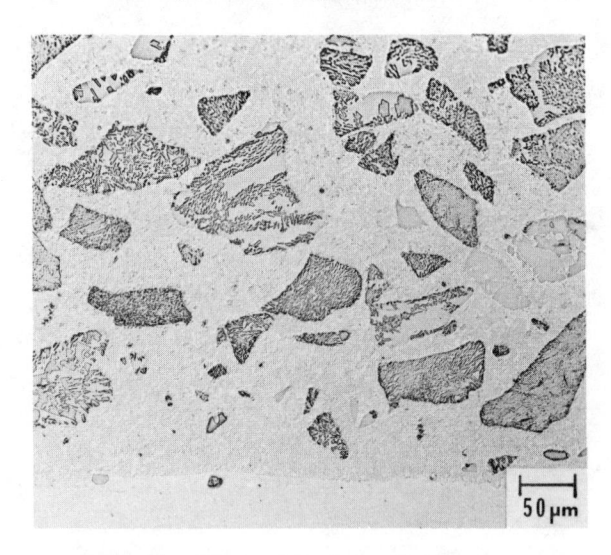

b) Metallographic Cross-Section

b) Metallographic Cross-Section

FIGURE 10. Laser Fused DELORO alloy No. 60
Powder Paste 1200 Watts
6 IPM Traverse, 75 Hz Oscillation

FIGURE 11. Laser Fused HAYSTELLITE Composite
Powder No. 1 Powder Paste 1200 Watts
6 IPM Traverse, 75 Hz Oscillation

Comments on Laser Fusing Applications, Advantages, Limitations

Figures 12 and 13 show the results of a "best effort" attempt to apply hardfacing powder pastes to actual industrial components by laser fusing. The results are interesting and show great promise, bearing in mind these were brief, isolated trials (i.e. no attempt was made to develop an optimized procedure). Figure 12 shows a diesel engine valve hardfaced with a STELLITE alloy SF 6 paste. An insufficient amount of paste was used to completely "fill" the premachined groove. However, the cross-section shows a homogeneous deposit thickness of about .020 inch (0.5mm). Figure 13 shows a laser hardfaced 1-1/2 inch (38mm) diameter gate valve seat. A deposit depression visible in Figure 13a is end of the fusion run, when the beam power was abruptly terminated. This tie-in region probably could be made smoother with proper "downslope" of beam power.

The potential utilization of laser for hardfacing of preplaced pastes is probably the greatest in applications characterized by high volume production. However, manufacturing problems of automatically preplacing powder paste would also have to be addressed. Typical deposition thicknesses up to .060-inch (1.5mm), and speeds of 4 to 12 IPM seem feasible, using a 1.2 Kw laser. A higher powered laser would probably allow thicker deposits and much faster fusing speeds to be used. Some advantages of laser fusing can be listed as follows:

1. smooth deposit of almost any hardfacing alloy composition with very little base metal dilution (zero to 5 percent).

2. very low heat input to base metal which translates into minimum heat-affected-zone problems and minimal distortion.

3. ability to reach deeply recessed areas inaccessible by conventional arc torches. Only line of sight is required.

a) Before and After Hardfacing

b) Cross-Section

FIGURE 12. Diesel Engine Valve Hardfaced by Laser Fusing STELLITE alloy No. SF6 Powder Paste

a) Before and After Hardfacing

b) Cross-Section

FIGURE 13. Fluid Flow Valve Seat (1-1/2-inch dia) Hardfaced by Laser Fusing DELORO 60 Type Powder Paste

There may be certain limitations associated with laser fusing of preplaced powder pastes and these should also be discussed. For instance, the problem of tie-in has not been fully investigated for situations requiring multiple passes. All of the work described in this paper involved single "bead-on-plate" type deposits. Cursory attempts to deposit multiple beads have met with only limited success. The problem involved keeping the paste "in place" immediately adjacent to a fused deposit. The thermal cycle associated with the initial deposit tended to "dry up" the adjacent powder rendering it very friable and prone to displacement by slight mechanical disturbances.

Another potential problem is the possibility of forming weld defects such as porosity, cracks, and lack of fusion. Rounded pores, some of considerable size (0.38mm diameter) appeared in some of the metallographic cross sections. Even experiments conducted at high values of N were not immune from porosity formation. It is not certain whether the porosity was caused by residual moisture in the paste or caused by metal/binder vaporization under the high energy density laser beam. It should be noted that porosity has also been observed in laser welding studies on wrought material suggesting that porosity formation may be an attendant defect possibility with any laser process and needs to be carefully investigated.

Some of the experimental deposits also cracked, but this is to be expected with any low ductility hardfacing overlay subjected to steep thermal gradients. In practice, stress cracking of this nature is usually eliminated by application of appropriate amounts of preheat. All of the fusing experiments were conducted on specimens at room temperature. It is believed that cracking can be controlled by proper preheat.

Finally, over most of the bond line areas observed, "metallurgical fusion" to the substrate was (albeit, metallographically) assumed to be good. It should be recognized that any process characterized by little or no dilution is also accompanied by the possibility of lack-of-fusion defects since by nature the amount of base metal melting is nil.

CONCLUSIONS

1. A 1200 watt CO_2 laser can be used for hardfacing by fusing of a preplaced powder paste onto a substrate producing a thin (1-1.5mm) fully solidified deposit with practically no base metal dilution.

2. A wide variety of hardfacing powders can be used including tungsten carbide composite powders. Unlike the oxyacetylene spray and fuse process, the laser fusing process is not limited to low melting point powder compositions.

3. Laser beam oscillation and specimen travel speed affect deposit smoothness and solidification microstructure. Good results in this study seemed to be achieved by traversing the specimen at 6 IPM (2.5mm/sec) under a beam oscillating frequency of 75 Hz.

4. While laser fusing of hardfacing powders shows great commercial promise, deposit defects such as cracking, porosity, and lack-of-fusion are possible, and proper parameters need to be developed in each application to minimize these defects.

REFERENCES

1. D. S. Gnanamuthu, "Cladding", US Patent 3942180, April 20, 1976.

2. J. Powell, and W. M. Steen, "Vibro Laser Cladding", from Lasers in Metallurgy, edited by K. Mukherjee and J. Mazumder, Conference proceedings published by the Metallurgical Society of AIME, 1981.

3. J. D. Ayers, R. J. Schaefer, and W. P. Robey, "A Laser Processing Technique for Improving the Wear Resistance of Metals", Journal of Metals, August 1981.

4. W. M. Steen and C. G. H. Courtney, "Hardfacing of Nimonic 75 using 2 Kw Continuous-Wave CO_2 Laser, Metals Technology, June 1980.

5. G. C. Irons, "Laser Fusing of Flame Spray Coatings", Welding Journal Research Supplement, December 1978.

6. J. D. Ayers and R. J. Schaefer, "Consolidation of Plasma-Sprayed Coatings by Laser Remelting", SPIE Vol. 198 Laser Applications in Materials Processing, (1979).

7. C. E. Witherall, and T. J. Ramos, "Laser Brazing" Welding Journal Research Supplement, October 1980.

8. A. V. La Rocca, "Laser Applications in Manufacturing" Scientific American, March 1982.

9. A. Belmondo, and C. Catagna, "Process for Coating a Metallic Surface With a Wear-Resistant Material", U. S. Patent 4218494, August 19, 1980.

A MODEL FOR SURFACE TENSION DRIVEN FLUID FLOW IN LASER SURFACE ALLOYING

C. Chan, J. Mazumder and M. M. Chen

Department of Mechanical and Industrial Engineering
1206 West Green Street
University of Illinois at Urbana-Champaign
Urbana, IL 61801

ABSTRACT

A model of the fluid flow and heat transfer of laser surface alloying is presented. The general three-dimensional governing equations are first considered. Their non-dimensional form is derived. A limiting case is considered--line source parallel to scanning direction. Governing parameters are derived and their significance is discussed. A numerical solution for a line source parallel to scanning direction is obtained. The flowfield within the molten pool and the shape of the molten pool thus obtained are presented. The predicted heat affected zone is shown. Characteristics of the process in the presence of the flowfield are discussed.

INTRODUCTION

While laser surface alloying (LSA) has been of interest for some time, its potential has not yet been fully explored. The main reason is that the process is not fully understood. The heat transfer process has been studied mainly by using a conduction model. Convective heat transfer has not been investigated. Moreover, fluid flow within the molten region has been neglected, although several authors [1-7] acknowledged its influence during LSA.

While most of the work on fluid flow has been qualitative [1-5], Anthony and Cline did the first quantitative work [6]. They proposed that the fluid flow is driven by the surface tension gradient on the surface. The model that they considered is essentially a one-dimensional case. The flowfield thus obtained is not coupled to the heat transfer process. Hence, no additional information can be obtained about the heat transfer process.

Recently, Chande and Mazumder [7] observed that an effective diffusivity of 10^6 cm^2/sec is required to explain, experimentally obtained uniform solute redistribution during LSA, on the basis of molecular diffusion. The implication, as pointed out by Chande and Mazumder [7], is that the fluid flow within the molten pool plays a significant role in the mechanism of solute redistribution and hence also in the heat transfer process. The understanding of convective heat transfer and fluid flow is therefore crucial in understanding the process of LSA.

In this paper, the convective heat transfer and fluid flow is analyzed. The general three-dimensional governing equations are first considered. Their non-dimensional form is then derived. Governing parameters thus arisen are discussed. The numerical solution of a limiting case, line source parallel to scanning direction, is obtained. Results are discussed and presented.

PHYSICS OF THE PROCESS

A laser beam having a defined power distribution strikes the surface of an opaque material of infinite width, thickness, and length moving along the x-axis with a uniform velocity (Fig. 1). Some of the incident radiation is reflected while the rest is absorbed. Surface heating is considered. The heat absorbed develops a molten pool. Owning to the high temperature gradient on the surface in the radial direction, surface tension gradient is developed. It is this mechanism that drives the flow. As the flow develops, energy transfer mechanism becomes a rather complicated process. In fact, it becomes a convection problem--convection which is driven by the surface tension.

The basic assumptions are:

1. The liquid metal is considered to be Newtonian so that the Navier-Stokes equation is applicable.

2. All the properties of the liquid metal and solid metal are constant, not a function of temperature (except the surface tension). This allows simplifications of the model, however, variable properties can be treated with slight modifications. The dependence of surface tension on temperature, the driving force of the flow, is assumed to be linear.

3. Latent heat of fusion is neglected since the energy liberated is small compared to conduction heat.

4. Thermal conductivity is assumed to be the same for both liquid and solid phases for the sake of simplicity of the model.

5. Surface of the melt pool is assumed to be flat so that the surface boundary condition can be coped with.

MATHEMATICAL MODEL

The appropriate governing equations are the energy equation

$$(\underline{u} \cdot \nabla)T = \kappa \nabla^2 T, \qquad (1)$$

LIST OF SYMBOLS

		UNIT	DIMENSION
C	constant which defines the interface	no unit	dimensionless
d_o	diameter of laser beam	mm	[L]
D_{eff}	effective diffusion coefficient	m²/sec	[L²/t]
k	thermal conductivity	kW/m°K	[ML/t²T]
ℓ	length of the rectangular heat source	mm	[L]
p	pressure	N/m²	[M/t²L]
Pe	Peclet Number $u_o d/\kappa$	no unit	dimensionless
q	net heat flux from laser	kW/mm²	[M/t²]
Re	Reynolds Number $u_o d/\nu$	no unit	dimensionless
r_o	radius of laser beam	mm	[L]
S	Surface Tension Number	no unit	dimensionless
T	temperature	°K	[T]
T_{melt}	melting temperature	°K	[T]
T_{metal}	temperature of metal when it is not heated	°K	[T]
$\underline{u}$	velocity vector	mm/sec	[L/t]
u	x-component of $\underline{u}$	mm/sec	[L/t]
u_o	scanning speed of the laser beam	mm/sec	[L/t]
v	y-component of $\underline{u}$	mm/sec	[L/t]
w	z-component of $\underline{u}$	mm/sec	[L/t]
x,y,z	Cartesian coordinate	mm	[L]

Greek

ρ	density	kg/m³	[M/L³]
κ	thermal diffusivity	m²/sec	[L²/t]
ν	kinematic viscosity	m²/sec	[L²/t]
μ	viscosity	kg/sec-m	[M/t-L]
σ	surface tension	N/m	[M/t²]

Superscript

*   dimensionless quantities

'   derivative with respect to its independent variable

Operator

∇	del operator	$\vec{i}\frac{\partial}{\partial x} + \vec{j}\frac{\partial}{\partial y} + \vec{k}\frac{\partial}{\partial z}$
$(\underline{u}\cdot\nabla)$	convective operator	$u\frac{\partial}{\partial x} + v\frac{\partial}{\partial y} + w\frac{\partial}{\partial z}$
∇^2	Laplacian operator	$\frac{\partial^2}{\partial x^2} + \frac{\partial^2}{\partial y^2} + \frac{\partial^2}{\partial z^2}$

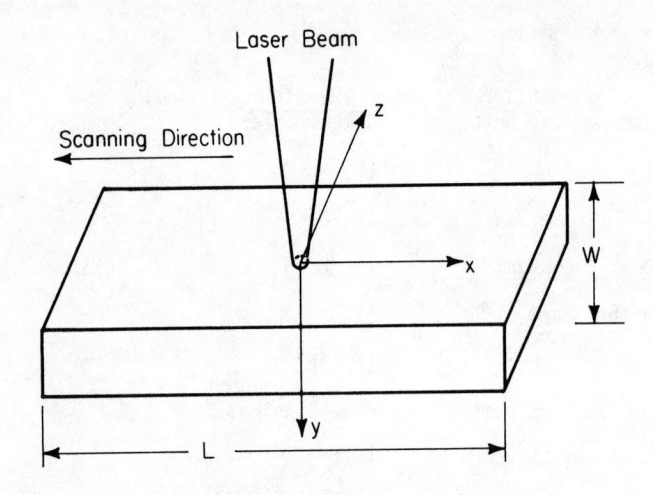

Fig. 1 Schematic Diagram of the Process

the continuity equations

$$\nabla \cdot \underline{u} = 0, \qquad (2)$$

and the momentum equations

$$(\underline{u} \cdot \nabla)\underline{u} = -\frac{\nabla p}{\rho} + \nu \nabla^2 \underline{u}. \qquad (3)$$

The boundary conditions are

$y = 0, \ v = 0$

$$k\frac{\partial T}{\partial y} = \begin{cases} -q & |r| \leq r_0, \\ 0 & \text{otherwise} \end{cases} \quad r = (x^2 + z^2)^{1/2}$$

$$\sigma'\frac{\partial T}{\partial x} = -\mu\frac{\partial u}{\partial y},$$

$$\sigma'\frac{\partial T}{\partial z} = -\mu\frac{\partial w}{\partial y}. \qquad (4)$$

And at the liquid-solid interface,

$$\begin{cases} f(x,y,z) = C, \text{ for some constant } C \\ u - u_0 = v = w = 0 \\ T = T_{melt}. \end{cases} \qquad (5)$$

Finally,

$$|x|, y, |z| \to \infty, \ T \to T_{metal}. \qquad (6)$$

Note $f(x,y,z)$ is also part of the problem to be solved.

To this end, we introduce the dimensionless variables

$$(x^*, y^*, z^*) = \frac{(x,y,z)}{d_0} \qquad (7a)$$

$$\underline{u}^* = \frac{\underline{u}}{u_0} \qquad (7b)$$

$$p^* = \frac{p}{\rho u_0^2} \qquad (7c)$$

$$T^* = \frac{T - T_{metal}}{d_0 q/k}. \qquad (7d)$$

The governing equations then become

$$(\underline{u}^* \cdot \nabla^*) T^* = \frac{1}{Pe} \nabla^{*2} T^* \qquad (8)$$

$$\nabla^* \cdot \underline{u} = 0 \qquad (9)$$

$$(\underline{u}^* \cdot \nabla^*) \underline{u}^* = -\nabla^* p^* + \frac{1}{Re} \nabla^{*2} \underline{u}^* \qquad (10)$$

and the boundary conditions become

$y^* = 0, \ v^* = 0$

$$S\frac{\partial T^*}{\partial x^*} = -\frac{\partial u^*}{\partial y^*} \qquad (11)$$

$$S\frac{\partial T^*}{\partial z^*} = -\frac{\partial w^*}{\partial y^*}$$

$$\frac{\partial T^*}{\partial y^*} = \begin{cases} -1 & |r^*| \leq 1/2 \\ 0 & \text{otherwise} \end{cases}.$$

and at the liquid-solid interface,

$f^*(x^*,y^*,z^*) = C^*,$

$$\underline{u}^* = (1,0,0), \ T^* = T^*_{melt} \qquad (12)$$

and

$$|x^*|, |y^*|, |z^*| \to \infty, \ T^* \to 0 \qquad (13)$$

where

$Re = u_0 d/\nu, \ Pr = \nu/\kappa,$

$$S = \frac{\sigma' qd}{\mu u_0 k}, \ T^*_{melt} = \frac{T_{melt} - T_{metal}}{q \, d_0/k} \qquad (14)$$

The three-dimensional problem is a very complicated one. Numerical solution would require huge amounts of computer time. To give more insight into the problem, we consider the limiting case, line source scanning in parallel to the x-axis.

LINE SOURCE SCANNING IN PARALLEL TO x-AXIS

This limiting case can be achieved by first considering a rectangular heat source $d \times \ell$, ℓ being the length coinciding with the x-axis, and then taking the limit as d/ℓ goes to zero. We introduce the dimensionless variables as follow

$$(x^*, y^*, z^*) = (\frac{x}{\ell}, \frac{y}{d}, \frac{z}{d}) \qquad (15a)$$

$$\underline{u}^* = \frac{\underline{u}}{u_0}, \qquad (15b)$$

$$p^* = \frac{p}{\rho u_0^2}, \qquad (15c)$$

$$T^* = \frac{T - T_{metal}}{q \, d_0/k}. \qquad (15d)$$

The final equations are obtained by substituting the dimensionless variables into Eqs. (6) through (11). Therefore, we have

$$v^* \frac{\partial T^*}{\partial y^*} + w^* \frac{\partial T^*}{\partial z^*} = \frac{1}{Pe}\left(\frac{\partial^2 T^*}{\partial y^{*2}} + \frac{\partial^2 T^*}{\partial z^{*2}}\right) \quad (16)$$

$$\frac{\partial v^*}{\partial y^*} + \frac{\partial w^*}{\partial z^*} = 0 \quad (17)$$

$$v^* \frac{\partial u^*}{\partial y^*} + w^* \frac{\partial u^*}{\partial z^*} = \frac{1}{Re}\left(\frac{\partial^2 u^*}{\partial y^{*2}} + \frac{\partial^2 u^*}{\partial z^{*2}}\right) \quad (18)$$

$$v^* \frac{\partial v^*}{\partial y^*} + w^* \frac{\partial v^*}{\partial z^*} = -\frac{\partial p^*}{\partial y^*} + \frac{1}{Re}\left(\frac{\partial^2 v^*}{\partial y^{*2}} + \frac{\partial^2 v^*}{\partial z^{*2}}\right) \quad (19)$$

$$v^* \frac{\partial w^*}{\partial y^*} + w^* \frac{\partial w^*}{\partial z^*} = -\frac{\partial p^*}{\partial z^*} + \frac{1}{Re}\left(\frac{\partial^2 w^*}{\partial y^{*2}} + \frac{\partial^2 w^*}{\partial z^{*2}}\right) \quad (20)$$

and the boundary conditions become

$y^* = 0, \; v^* = 0$

$$\frac{\partial u^*}{\partial y^*} = 0, \quad \frac{\partial w^*}{\partial y^*} = -S \frac{\partial T^*}{\partial z^*} \quad (21)$$

$$\frac{\partial T^*}{\partial y^*} = \begin{cases} -1 & |z^*| \leq 1/2, \; 0 \leq x^* \leq 1 \\ 0 & \text{otherwise.} \end{cases}$$

Interface $f^*(x^*, y^*, z^*) = C^*$,

$u^* - 1 = v^* = w^* = 0$,

$$T^* = T^*_{melt}. \quad (22)$$

$|x^*|, y^*, |z^*| \to \infty$
$T^* \to 0 \quad (23)$

The problem then reduces to a two-dimensional case and the solution is comparatively much easier to reach than for the three-dimensional case.

NUMERICAL SOLUTION

The computer program SOLA [8] is employed. The basic method of the algorithm is presented in Hirt's report [8]. It would not be repeated here; however, the modification is presented.

The first major modification is the addition of the energy equation. This equation is approximated by the same way as the y-direction momentum equation. That is,

$$T_{i,j}^{n+1} = T_{i,j}^n + \delta t \; (-FTX - FTY + TIS) \quad (24)$$

where

$$FTX = \frac{1}{4\delta z} \{(w_{i,j}^n + w_{i,j+1}^n)(T_{i,j}^n + T_{i+1,j}^n)$$

$$+ \alpha |w_{i,j}^n + w_{i,j+1}^n|(T_{i,j}^n - T_{i+1,j}^n)$$

$$- (w_{i-1,j}^n + w_{i-1,j+1}^n)(T_{i-1,j}^n + T_{i,j}^n)$$

$$- \alpha |w_{i-1,j}^n + w_{i-1,j+1}^n|(T_{i-1,j}^n - T_{i,j}^n)\}$$

$$(25)$$

$$FTY = \frac{1}{4\delta y}\{(v_{i,j}^n + v_{i,j+1}^n)(T_{i,j}^n + T_{i,j+1}^n)$$

$$+ \alpha |v_{i,j}^n + v_{i,j+1}^n|(T_{i,j}^n - T_{i,j+1}^n)$$

$$- (v_{i,j-1}^n + v_{i,j}^n)(T_{i,j-1}^n + T_{i,j}^n)$$

$$- \alpha |v_{i,j-1}^n + v_{i,j}^n|(T_{i,j-1}^n - T_{i,j}^n)\} \quad (26)$$

$$TIS = \frac{1}{Re \; Pr}\{\frac{1}{\partial x^2}(T_{i+1,j}^n - 2T_{i,j}^n + T_{i-1,j}^n)$$

$$+ \frac{1}{\delta y^2}(T_{i,j+1}^n - 2T_{i,j}^n + T_{i,j-1}^n)\}. \quad (27)$$

The second modification is the irregular boundary of the interface. It is approximated by steps of grids that are closest to the interface (see Fig. 2).

Finally, the surface boundary condition as in Eq. (21) is approximated by the finite difference equation,

$$\frac{w_{i,1} - w_{i,2}}{\delta y} = S \cdot \frac{T_{i+1,1} - T_{i,1}}{\delta z}. \quad (28)$$

It is important to point out that the dependence of the momentum equation on the energy equation is only through the surface boundary condition. The algorithm that we choose in effect decouples the two. At each time step t, the momentum equations are solved based on the previous time step temperature field. The marching in time is then repeated.

The numerical solution for the energy equation with zero velocity is first carried out to x = 0.1 d before the aforementioned procedure is followed. This step is necessary since velocity field does not exist without the molten pool.

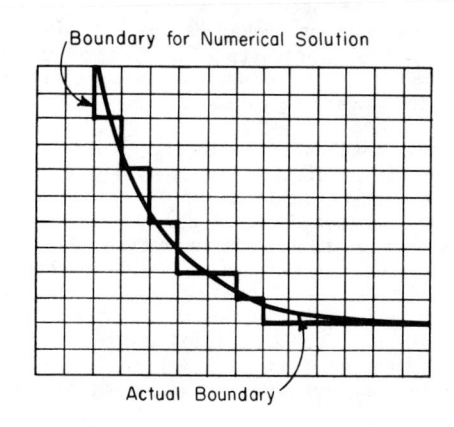

Fig. 2 Numerical Solution Boundary of Solid-liquid Interface

RESULTS AND DISCUSSIONS

The governing parameters arising from the non-dimensionalized equations are Reynolds number (Re), Prandtl number (Pr), surface tension number (s), and the melting temperature (T^*_{melt}). Each of these would govern the characteristics of the problem and each has its own physical interpretation.

Reynolds number can be interpreted as the ratio of the inertia force and viscous force. In our case, the inertia force is due to the scanning while the viscous force is due to the shear of the molten material. Prandtl number is a property of the molten material. It is a ratio of momentum diffusion and thermal diffusion. Surface tension number is a ratio of the characteristic velocity of the surface tension driven flow to the scanning velocity. Melting temperature is the dimensionless form of the melting temperature.

The surface tension number governs the existence of convection flow in the molten pool. For surface tension number to be very small, convection flow does not exist and so the problem becomes conduction dominated. On the other hand, if the number is very large, convection flow would be very strong and the physics of the problem is quite different. The surface tension number is also a measure of the propogation time of the convection flow compared to the scanning interaction time. For small surface tension numbers, the propogation time of the convection is so large compared to the scanning time that convection does not make a significant contribution to the heat transfer; in contrast, for large surface tension numbers, convection would make a significant contribution to the heat transfer determining the size of the molten pool. The larger the dimensionless melting temperature, the smaller its size.

The shape of the molten pool would depend on the surface tension number and the Prandtl number as well as the Reynolds number. It is of interest to point out that in the case of line source scanning in parallel to x-axis, the flow field is symmetrical along the y-axis. Even more interesting is the fact that the Reynolds number and the surface tension number can be grouped into one parameter--their product. The consequence of such a degeneracy is that the scanning speed is not part of the governing parameter.

Numerical results of velocity field, at a cross-section perpendicular to the direction of travel, at different stations of x, the scanning direction, are presented in Figs. 4 through 10. The corresponding surface temperature is also presented in Figs. 11 through 17. The governing parameters match those of one of our experimental results, namely s = 55,000, Re = 12.5, Pr = 0.4, and T^*_{melt} = 0.05. The grid size is presented in Fig. 3. A finer grid is used within the molten region, while coarser grid is used outside the region.

The recirculating flow can be seen clearly. The magnitude of the recirculating flow velocity is one or two orders of magnitude higher than that of the scanning speed. Consequently, the heat transfer process is convection dominated.

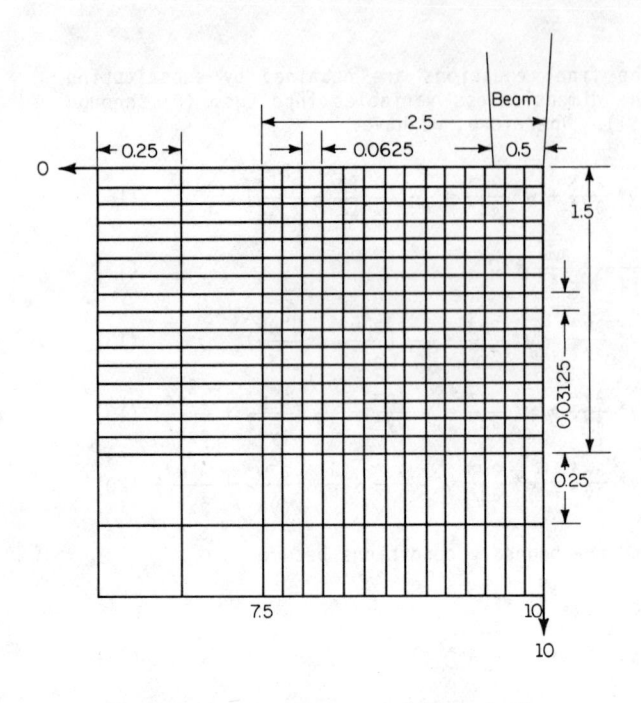

Fig. 3 Grid Distribution of Numerical Solution (not to scale)

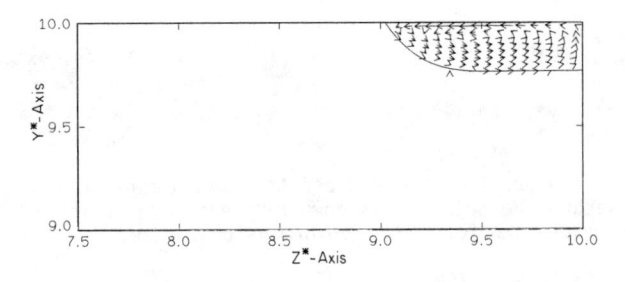

Fig. 4 Velocity Field in Molten Pool at x* = 0.25

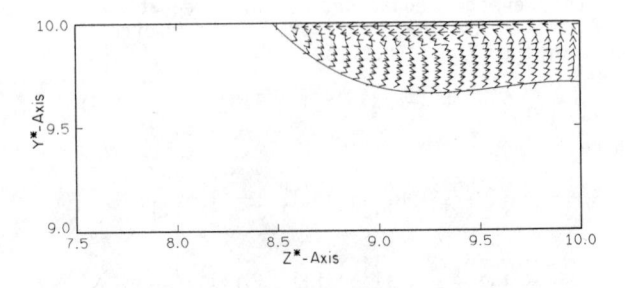

Fig. 5 Velocity Field in Molten Pool at x* = 0.5

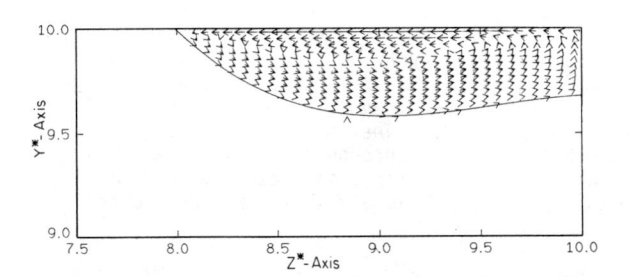

Fig. 6 Velocity Field in Molten Pool at x* = 0.8

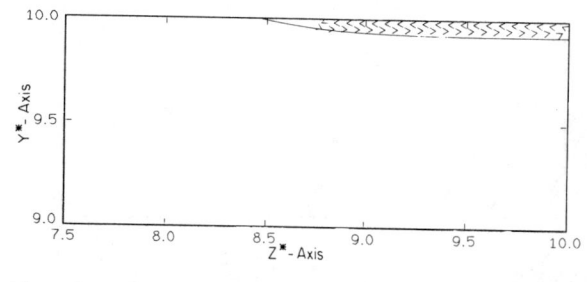

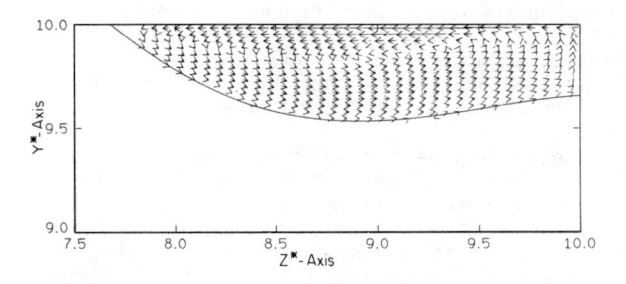

Fig. 7 Velocity Field in Molten Pool at x* = 1.0

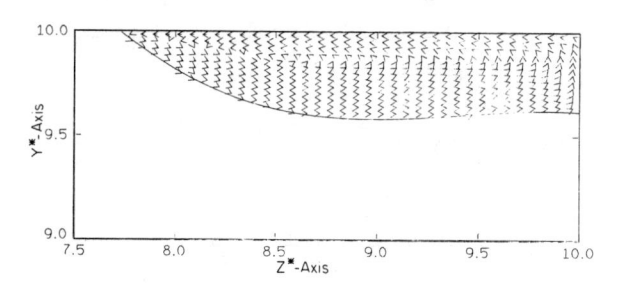

Fig. 8 Velocity Field in Molten Pool at x* = 1.25

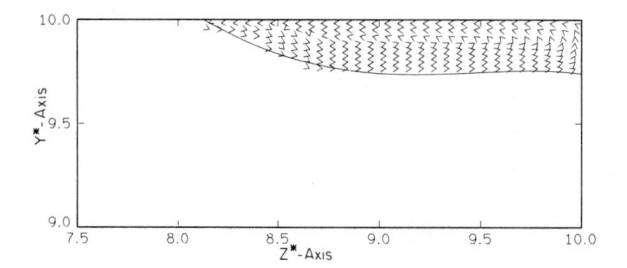

Fig. 9 Velocity Field in Molten Pool at x* = 1.50

Fig. 10 Velocity Field in Molten Pool at x* = 1.6216

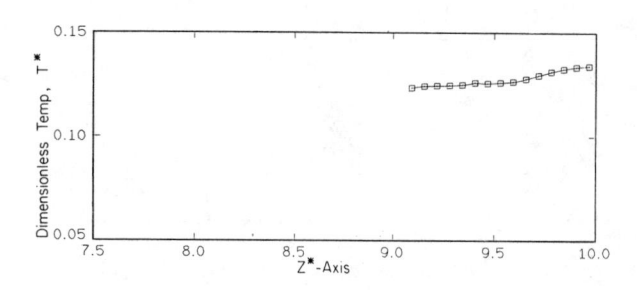

Fig. 11 Surface Temperature at x* = 0.25

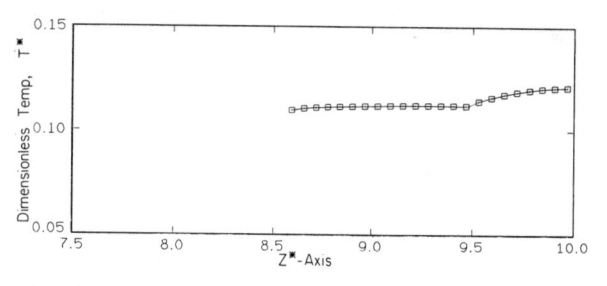

Fig. 12 Surface Temperature at x* = 0.5

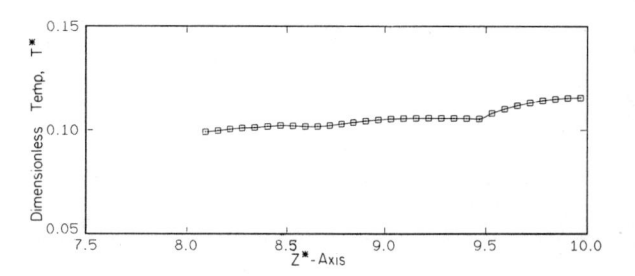

Fig. 13 Surface Temperature at x* = 0.8

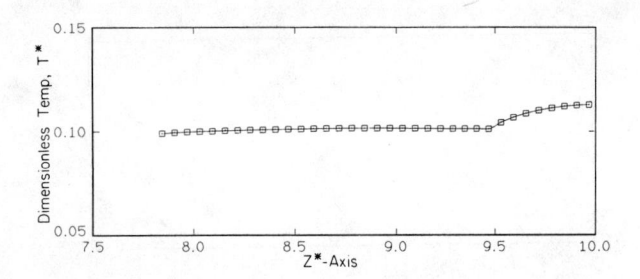

Fig. 14 Surface Temperature at x* = 1.0

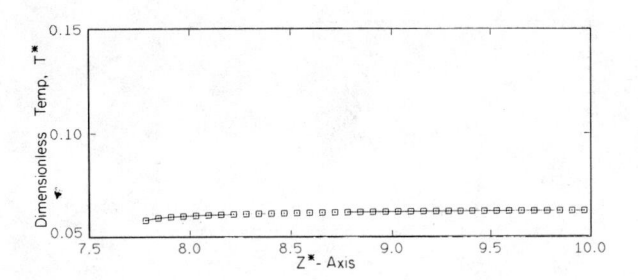

Fig. 15 Surface Temperature at x* = 1.25

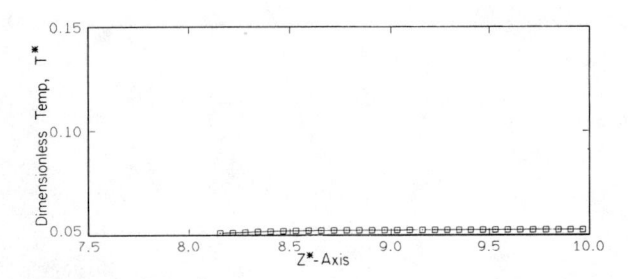

Fig. 16 Surface Temperature at x* = 1.5

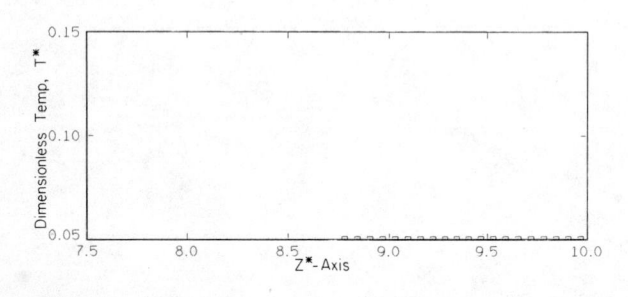

Fig. 17 Surface Temperature at x* = 1.6216

Widening of the molten pool is due to the fact that the heat transfer is mainly by convection. The shape of the molten pool is not spherical like as can be seen in Figs. 4 through 10. Such a shape is due to the flow pattern of the recirculation. The freezing process can be observed in Figs. 7 through 10. The corresponding surface temperature are also presented in Figs. 11 through 17. The heat affected zone predicted is shown in Fig. 18.

Because of the fact that the recirculating flow velocity is of one or two orders of magnitude higher than that of the scanning speed, a fluid particle would recirculate many times before it solidifies. This can account for the highly disperse and uniform distribution of solute within the molten region. The transport process, both heat and mass, are convection dominated. The transport phenomena are actually by the carriage of the heat or mass by the flowfield; molecular diffusion plays a rather subordinate role.

The centerline surface temperature versus x after the beam is plotted in Fig. 19. The cooling rate can then be estimated from the plot. It is found to be of the order of 10^7 °K/sec. Consequently, the freezing process occurs rapidly; in fact, the whole molten region is frozen within one beam diameter after the beam. The magnitude of the recirculating flow also decreases very rapidly due to the fact that the kinetic energy is dissipated through the viscous term.

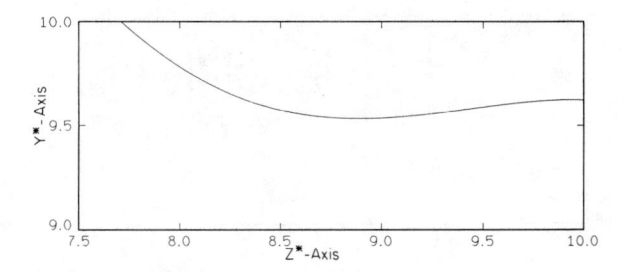

Fig. 18 Heat Affected Zone (HAZ)

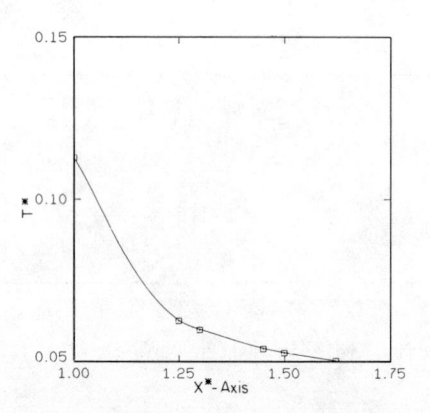

Fig. 19 Centerline Surface Temperature versus x* after Beam

CONCLUSIONS

The general three-dimensional model is presented. Numerical solution of its limiting case, line source parallel to scanning direction is also presented. The importance of convection in terms of heat transfer process is demonstrated. The characteristics of process are discussed. The cooling rate thus predicted is of the order of 10^7 °K/sec. The over-prediction is due to the fact that the model is two dimensional. As a result, the length of the molten pool is underpredicted while the width and depth of it are overpredicted. It is evident that the flowfield in the scanning direction also plays a role in the process. It, therefore, suggests that a three-dimensional model should be considered to simulate the process.

REFERENCES

1. Gnanamuthu, D. S., "Applications of Lasers in Materials Processing," E. A. Metzbower, ed., pp. 177-289, ASM, 1979
2. Moore, P. M., and L. S. Weiman, SPIE, Vol. 198, p. 120, 1979.
3. Weinman, L. S., J. H. Devault, and P. Moore, Applications of Lasers in Materials Processing, E. A. Metzbower, ed., pp. 245-259, 1979.
4. Draper, C. W., Proc. Conf. of Lasers in Metallurgy, Chicago, Ill., K. Mukherji and J. Mazumder, eds., AIME, 1981.
5. Weinman, L. S., and J. H. Devault, AIP Conference Proc. No. 50, Symp. on Laser Solid Interactions and Laser Processing, Boston, Mass., p. 239, 1978.
6. Anthony, T. R., and H. F. Cline, J. Appl. Phys., Vol. 48, No. 9, pp. 3888-3894, 1977.
7. Chande, T., and J. Mazumder, Appl. Phys. Letter, Vol. 41, No. 1, p. 42, 1982.
8. Hirt, C. W., B. D. Nichols, and N. C. Romero, "A Numerical Solution Algorithm for Transient Fluid Flows," UC-34 and UC-79d, April 1975.

BASIC COMPUTER MODEL OF THE PULSED LASER DRILLING PROCESS WITH A NEODYMIUM LASER

M. G. Jones, G. Georgalas, and A. Brutus

INTRODUCTION

A model of the laser drilling process gives additional insight into understanding the interactions between a neodymium laser and materials. The laser drilling process occurs when a material undergoes an alteration in physical state due to the absorption of laser light. Since lasers can be optically focused to incident power densities, they can produce intense pulses of energy sufficient to melt and vaporize most metals. This interaction of laser energy with materials can be described as a conversion of electromagnetic energy into intense localized heating of a workpiece. The reaction is at best quasi-steady. The process occurs in fractions of a millisecond and in an irreversible manner. Due to the rapid localized heating, a reversible thermal balance on the region of treatment is a good first approximation of the process.

THERMAL BALANCE CONCEPT

The thermal balance concept concentrates on coupling between incident laser radiation and a reversible thermal response of a localized region on the metal.[1] This process is best described as a constant pressure heat addition phenomena. A material being heated at constant pressure follows a thermal response curve as shown in Figure 1. This temperature versus time relationship is characterized by four principal zones. The first (A-B) represents a constant pressure heat addition needed to bring the metal from room temperature up to melting (Tm). The second zone (B-C), represents the addition of heat needed to bring the sample completely from solid to the molten phase. This is an isothermal heat addition process commonly termed fusion. The third critical zone (C-D), is a constant pressure heat addition which brings the molten material up to its boiling temperature (Tb).

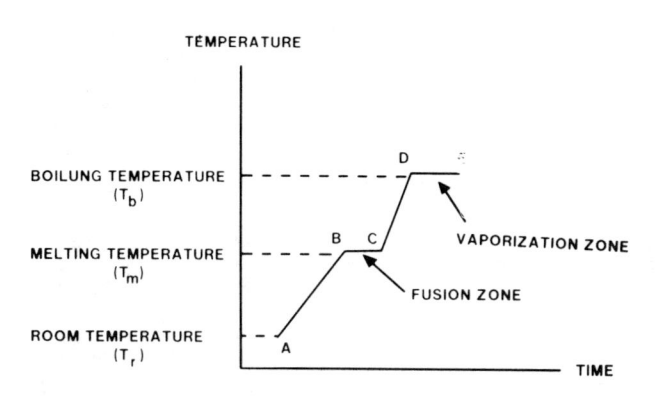

Figure 1. Typical Thermal Response Curve for Pure Metals

At the boiling point, the material undergoes another isothermal transformation (D-E), termed vaporization. At this point, the material reaches a gaseous or vapor state.

The total heat applied to bring the sample from room temperature (Tr) to vaporization can be written as:

$$Q_{total} = M_t C (T_m - T_r) + M_t H_f + M_t C (T_b - T_m) + M_t H_v \quad (1)$$

where
M_t = total mass of material in the laser drilled zone (gram)
C = specific heat of the material (cal/gm°C)
H_F = heat of fusion (cal/gm)
H_v = heat of vaporization (cal/gm)
T_m = melting temperature (°C)
T_r = room temperature (°C)
T_b = boiling temperature (°C)

With the thermal response of the material adequately described, a heat balance between the incident laser radiation and the heat needed for these transformations can be developed.

MODIFICATIONS TO THE THERMAL BALANCE EQUATIONS TO COMPENSATE FOR LASER HEATING

Reflectivity Considerations: The heat supplied to bring the material through its transformation is supplied by the incident laser pulse. This energy is wave-like in nature and susceptible to reflectivity[2] concerns. The thermal balance model has been adjusted to address different reflectivity parameters for each portion of the thermal cycle. Our first order approximation of reflectivity losses occurs only during the initial portion of the thermal cycle, Q (A-B). This means that the reflectivity parameters will dominate only as the sample approaches its melting temperature.

Expulsion Considerations: In hole drilling with lasers, not all material leaves by vaporization.[3] Part of the removed material leaves in the liquid state. The molten material is ejected from the hole by the pressure gradient developed during the vaporization of a portion of the material. This process results in a mass removal greater than if all mass was vaporized. The thermal balance model addresses this concept directly in the heat transfer equations. Since the ejected material has to leave in molten form, our assessment accounts for all material ejected to leave when the localized drilling zone reaches the fully melted state.

Final Thermal Response Equation

Considering the reflectivity and expulsion parameters, the original thermal balance equation was modified. The first order approximations used in this analysis transformed equation (1) into:

$$Q_{total} = (\frac{1}{1-r})[M_t C_p (T_m-T_r)] + M_t H_f$$
$$= (1-E)[M_t C_p (T_v-T_r)] + (1-E)(M_t H_v), \quad (2)$$

where

R = reflectivity fraction

E = expulsion fraction of material removed

This form of the thermal balance equation explicitly accounts for the modifications needed to accommodate reflectivity and expulsion losses.

ANALYSIS

Testing Plan

With the thermal balance analytically assessed, actual material removal tests were performed. These tests considered the interaction of 1.06 micrometer wavelength energy from a neodymium-glass laser with samples of carbon steel, stainless steel and aluminum. The laser parameters and optical transmission equipment used in all tests are shown in Table 1. A schematic of the optical system, with laser related hardware, is shown in Figure 2. All tests were performed with interactions of single pulse duration. Efforts to correlate material removed (determined by direct weight measurement or hole geometry approximations) and delivered laser energy (determined by pulse energy measurements made through optical hardware) enhanced confidence in the accuracy of the model working parameters.

TABLE 1

LASER AND OPTICAL PARAMETERS USED DURING ALL LASER DRILLING TESTS

LASER TYPE - PULSED NEODYMIUM-GLASS
PULSE RATE (ppm) - 15
PULSE ENERGY (joules) - 10-40
PULSE WIDTH (msec) - 0.63
BEAM QUALITY (times diffraction limit) - 100.0-165.0
FOCUSING OPTICS (focal length) - 51 mm DOUBLE CONVEX LENS
LENS PROTECTION (mm x mm) - 50 x 75 GLASS SLIDE

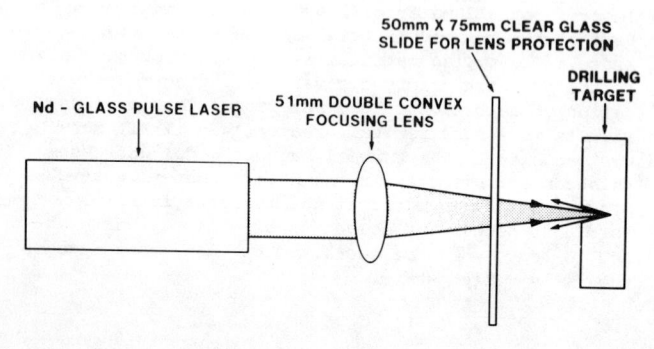

Figure 2. Laboratory Setup Used in Pulsed Laser Drilling Tests

Expulsion considerations were also of principal interest to the model. In an effort to gather information on the relative amount of material removed, an expulsion collection scheme was put together. This scheme consolidated a large portion of the material removed in the molten state and allowed us to directly measure it. Comparing the amount of ejected material to that of total material removed allowed us to put Table 2 together. Examination of the expulsion spray zone indicated an increase in expansion area with laser energy. Due to limitations in the collection scheme, it would be difficult to support the trend of expulsed material decreasing with increasing pulse energy.

TABLE 2

RELATIVE PERCENTAGE OF MATERIAL THAT IS REMOVED FROM SAMPLE IN THE LIQUID FORM

INCIDENT ENERGY (joules)	CARBON STEEL	STAINLESS STEEEL	ALUMINUM
9.0	n.a.	69.20	54.20
11.75	41.42	66.75	57.50
15.00	n.a.	54.14	55.80
19.50	38.00	59.23	58.60
23.00	39.10	52.40	42.90
26.50	n.a.	53.00	44.70

n.a. = data not available

MODEL RESULTS

Carbon Steel: Two series of tests were performed with carbon steel. The first (Test Case 1) considered hole geometry approximations for material removed. The second (Test Case 2) used direct weight measurements for this parameter. Initial examination of the material removed versus input energy profile showed a close linear relationship between the two parameters. Putting the raw data through the heat balance model yielded two combinations of reflectivity and expulsion factors which best fit the data. The first set of parameters (CRS-Method A) was determined by analytically minimizing the expulsion factor and varying the reflectivity to compensate for it. The second set of parameters (CRS-Method B) set the reflectivity[4] at a level of 50% and optimized the expulsion until the error in predicted pulse energy was minimized. Both methods, with relative error percentages, are shown in Table 3. The relative

TABLE 3

RELATIVE PERCENTAGES IN CALCULATED ENERGIES FROM THE THERMAL BALANCE MODEL FOR TESTS WITH CARBON STEEL

MATL. REMOVED (mgrams)	ACTUAL ENERGY (joules)	CRS - METHOD A refl.=0.30 expl.=0.75	CRS - METHOD B refl.=0.50 expl.=0.80
TEST CASE 1:			
2.153	9.00	20.80	24.20
3.745	12.00	-3.30	1.10
3.780	15.00	16.60	20.10
6.038	20.00	0.05	4.30
7.053	24.00	2.70	6.90
8.278	29.50	7.10	11.10
TEST CASE 2:			
3.930	11.75	-10.70	-6.00
5.600	17.25	-7.50	-2.90
7.060	23.00	-1.60	2.70

model performance is graphically shown in Figures 3 and 4. Good agreement for both assessment methods follows for Test Case 1 and Test Case 2. A measure of the relative error in the model predictions is best shown in Figures 5 and 6. These two plots assess the relative percentage error between the actual laser energy and the predicted laser energy needed to process discrete amounts of material.

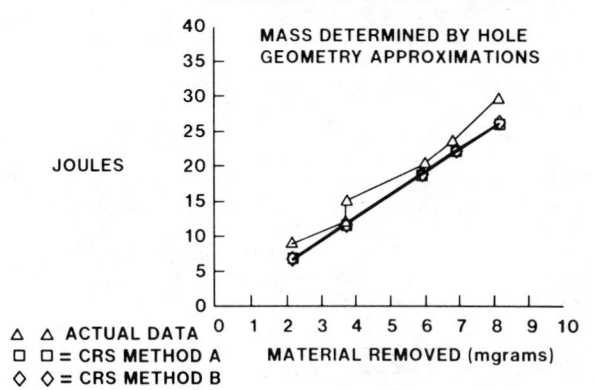

Figure 3. Comparison of the Predicted and Actual Energy Levels Resulting from the Thermal Balance Model of the Laser Drilling Process on Carbon Steel

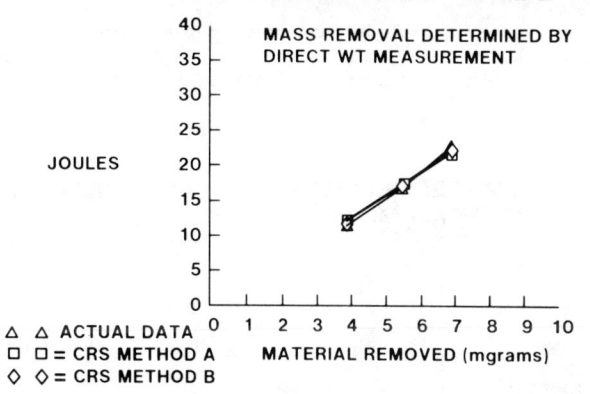

Figure 4. Comparison of the Predicted and Actual Energy Levels Resulting from the Thermal Balance Model of the Laser Drilling Process on Carbon Steel

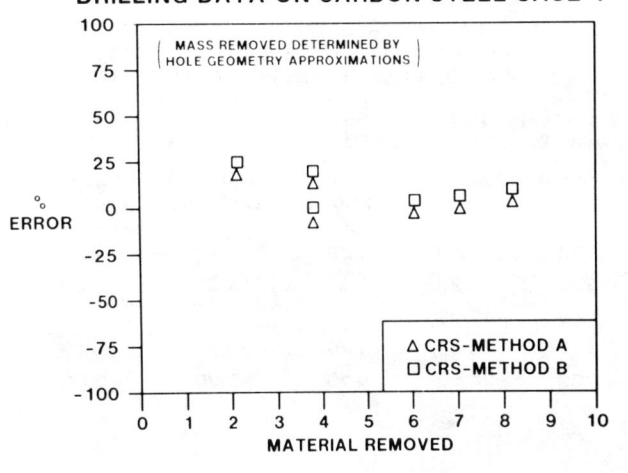

Figure 5. Measurement of the Relative Accuracy Between the Actual and Predicted Laser Energy Levels Needed to Process Holes in Carbon Steel

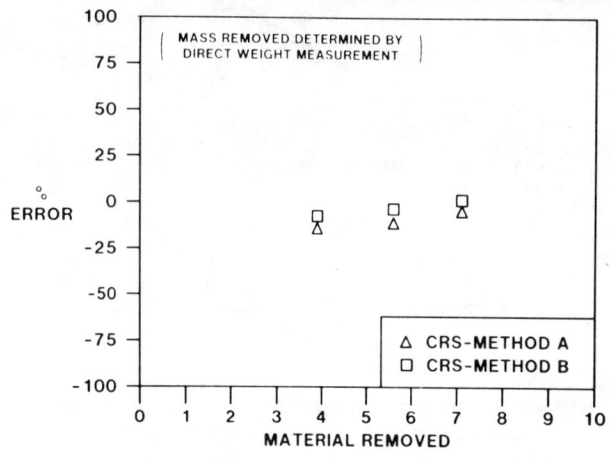

Figure 6. Measurement of the Relative Accuracy Between the Actual and Predicted Laser Energy Levels Needed to Process Holes in Carbon Steel

Stainless Steel: Drilling tests on stainless steel have followed good linear correlation between material removal and input pulse energy. With this as a basis, the thermal model with relative material characteristics and drilling performance levels was exercised. Two approaches were taken to best fit the data. The first concentrated on minimizing the expulsion factor while the second tried to match an expulsion level to a reflectivity of 50%. The first attempt yielded a narrow range of minimum error in predicted pulse energies (see SS-Method A and SS-Method B). The second approach resulted in a tight fit for all the data (SS-Method C). All three methods and their respective percentage error levels are detailed in Table 4. Representation of the model response as compared to actual material removal is shown in Figure 7. Excellent agreement between these levels is seen over the entire range of laser energy levels. A measure of the relative error of these predictions is shown in Figure 8. All three assessment methods fall within a $\pm 10\%$ error level.

TABLE 4

RELATIVE PERCENTAGE OF ERROR IN
PREDICTED ENERGY LEVELS TO PROCESS
LASER DRILLED HOLES IN STAINLESS STEEL

MATL. REMOVED (mgrams)	ACT. ENERGY (joules)	SS - METHOD A refl.=0.15 expl.=0.80	SS - METHOD B refl.=.35 expl.=.80	SS - METHOD C refl.=.50 expl.=.85
3.463	9.0	1.30	-7.70	-4.60
4.333	12.25	9.30	1.00	3.80
6.083	16.50	5.40	-3.10	-0.25
7.517	19.50	1.10	-7.80	-4.80
8.682	24.50	9.10	0.80	3.60

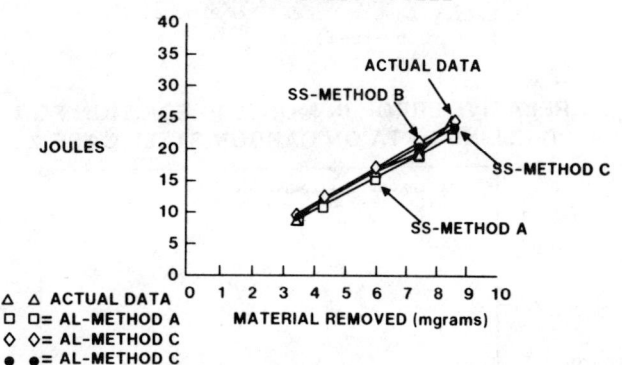

Figure 7. Comparison of the Predicted and Actual Energy Levels Resulting from the Thermal Balance Model of the Laser Drilling Process on Stainless Steel

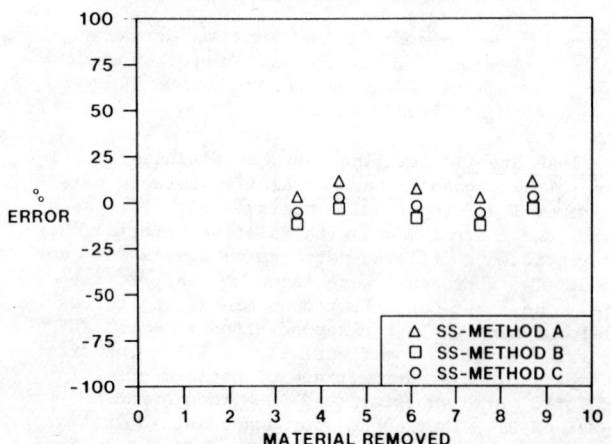

Figure 8. Measurement of the Relative Accuracy Between the Actual and Predicted Laser Energy Levels Needed to Process Holes in Stainless Steel

Aluminum: Drilling tests on aluminum have shown a non-linear relationship between pulse energy and material removed. Since our model is a first order linear approximation, results had to concentrate on either the low power or high power end of the response curve. Two attempts have been made to determine the best set of parameters that could be used to describe the response of the metal. The first attempt minimized the expulsion factor while the second concentrated on higher reflectivities (70-80%).

Addressing the expulsion minimization scheme, AL-Method A and AL-Method B were derived. These two plans concentrated on a minimization of the percentage error in the predicted energy levels at the low power and high power ends of the response curves, respectively.

Considering the high reflectivity approach, AL-Method C and AL-Method D were developed. These two concentrated on developing the best combination of expulsion factor and reflectivity level to yield the minimum percentage error between actual and predicted energy levels for both the low power and high power regimes of the laser incidence profile.

The relative performance of these methods is shown in Figures 9 and 10. Typical data are also shown in Table 5. A better fit results from the expulsion minimization approach. This could be a result of the low melting temperature (Tm = 660°C) and relative expulsion levels shown in Table 2. Available data indicate that aluminum is a high reflector, but our data tend to support that it is more of a material response phenomena than a surface reflectivity dominated one.

A relative measure of the accuracy of these predictions is shown in Figures 11 and 12. Once again the superior fit of AL-Method A and AL-Method B can be seen.

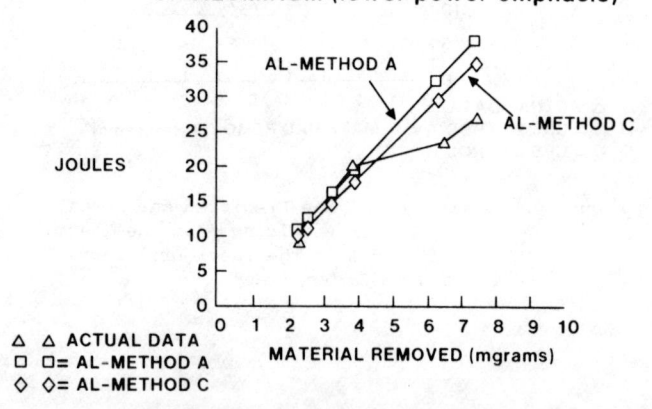

Figure 9. Comparison of the Predicted and Actual Energy Levels Resulting from the Thermal Balance Model of the Laser Drilling Process on Aluminum

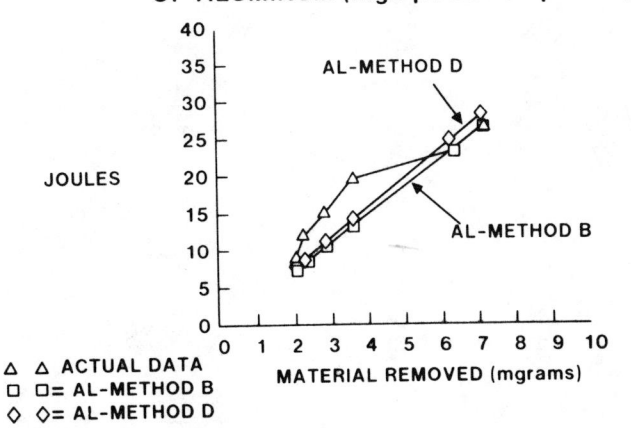

Figure 10. Comparison of the Predicted and Actual Energy Levels Resulting from the Thermal Balance Model of the Laser Drilling Process on Aluminum

TABLE 5

RELATIVE PERCENTAGE OF ERROR IN PREDICTED ENERGY LEVELS TO PROCESS LASER DRILLED HOLES IN ALUMINUM

MATL. REMOVED (mgrams)	ACT. ENERGY (joules)	AL - METHOD A refl.=.50 expl.=.70	AL - METHOD B refl.=.30 expl.=.80	AL - METHOD C refl.=.70 expl.=.80	AL - METHOD D refl.=.80 expl.=.95
2.030	9.0	-17.00	17.70	-7.10	12.45
2.260	11.75	0.10	29.80	8.60	25.40
2.860	15.00	0.70	30.20	9.20	25.80
3.630	19.50	3.20	32.00	11.60	27.70
6.300	23.00	-42.00	0.04	-29.90	6.30
7.233	26.50	-41.00	0.40	-29.50	5.60

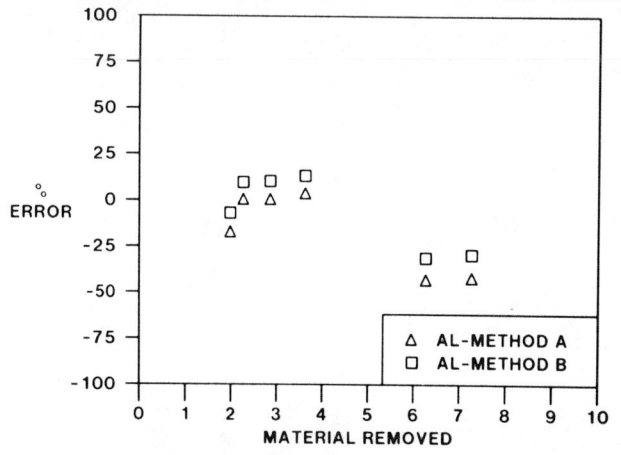

Figure 11. Measurement of the Relative Accuracy Between the Actual and Predicted Laser Energy Levels Needed to Process Holes in Aluminum

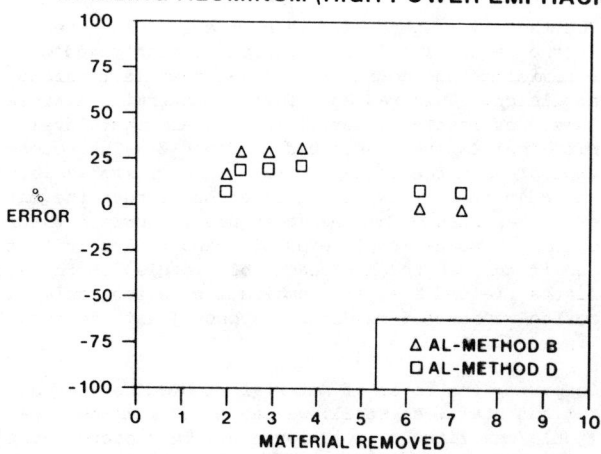

Figure 12. Measurement of the Relative Accuracy Between the Actual and Predicted Laser Energy Levels Needed to Process Holes in Aluminum

DISCUSSION

Validity of the Thermal Balance Approach

Aluminum: The thermal balance approach to describe the laser drilling process is heavily dependent on localized heat concentration. The fundamental balance between input laser energy and thermal response of the material can break down if the heat transferred to the sample is quickly conducted away from the drilling zone. This thermal transport mechanism will result in a larger zone between being heated but not necessarily being vaporized. As a result, the laser interaction must consider a larger zone which may not reach the same upper temperature limits. Comparing the material response curves of carbon steel (Figure 3) and aluminum (Figure 9), we see an initial slower removal rate (joules/mgm removed) for the aluminum. This can best be explained by the ratio of their thermal diffusivities (0.21 cm^2/sec for carbon steel and 0.91 cm^2/sec for aluminum). The aluminum thermal rate mechanism takes the laser energy into the material body faster, thus not allowing the localized region to reach the critical temperatures (Tmelt and Tvap). This prohibits drill zone heating and results in a lower material removal rate. However, looking at the upper end of aluminum response curve (higher beam energies) a change in material removal rate occurs. This is because the input beam has sufficient energy density to offset the thermal transport and thus remove material at a higher rate. The rate of material removed in the higher power regime seems close to that for the carbon steel. This thermal transport match even shows in the comparison of expulsion material (see Table 2) at the higher energy levels.

Carbon and Stainless: Turning to the model performance for carbon steel and stainless steel, an excellent correlation between predictions and actual material response is noted. This agreement helps to support the localized heat approximation and the thermal balance concept. As indicated, this laser drilling model will work well with metals that are primarily not good thermal conductors.

Expulsion Minimization vs High Reflectivity

The thermal response for each material was assessed in two ways. The first technique concentrated on minimizing the amount of material that left through expulsion. This was an effort to determine minimum levels of ejected material and compare that level with the response described in Table 2. The second approach set the reflectivity at a comfortable level and adjusted the expulsion fraction to meet the material response. This was designed to compare the resulting combination of expulsion and reflectivity to see if they approached reasonable levels. Both techniques yielded best fit combinations which truly gave a fine measure of material response (best shown in Figures 5, 6 and 8).

Comparing the rates of material removed between carbon and stainless steel, we see an almost identical (joule/mgm removed) relationship. This occurs even with different expulsion and reflectivity combinations because of the close similarity in the material characteristics.

FUTURE CONSIDERATIONS

The fundamental equations used in this first order approximation use material properties which are not temperature dependent. Since these thermal characteristics vary directly with temperature, our next approach will be to incorporate this response into our model.

In addition to our efforts of matching thermal properties, more materials will be considered. This will allow us to generate a broad database regarding rates of material removal and give a higher confidence in our model performance.

Once the broad database for material removal is available, our model will look towards drilling and cutting on neodymium-YAG lasers. The repetitive pulse phenomena, associated with our understanding of expulsion and reflectivity, will help to formulate a reasonable laser cutting model.

REFERENCES

1. Charschan, S. S., Ed., Lasers in Industry, Van Nostrand Reinhold, New York (1972)

2. Mukherjee, K. and Mazumder, J., Ed., Lasers in Metallurgy, Metallurgical Society of AIME (1981)

3. Ready, J. F., Ed., Lasers in Modern Industry, Society of Manufacturing Engineers (1979)

4. Charschan, S. S., Ed., Guide for Material Processing by Lasers, Paul M. Harrod (1977)

LASER CLADDING WITH PNEUMATIC POWDER DELIVERY

V. M. Weerasinghe & W. M. Steen
Metallurgy Dept. Imperial College, London

ABSTRACT

A process of laser cladding using argon blown powder delivery is described. The effect of the process parameters on the quality of the clad layer is discussed. The flexibility and versatility of this process are noted together with the significant process improvement obtained using a reflective shroud and shot blasted specimens.

INTRODUCTION

Cladding and surface alloying are two of the many material processing applications for which a laser has been and is being used[1-6]. It could be argued that they are the two uses with the greatest industrial potential in the future because of the almost infinite material possibilities involved in coating one metal with another and the frequent design problem which requires one set of properties for the surface and another for the bulk, all at minimum cost.

Cladding is not new, but laser cladding offers some unique advantages:

a) controlled dilution levels.
b) localised heating, which reduces thermal distortion
c) controlled shape
d) good thermal bonding
e) fine quench microstructures

These advantages usually lead to fundamental economic advantage over competing processes due to reduced after machining costs, reduced powder or clad material costs, and reduced product rejection due to unacceptable distortion. However added to these advantages are the improved metallurgy of the clad layer due to its moderately rapid quench and, in the process described here, its ease of application since the process can be a non contact one.

The basic technique for producing uniform clad layers by laser is to overlap single clad tracks. Intuitively this would seem tedious when compared to established methods such as roll bonding, explosion cladding or powder casting[7-11]. One may conclude that the laser is best suited for coating small confined areas. However cladding rates presented here indicate that the laser may well compete favourably in covering rates with conventional cladding processes. Three methods reported here help to make this possible:

a) high laser powers
b) reflective shronding
c) shot blasted surfaces

Most metal combinations can be clad provided the clad material does not have a significantly higher melting point than the substrate as for example has aluminium clad with stainless steel. The mixed solidification direction in these cases leads to interface porosity.

Discussed here are the effects of the major process parameters on the quality of the clad layer using a pneumatic powder delivery system for laser cladding.

EXPERIMENTAL

The process used here, is similar to that currently used in production by Rolls Royce[12] namely to blow the powdered clad material into the laser generated melt pool as illustrated in fig. 1.

Fig. 1. Laser cladding by powder injection.

The powder continuously impinging on the melt pool appears to give enhance energy absorption.

The process is fundamentally different from the fusion of preplaced powder beds. In this process control of dilution is by the powder feed rate and melting occurs first at the substrate surface, whereas in preplaced powder bed processes dilution is controlled by power density and melting occurs first at the powder bed surface and so on through the bed to the substrate surface.

The independent process variables are related to the three components of the system; the laser, powder delivery system; substrate/clad combination. These variables and how they were monitored are listed in table 1.

Table 1 main operating variables

System variable	method of monitoring
Laser	
power	flowing cone calorimeter; up to 3 kw used.
Beam diameter and mode structure	rotating rod[13,14], acrylic prints, fluorescent screens (0.2-20mm used)
traverse speed	electronic timer
Powder feed	
mass flow rate	load cells on feed hopper, acoustic emmission on feed line[15]
velocity of feed	rotameters and photography
powder shape & size	photography (not a relevant parameter with the screw feeder used here).
Substrate/clad combination	
materials used	most of the work reported here used powdered 316 stainless steel on En 3 mild steel substrate. Other systems reported here are: stellite/mild steel stellite/powder formed tool steel TiC & stainless steel/mild steel

The rest of this paper discusses the effect of these variables on the dependent variables of cladding rate, clad thickness and width, surface finish, powder utilisation, heat utilisation, substrate dilution, segregation, porosity, residual stress, cracking, microstructure and adhesion.

DISCUSSION & RESULTS

a) <u>Single track profiles</u>. Three basic cross section profiles of single tracks were observed. They are illustrated in fig. 2. Profile (c) is the preferred section for overlapping tracks to cover an area. At higher powder feed rates or lower power densities profile (a) is produced while at lower powder feed rates or higher power densities profile (b) is produced. For a given traverse speed and powder flow there is an optimum spot diameter for maximising the cladding rate.

b) <u>Dilution</u>. For a given specific energy (power/(beam diameter x traverse speed), J/mm²) the injected powder mass flow controls the level of dilution. This is illustrated in fig. 3. Unlike preplaced powder cladding dilution is independent

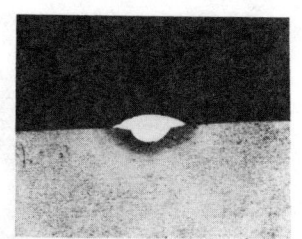

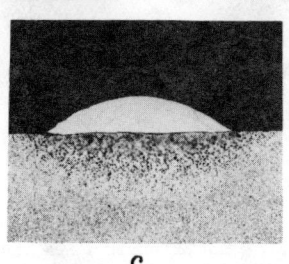

Fig. 2. Three basic clad section profiles: Clad-stainless substrate. Mild steel

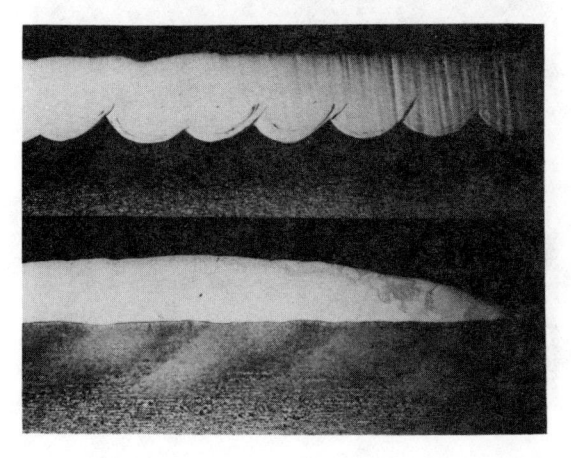

Fig. 3. Control of dilution by increasing the powder mass flow.
Powder mass flow 0.090 g/s top
0.212 g/s bottom
Specific energy input - 54 J/mm²

of the cladding speed.

Control of the powder feed is of prime importance. The ability to set a precise value allows control of dilution whereas any irregularity in the feed shows as a corresponding surface irregularity in the clad track. The sensitivity is sufficient to allow surface contouring by deliberate variations in powder feed rate.

If defocussed beams are used, such as 5mm in diameter, a symmetric mode structure is needed to avoid local power spikes causing dilution as illustrated in fig. 4.

c) <u>Porosity</u>. Porosity in laser clad layers may be caused by one or a combination of the following: cavities between tracks (interrun porosity), solidification cavities, gas evolution.

Interrun porosity, illustrated in fig. 5, occurs when two tracks of section profile similar to fig. 2a are laid side by side with insufficient overlap.

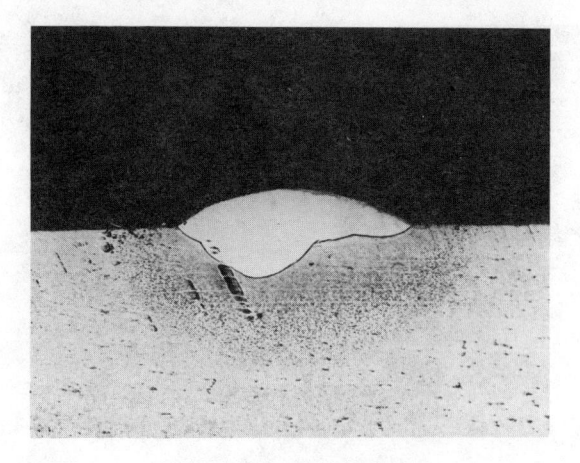

Fig. 4. Effect of unsymmetric beam intensity distribution

Fig. 5. Inter-run porosity.
clad - s/s substrate m/s

It can be eliminated totally by operating in such a way to produce section profiles as in fig. 2. It can also be eliminated as shown by Powell[15] by angling the powder feed around 10° to the line of the clad track or ultrasonically vibrating the substrate during cladding.

Solidification cavities occur as porosity at or near the interface between the cladding and substrate if the clad layer has a significantly higher melting point than the substrate. Under such conditions the solidification front finishes at the interface causing shrinkage cavities. For sound cladding the solification front should finish at the surface. Fig. 6 is an example of this for the system stainless steel cladding on aluminium.

d) Cracking. Cracking occurs due to tensile solidification stresses which result from differential expansion in the steep thermal gradients associated with laser processing. These tensile stresses result in cracking of non ductile clad layers or layers with stress raisers due usually to porosity. Thus it is observed in hard coatings such as iron/boron (>1000 VHN) but not on coatings of stainless steel (220 VHN) or stellite (580 VHN) in mild steel.

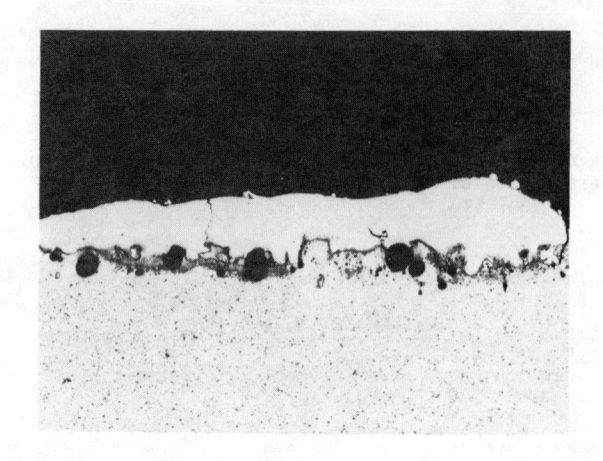

Fig. 6. Solidification void cracking.
clad - s/s substrate - Al.

While coatings of stainless steel on aluminium do show cracks, associated with pores.

Cracking of difficult systems such as Fe/B have been successfully eliminated by preheating the substrate as shown in fig. 7.

room temperature   500°C   700°C

Fig. 7. Progressive ellimination of solidification stress cracking by increasing substrate pre-heat temperature.
clad - Iron Boron (1100 HV)
Substrate - Mild steel

c) Surface Finish. The degree of overlap of adjacent tracks, and the section profile of each track govern the surface finish of the clad layer. With sufficient percentage overlap (usually about 50-60%) a good surface finish can be obtained as illustrated in fig. 8 and 9.

f) Homogeneity of clad layer. Fig. 10, 11 illustrate the distribution of alloying elements in the clad layer of 316 stainless and interface regions by an EDX scan. The almost negligeable dilution and lack of macro segregation are apparent.

Fig. 8. Stainless clad surface
Laser Power - 1700 W
With utilization of reflected energy
Spot Din - 5 mm
1st coating - Cladding rate 18·8 mm²/s
Speed   12·5 mm/s
2nd cross clad coating - cladding rate
10 mm²/s
speed   6·7 mm/s
Powder flow - 0·2 g/s

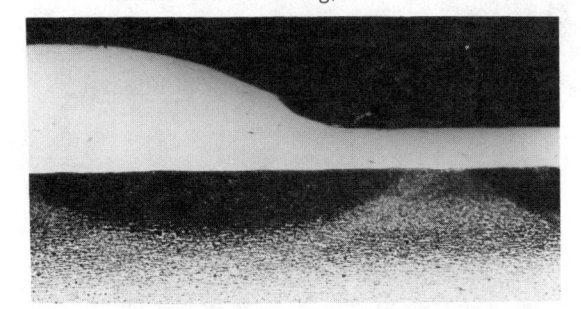

Fig. 9. Section of the clad shown in Fig. 8
Etchant - Nital 10%

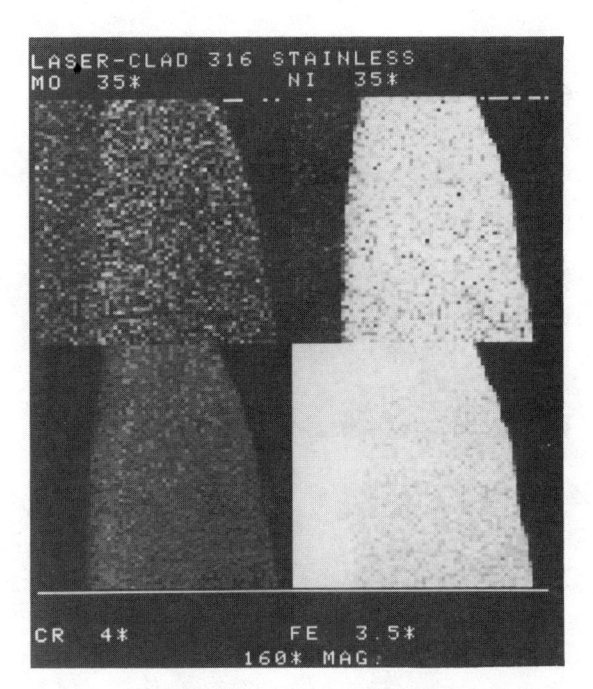

Fig. 10. X-ray digimap off alloying elements in a stainless 316 clad showing no macro-segregation.

Fig. 11. Line scan of Cr across the clad/substrate interface.

Layers which were heavily diluted also showed an almost uniform composition, suggesting that there is considerable stirring in the weld pool.

The solidification mechanism is usually dendritic along the steepest thermal gradients fig. 12.

g) <u>Substrate distortion</u>. Measurements on the heat affected zone (HAZ) indicate that there is minimal thermal penetration of the substrate (of the order of 1-2mm). Distortion from this source is therefore slight. However, there is some distortion due to the contractural stress set up by the solidifying clad layer.

h) <u>Substrate adhesion</u>. 180° bend tests on stainless/En3 clad layers only revealed vertical cracks through the clad layer or none at all. There were no examples of delaminating or interfacial cracks.

On other systems (stellite/steel, monel/steel, bronze/steel) a side blow with a chisel, very rarely removed the clad at the interface.

From these examples it can be concluded that the adhesion of laser clad layers is unusually good. The process produces, as expected, a firm fusion bond.

i) <u>Cladding rates</u>. Fig. 13, 14 show the relationship between the cladding speed and the width and height of a single track for a fixed beam diameter, laser power, and powder flow conditions (mass flow, velocity, nozzle shape etc).
By varying the power, or beam diameter it can be shown that the edge of the track has a constant energy density. This suggests that successful cladding only occurs if certain thermal conditions in the substrate are satisfied.

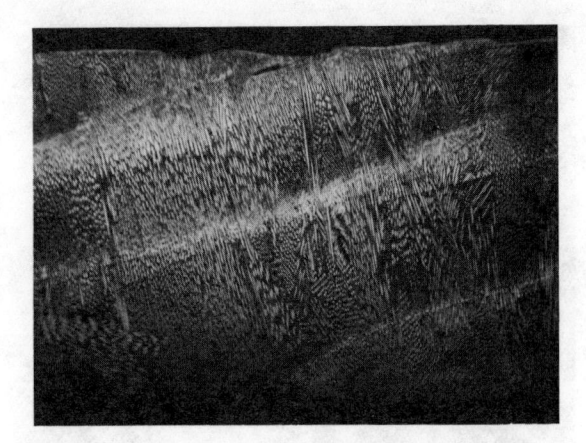

Fig. 12. Dendritic micro-structure of a stainless 316 clad.
Electrolytic etch in Oxalic Acid.

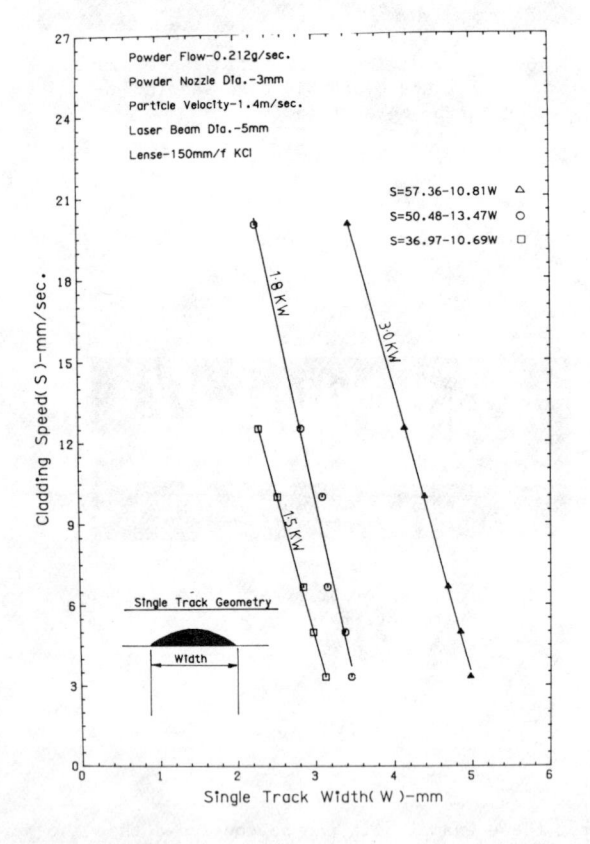

Fig. 13. Effect of cladding speed on single track width.

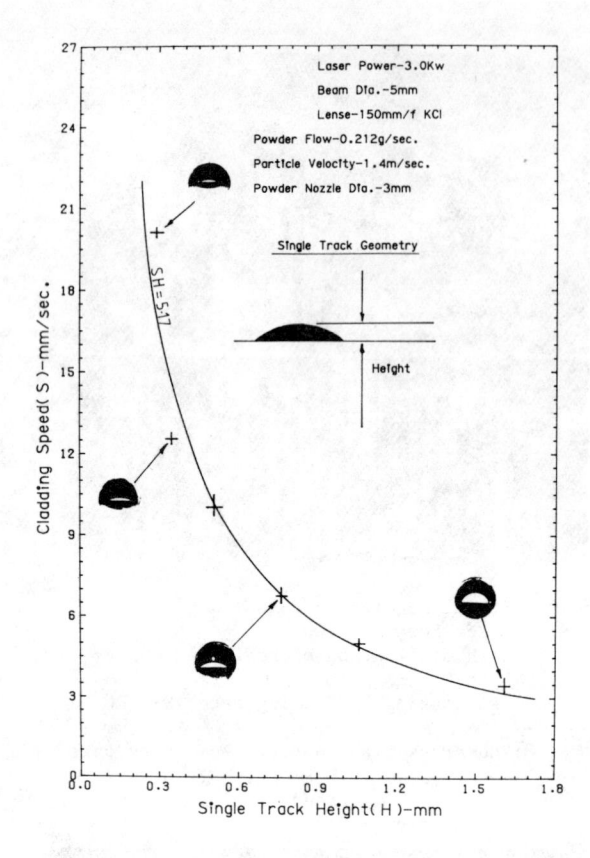

Fig. 14. Effect of cladding speed on single track height

Currently we believe this to be superficial melting of the substrate. The powder particles need not necessarily arrive at the surface molten for adhesion.

The effect of overlap on the minimum clad layer thickness is shown in fig. 15. Where the percent overlap 'k' is plotted against the dimensionless group (T/H) where T is the minimum or uniform clad thickness and H is the height of a single clad track.

From these graphs, all at constant laser power, beam diameter and powder flow conditions, it is possible to draw up relationships of the form:

$$S = a - bW \quad (1) \text{ (see fig. 13)}$$
$$k = \exp(-T/1\cdot 8H) \quad (2) \text{ (see fig. 15)}$$
$$SH = d \quad (3) \text{ (see fig. 14)}$$
$$C = kWS \quad (4)$$

where
S = cladding speed mm/s
W = single track width mm
H =   "   " height mm
T = multiple track minimum clad thickness mm
k = overlap factor
C = coverage rate mm²/S (not allowing for indexing time)
a,b,c,d = constants

From the equations the cladding rate as a function of speed and thickness can be obtained. By differentiating this expression the speed giving the

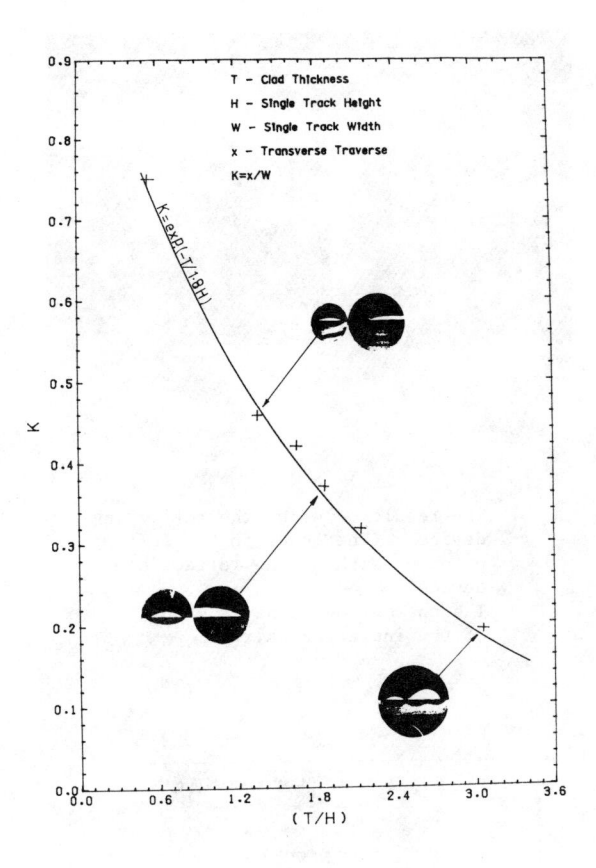

Fig. 15. Effect of overlap on clad thickness

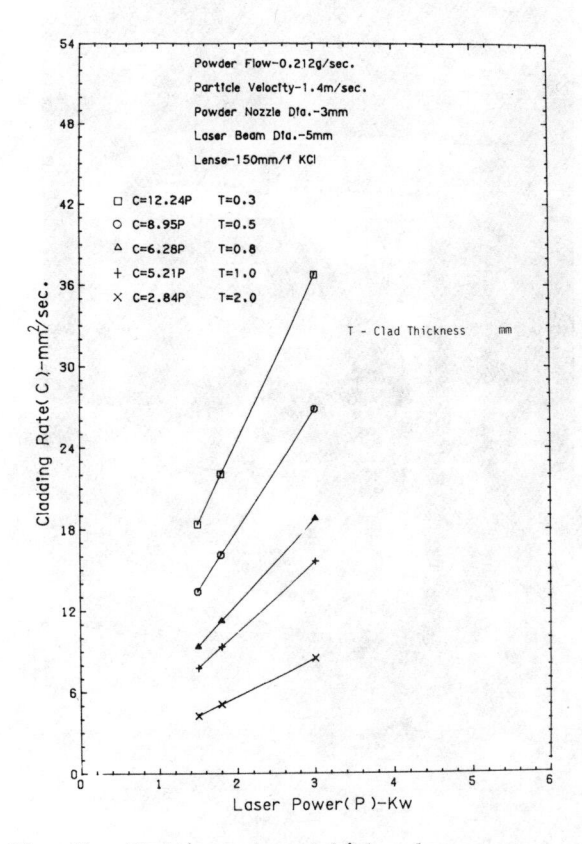

Fig. 16. Cladding rates at higher laser powers

maximum cladding rate can be found.

From equation 2 & 3 this speed gives the optimum overlap required. Such calculations are presented in fig. 16, showing the maximum cladding rate for particular clad thicknesses and laser powers using a fixed powder feed system. The lines in fig. 16 can not be safely extrapolated into the lower power region since powers less than around 1kW are unable to produce continuous clad tracks. Also, operating at high traverse speeds means the process is even more sensitive to variations in powder feed. Speeds between 4-18 mm/s are found good for laser powers $1.5 < P < 3.0$ kW.

j) <u>Utilisation of reflected energy</u>. The reflected energy consists of the reradiated energy from the hot molten pool and the reflected incident laser radiation.

This radiation was partially recovered by a hemi spherical reflecting device fig. 17, placed so that the molten pool was at its centre.
A remarkable improvement in cladding rate was obtained. Fig. 18-20. For example the mass deposition rate for 1700 W laser power improved to that equivalent to a 2600 W laser power. This suggests around 900 W of energy were recovered. Without the reflective dome the melting efficiency was around 6% with it the efficiency is around 15%.

Using a Milletron thermo scope at two frequencies the molten pool temperature was measured as 1900°C. For the known pool dimensions and assumed emissivity of around 0.6 the expected reradiated power is around 15 W. Thus the reflecting dome must be principally recycling reflected laser radiation.

An effective gas shroud was incorporated into the dome, allowing cladding of such sensitive metals as titanium.

k) <u>Substrate surface condition</u>. Fig. 20, illustrates the effect of shot blasting the substrate surface prior to cladding. The upper part was shot blasted, while the bottom half retained the original ground finish. The track on the right was produced without the reflecting dome and is seen to cease on reaching the more reflective surface. The track on the left was produced using the reflective dome, and is continuous on both surfaces.

l) <u>Some special applications</u>. Fig. 21 illustrates the hardfacing of surfaces of small area and complex shape. Fig. 22 shows a totally new clad style of alternate ribs of stellite and stainless steel.

Fig. 17. Hemispherical reflecting device 4" sq.

Fig. 18. The result of using the reflecting device. (The track on the left was produced without the reflecting device).
The one on the right is with the use of the inert gas skirt.

Fig. 23, 24, shows a clad layer made from a mixed powder feed of stainless steel and TiC particles. These particles tend to float as shown in fig. 25. Ayers[16,17] reported a similar effect with his particle injection process.

CONCLUSIONS

1. Metallurgically sound clad layers of low dilution, good adherence and no porosity can be produced by a non contact process based on pneumatic powder delivery.

2. The process has low thermal distortion and includes the possibility of making shaped deposits.

3. The cladding rates are industrially attractive at laser powers over 3kW.

4. A considerable improvement in process efficiency is obtained by using a reflective dome.

5. The process is easily adapted to an industrial line and is versatile in treatment as illustrated by some novel clad layers.

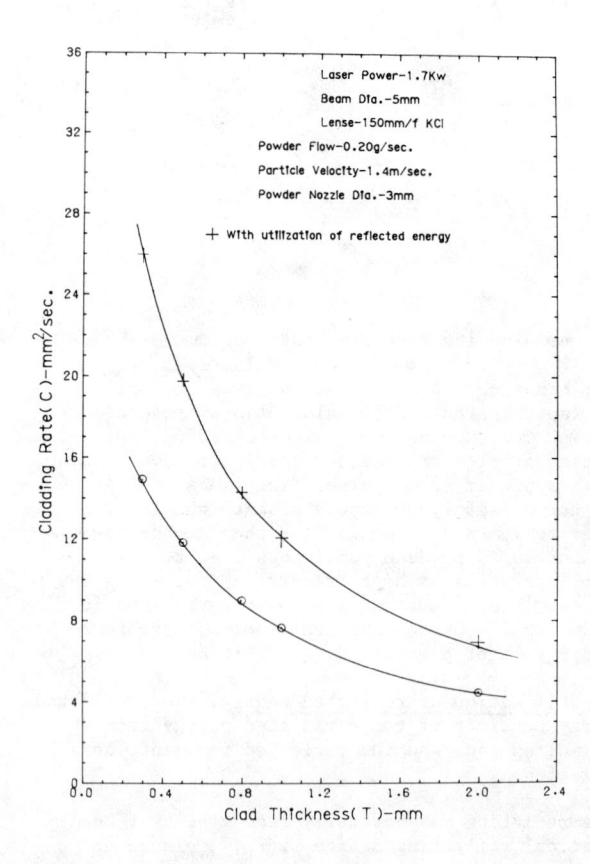

Fig. 19. Effect of utilizing the reflected energy on the cladding rate.

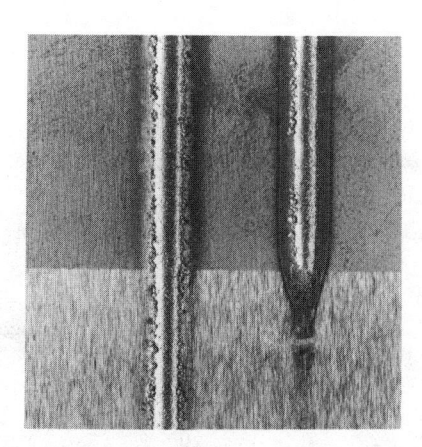

Fig. 20. Effect of shot blasting the substrate surface.

REFERENCES

1. F.D. Seaman, D.S. Gnanamuthu, "Using the Industrial laser to surface harden and alloy", - Metal Progress, Vol. 108, No.3, 1975, P.67

2. W.M. Steen, C.G.H. Courtney, "Hardfacing of Nimonic 75 Using a 2kW continuous wave CO_2 laser", - Metals Technology, June 1980 PP 232-237.

3. W.M. Steen "Surface Coating Using a laser", Paper 3 Conf. Advances in Surface Coating Technology, Welding Int. London, Feb. 1978.

4. J. Powell, W.M. Steen "Vibro Laser Cladding" - Conf. Proceedings, The Metallurgical Society of AIME, Feb. 1981.

5. A. Belmondo, M. Castagna, "Wear-resistant coatings by laser processing" - Thin solid Films, 64 (1979) p. 249-256.

6. E.M. Breinan, B.H. Kear & C.M. Banas" Processing Materials with lasers" - Physics Today, Nov. 1976 P. 44-50.

7. B.B. Moreton, "Copper-Nickel clad steel for Marine Use" - The Metallurgist, May 1981, P. 247-252.

8. H. Thielsch, "Stainless Clad Steels" - Welding Journal, Welding research Suppl. Vol. 31; March 1952, P. 142-150.

9. R.S. Zuchowski, E. Garrabrant, "New developments in Plasma arc Weld Surfacing" - Welding Journal (43) Jan. 1964 P.13-20.

10. M. Riddinhough, "Hardfacing by Welding" Pub. The Louis Cassier Co. Ltd. 1949

11. J. Hinde, "The Welding of clad steels" Pub. The Mond Nickel Co. Ltd. 1956

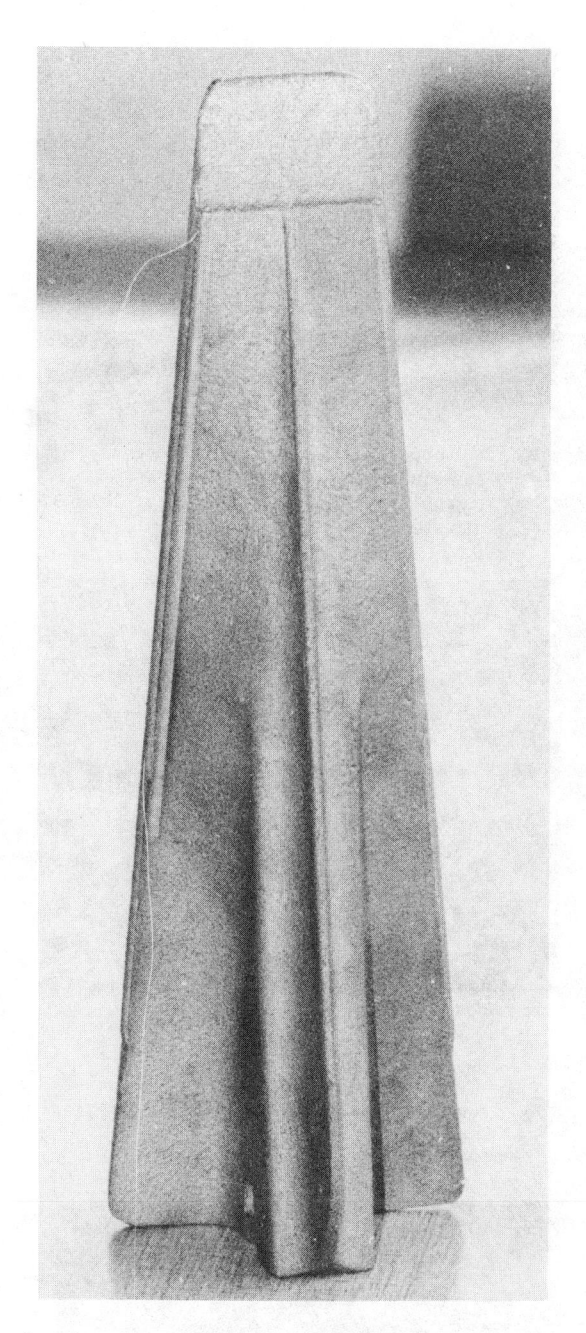

Fig. 21. An agricultural plough blade, laser hardfaced with stellite.
Covering rate 12·5 mm²/s
Laser Power 1700 W.

12. M. McIntyre "Laser cladding at Rolls Royce" This conference, paper

13. G.C. Lim & W.M. Steen, "Measurement of temporal & spatial power distribution of a high powered CO_2 laser beam" Optics & Laser Tech. June 1982 p. 149-153.

14. A.L.L. brochure. Hans Gressel weg München - 8000, W. Germany. 1982

15. V.M. Weerasinghe, W.M. Steen, "Monitoring of gas borne powder mass flow using an acoustic

Fig. 22. 'Ribbed' clad layer on mild steel consisting of alternate tracks of hard (stellite) and soft (stainless 316) material.

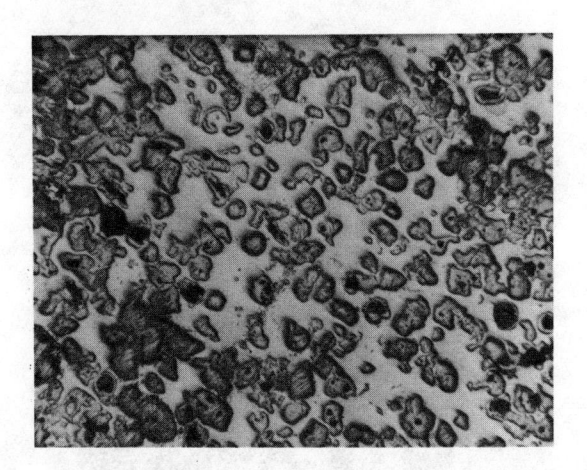

Fig. 23. TiC particles in a stainless 316 clad matrix.

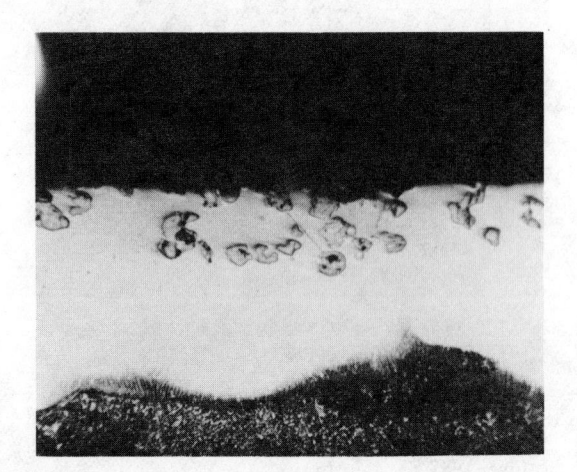

Fig. 24. Section of the clad surface shown in Fig. 25.
Etchant - 10% Nital.

emission technique" to be published.

16. W.M. Steen, J. Powell. Proc. Laser 81 conf. Munich p. 200-204 publ. Springer Verlag, Berlin

17. J.D. Ayers, "Particulate - TiC - Hardened Steel Surfaces by Laser Melt injection, - Thin Solid Films, 73, (1980), P. 201.

18. J.D. Ayers, "Particulate Composite Surfaces by laser Processing" - Conf. Proceedings, The Metallurgical Society of AIME Feb. 1981.

Acknowledgements.

Both the authors would like to thank the British Steel Corp., Scottish Laboratories for encouragement and financial support with this work. Weerasinghe would also like to thank the Science & Engineering Research Council for a grant during his research studies.

LASER PROCESSING OF PLASMA-SPRAYED NiCr COATINGS

H. Bhat, H. Herman
Dept. of Materials Science & Engineering
State University of New York
Stony Brook, N.Y. 11794

and

R. J. Coyle
Engineering Research Center
Western Electric Co.
P.O. Box 900
Princeton, N.J. 08540

ABSTRACT

Plasma spraying is a high velocity melt deposition process yielding thick (>75 μm) coatings for protecting substrates in extreme environments. While plasma spraying technology has grown rapidly, there remain deficiencies associated with such coatings, the principal one being porosity. Pores which extend from the outer surface of the coating to the substrate can permit entry of corrodents, giving rise to rapid deterioration of the coating system. Laser melting of the outer regions of the plasma-sprayed coating can seal the pores, giving significantly improved corrosion resistant behavior. In this study the corrosion behavior of plasma-sprayed NiCr was examined as affected by post-spray laser treatment.

Ni-20 w/o Cr coatings (250-300 μm thick) were plasma sprayed onto properly prepared low carbon steel substrates, the spray process parameters being controlled in order to obtain varying degrees of porosity. The porosity was characterized in cross-section using scanning electron microscopy. The surface of these coatings was melted to a depth of 20-25 μm employing two passes of a continuous wave CO_2 laser operating at 200 and 300 watts. Metallographic examination of the cross-section of the thus treated coating revealed no surface connected porosity, a dense, continuous layer of alloy having been formed at the surface.

The laser-treated plasma-sprayed specimens were subjected to corrosion tests in concentrated HCl solutions. The nature of corrosion was monitored through metallography, analytical chemistry techniques, and by way of acoustic emission (AE) emanating from the corroding surface. For as-sprayed coatings, bubble evolution from the corroding surface gave rise to recordable AE events, the number of which increased linearly with time for a given concentration of HCl.

Laser-treated coated specimens showed no corrosive attack up to 6 volume percent HCl. Thus, laser sealing of pores results in a highly corrosion resistant plasma-sprayed coating. This duplex process, it is suggested, could significantly extend the use of plasma spraying to extreme application areas where porosity cannot be tolerated.

INTRODUCTION

A major deficiency of plasma-sprayed corrosion-resistant coatings is the surface-connected porosity which permits the entry of corroding constituents, thereby degrading the coating-substrate interface. Laser fusing is being employed to enhance the corrosion-resistant features of such coatings (1,2). Laser-induced surface melting and rapid solidification of the top layer (25 μm thick) in the sprayed coatings results in pore closure and homogenized microstructures.

Bhat et al (2) have carried out a preliminary study of the potential utility of laser treating of sprayed coatings for corrosive environments, and Dallaire and Ceilo have optimized the laser processing parameters in terms of the heat-transfer characteristics of Ni-coated steel substrates (3). In the present work, acoustic emission (AE) methodology has been employed to study ambient temperature aqueous corrosion behavior of the as-sprayed coatings with and without post-spray laser treatment. Of central interest in this study has been the effect of laser treatment on coating-substrate interface attack.

Mansfeld and Stocker have reported a qualitative relationship between AE and corrosion rate, having reported that the acoustic emission detected rate of noise generation accurately reflects pitting behavior in aluminum alloys and stainless steel (4). Zhakarov, et al, employed AE to monitor the corrosion rate of steel in acid environments, concluding that AE intensity depends on the concentration and nature of the acid (5). AE techniques have been employed to detect the selective phase corrosive attack occurring in cast nickel aluminum bronze (6). The application of this technique has been more versatile in stress corrosion cracking and hydrogen embrittlement studies (7,8,9). The spectral study of noise generating from corroding nodular cast iron has been used to differentiate the origin of AE (10). AE studies can also be carried out in the presence of turbulent flux and cavitation noise, by employing a transducer in the range of 1.5 to 2.5 MHz (11).

EXPERIMENTAL PROCEDURE

Ni-20 wt.% Cr powder (Metco 43C) was plasma sprayed onto a grit-blasted mild steel substrate measuring 12cm X 1.5cm. One half of the surface area of one face of the substrate was coated. Coatings with various characteristics were obtained using the spray process parameters listed in Table 1.

Table 1. Plasma-Spray Parameters

Specimen	Feed Rate (kg/hr)	Spray distance (mm)	Current (ampere)	Voltage (Volt)	Thickness (mm)	Characertistics*
A	6.5	150-175	500	65	0.3	Porous (10-11%)
B	4.8	100-125	500	65	0.3	Dense (5% porosity)
C	3.2	75-100	500	60	0.3	Unmolten particles 8% porosity

*Determined by optical metallography and SEM examination of cross-sections.

The as-sprayed coatings with the highest porosity (Specimen A) was laser treated employing a continuous wave CO_2 laser. Two mutually perpendicular passes of the laser, operating at 200 and 300 watts, were employed (Table 2). The detailed description of the laser treatment procedure has been given elsewhere (12). The thickness of the laser-treated zone so obtained was 20-25 µm.

Table 2. Comparison of surface-roughness

a) Laser treated surface

Power (watts) Pass 1	Pass 2	Mutual direction of passes	Surface Roughness* AA (µm)	Maximum peak to volley distance (µm)
300	200	Parallel	3.05-3.55	40.6
200	300	Parallel	2.54	35.6
300	200	Perpendicular	1.78-1.9	20.3-22.9
200	300	Perpendicular	1.75-1.9	20.3

b) As-sprayed surface - 6.1 to 7.1 µm (AA).

*Determined by electro-mechanical profilometer.

The as-sprayed and sprayed and laser-treated coatings were subjected to AE tests. The schematic of the AE test setup is shown in Fig. 1. A thin film of couplant fluid was applied to the contact surface of the transducer (150-300 KHz), which was clipped onto the unsprayed portion of the substrate by a constant pressure clip. The edges and the backside of the specimen (i.e., the uncoated face) were sealed using epoxy so that only the top surface of the coating was exposed to the HCl corrodent, the exposed area being the same in all cases. Preliminary tests were carried out to establish the necessary gain and threshold level voltage in order to monitor corrosion and to exclude background counts. All of the experiments were carried out by employing a threshold level of 1.05V and a gain of 98 dB.

The standard dimethyl glyoxime and NH_4OH tests were employed for the detection of nickel and iron radicals, present in the corrodent respectively (13).

RESULTS AND DISCUSSION

1. Defects in the laser-treated zone

A microstructural examination was carried out to determine the fate of the as-sprayed features of the coatings following laser treatment. These features include porosity, oxide inclusions, interlamellar separation, etc. The most commonly occuring defect is the formation of vertical cracks confined to the regions of oxide inclusions in the laser treated zone. Such a crack through the oxide phase, as shown in Fig. 2a, extends to the top surface of the coating. The other branch of the crack has stopped where it intersects metal. The use of oxygen-free powder for plasma spraying and proper spray practice might obviate this problem.

A major difficulty encountered in laser-treated coatings is edge lift-off, Fig. 2b, which can be attributed to the presence of stress gradients through the cross-section of the sprayed coatings (14). This problem can be overcome by proper spraying and laser treatment practice: e.g., i) bevelling the edges of the substrate before spraying; ii) the use of lower power for initial passes and using higher power for subsequent passes.

2. Surface roughness of laser-treated coatings

The effect of laser power and the mutual directionality of the passes on the surface roughness of laser-treated coatings is given in Table 2. It is evident from the table that mutually perpendicular passes yield a much smoother surface than that obtained from parallel passes. Also, the surface roughness of laser-treated coatings is one half the value of the as-sprayed coatings. In corrosion-erosion applications, the higher the surface roughness, the greater

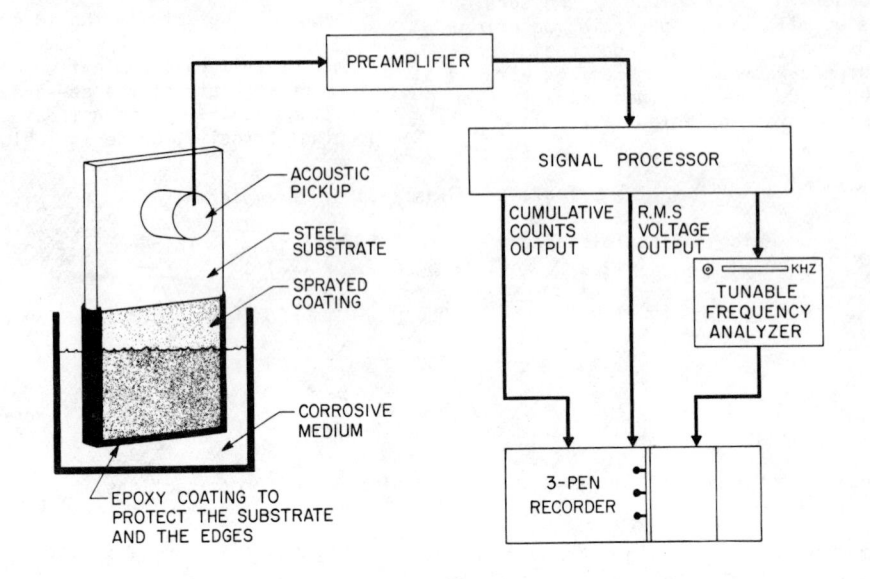

Fig. 1   Schematic of acoustic emission set-up

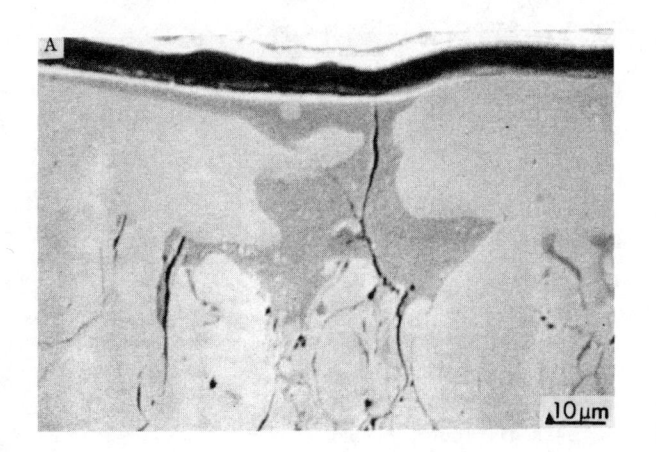

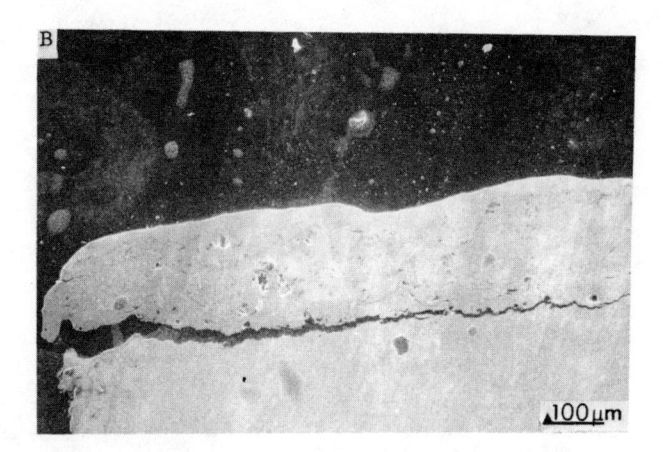

Fig. 2   Defects occuring in laser treated coatings.
a) A vertical crack through the oxide phase in laser treated zone.
b) Edge lift-off.

will be coating degradation (15). No significant difference in the thickness of the laser-treated zone was detected by employing the variables listed in Table 2.

3. Corrosion monitoring using AE techniques

a. As-sprayed coatings:

The sprayed coatings with different porosities, as listed in Table 1, showed a linear relationship between AE counts and time during corrosion in a given HCl solution. The slope of this relation is tabulated as a function of different HCl concentrations in Tables 3, 4 and 5 for Specimens A, B and C, respectively. In all of these cases, at low concentrations of HCl, the acid penetrated through the pores of the coating, attacking and dissolving iron from the coating-substrate interface. A typical cross-section of the as-sprayed coating before and after corrosion is shown in Figs. 3a and 3b, respectively. The excellent as-sprayed coating-substrate interface separates out completely after undergoing corrosion. The fact that iron corrodes when the as-sprayed

Table 3. AE Count Rate as a Function of HCl Concentration for As-Sprayed Specimen A.

Concentration of Corrodent (%)	Time (min.)	AE Count Rate (counts/min.)	Correlation* Co-efficient	Test For Fe	Ni
1.0	15	36	0.992	Yes	No
1.5	10	148	0.998	Yes	No
2.0	10	210	0.999	Yes	No
3.0	10	327	0.999	Yes	No
4.0	10	426	0.998	Yes	No
5.0	10	511	0.998	Yes	No

*Refers to the linearity of relationship existing between AE counts and time at a given corrodent concentration

Table 4. AE Count Rate as a Function of HCl Concentration for As-Sprayed Specimen B.

Corrodent Concentration (%)	Time (min.)	AE Count Rate (counts/min.)	Correlation Co-efficient	Test For Fe	Ni
1	10	None	——	No	No
2	10	35	0.990	Yes	No
3	10	78	0.992	Yes	No
4	5	186	0.993	Yes	No
	5	54	0.984		
5	5	103	0.994	Yes	No

Table 5. AE Count Rate as a Function of HCl Concentration for As-Sprayed Specimen C.

Corrodent Concentration (%)	Time (min.)	AE Count Rate (counts/min.)	Correlation Co-efficient	Test For Fe	Ni
0.3	10	300	0.993	Yes	No
0.8	10	2665	0.996	Yes	No
2.1	10	3940	1.0	Yes	No
4.3	10	6208	0.999	Yes	No
7.14	10	37900	0.999	Yes	Yes
11.43	10	58740	0.999	Yes	Yes

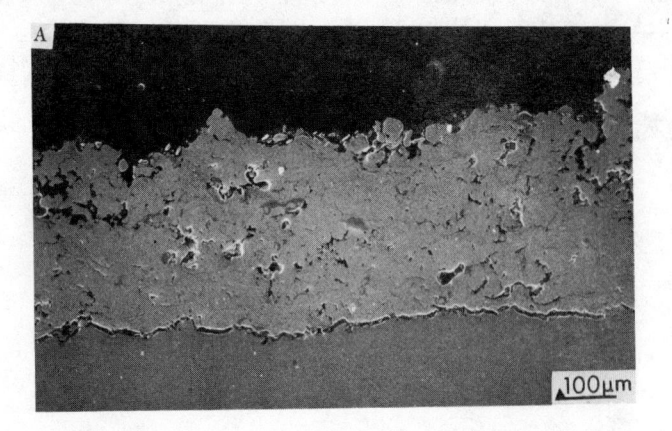

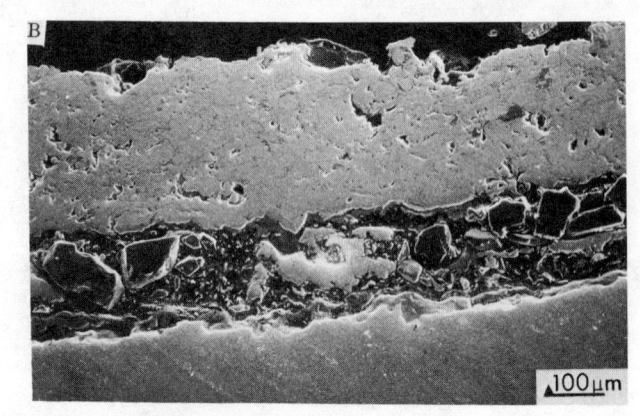

Fig. 3. SEM Cross-Section of the As-Sprayed Coating.
A. Before corrosion.
B. After corrosion.

coating is exposed to the corrosive environment is further confirmed from the standard analytical chemistry tests for iron and nickel in the corrodent (Tables 3, 4, and 5). It is also clear from these tables that AE activity for the case of Specimen B (lowest porosity), is lower than for the case of Specimen A and C. Also, the dissolution of iron does not occur below 2% HCl concentration for the case of Specimen B, which may be attributed to the longer time taken for the corrodent to penetrate to the coating-substrate interface.

An abnormally high rate of acoustic activity was observed for the 7.14% HCl concentration for Specimen C. At this concentration, the chemical test showed nickel to be present in the corrodent.

The spectral analysis of the AE signals generated during the corrosion in HCl of the as-sprayed samples showed peaks occuring at 92 KHz and 110 KHz. Earlier workers have reported such low frequency peaks occur when the yield point of a magnesium alloy is approached (16). It is expected that as fracture progresses, the specimen becomes less stiff and, consequently, emission frequencies should fall (17). In the present case, the corrosion environment present at coating-substrate interface may be playing a role in decreasing the apparent stiffness, thus enabling the stresses present to fracture the interface, giving rise to low frequency emissions.

b. Laser-treated coatings:

The beneficial effect on corrosion behavior of laser treating of the sprayed coatings is evident from the AE measurements; Table 6. In 6% HCl solution, nickel starts to dissolve, as indicated by the AE test and confirmed by analytical chemistry. However, for 6% HCl, no iron was detected within the corrodent. At this concentration after approximately 10 mins., the AE activity begins to fall off exponentially due to polarization of the coating resulting from H_2 bubbles adsorbed to the coating surface; Fig. 4. Thus, laser treatment has completely sealed the pores present in the as-sprayed coating. This fact is further confirmed by Fig. 5, which shows from the cross-section of the laser-treated coating, following corrosion, that the coating-substrate interface is intact (c.f. Fig. 3b). Figs. 6a and 6b show a comparison of the corrosion products formed on the surface of the as-sprayed coating and the laser-treated coating. These specimens were removed from the 25% HCl solution after 15 min. and dried in air. A yellow rust has formed for the as-sprayed case, which was absent for laser treatment.

4. Origin of AE Activity

A firm understanding of the origins of corrosion-induced AE has been limited, as discussed in the review by Wadley, et al (18). The sources of AE activity can be classified into two categories: i) physical processes of bubble formation and dissipation; ii) electrochemical origin of bubble formation, and associated changes in local stress and strain fields.

There is considerable evidence to support (ii): AE count rate is not proportional to bubble size; i.e., small and large bubbles produced at the same concentration of the corrodent do not give rise to different AE counts, and, an amplitude analysis of the plot of AE count rate vs. time at a given concentration of the corrodent shows that the amplitude remains essentially constant. Hence, the origin of these signals must be very uniform. If AE counts are generated due to the physical process of bubble formation and dissipation, such processes should be a function of contact angle, or the roughness of the coating-substrate interface, which does not appear to be the case.

Hence, the continuous release of stored energy (due to stresses present at the coating-substrate interface), on a micro-level (i.e., atomistic scale) as a result of electrochemical action we believe to be the principal origin of the AE counts generated in the corrosion of Ni-Cr coatings sprayed onto steel substrates.

Table 6. AE Activity as a Function of HCl Concentration for Laser Treated Coating.

Concentration of Corrodent (%)	Time (min.)	AE Count Rate (Counts/min.)	Correlation Co-efficient	Test For Fe	Test For Ni
0.7	10	None	———	No	No
2.1	10	None	———	No	No
3.6	10	None	———	No	No
4.3	10	None	———	No	No
6.0	10	794	0.997	No	Yes

*After 10 mins. polarization of the coating due to H_2 bubble evolution occurs.

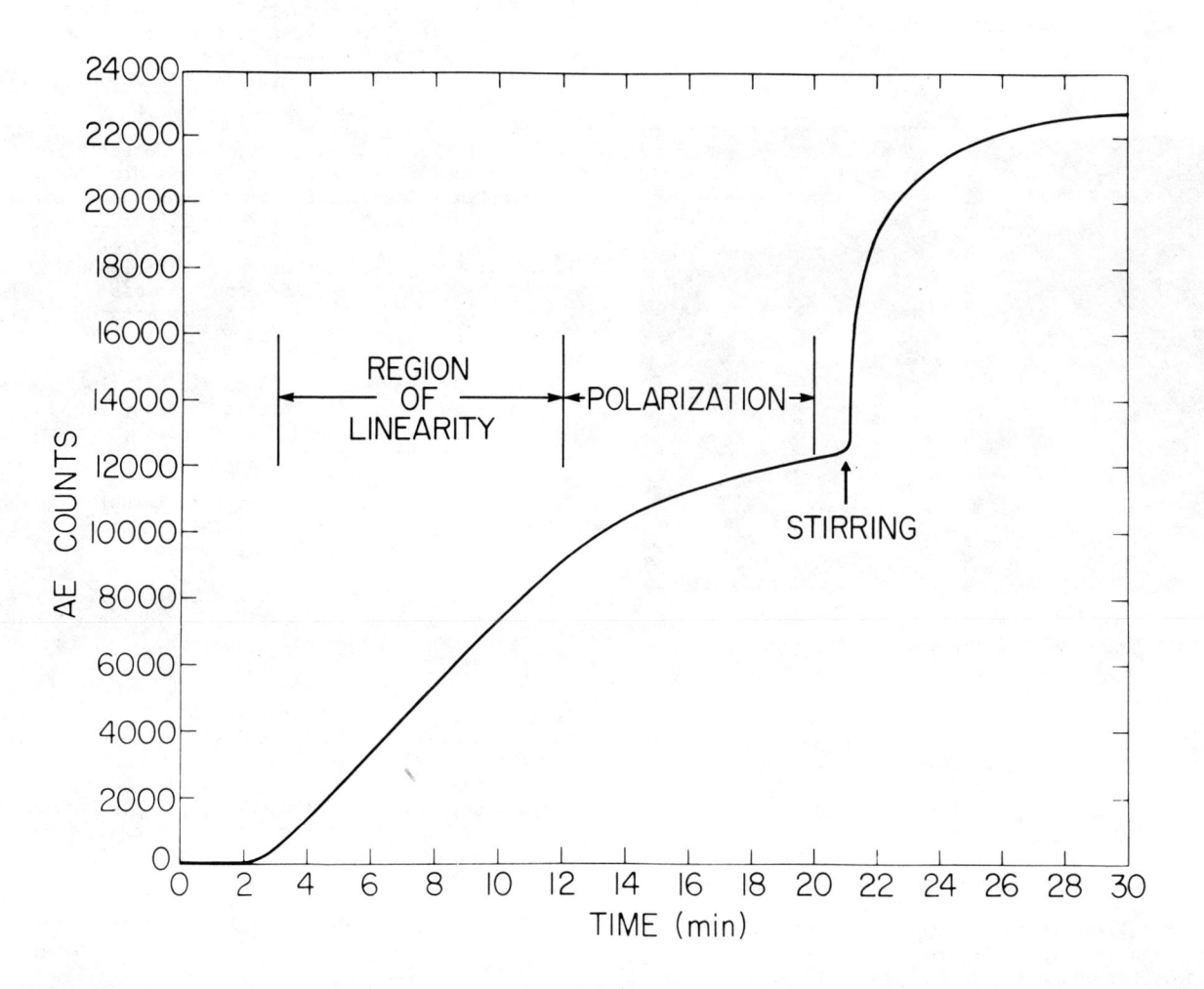

Fig. 4. Polarization of laser treated sample in 6% HCl solution.

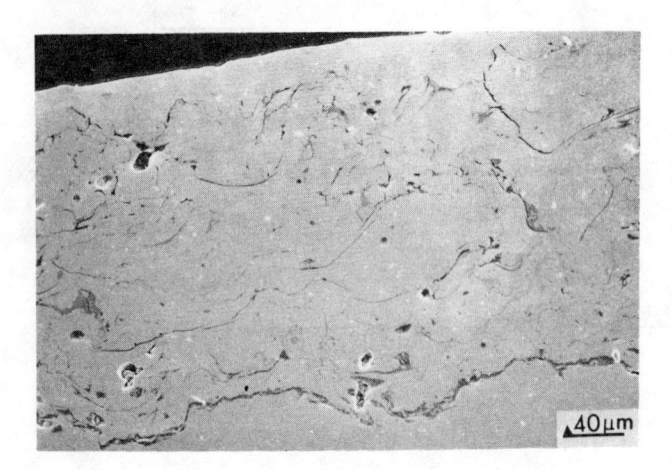

Fig. 5. SEM cross-section of the laser-treated coating after corrosion.

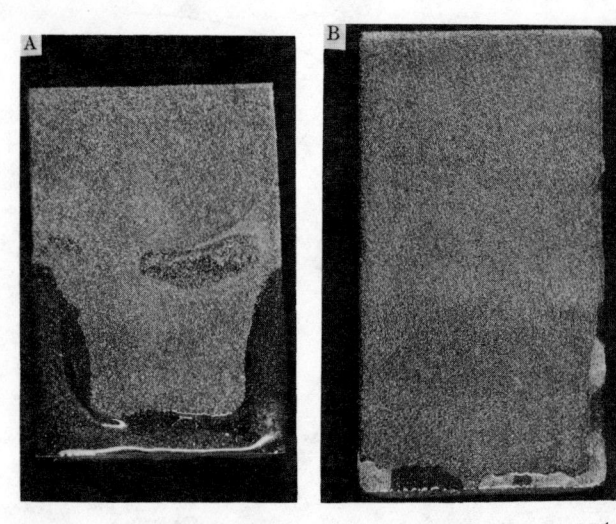

Fig. 6. Nature of corrosion products formed in 25% HCl; A. As-sprayed coating surface. B. Laser treated coating surface.

CONCLUSION

Laser treatment of the plasma-sprayed Ni-Cr coatings results in the closure of surface pores, thus significantly improving the corrosion resistance of the coating. The AE technique has been used to differentiate the corrosion characteristics of the Ni-Cr coating vis-a-vis coating-substrate interface, i.e., corrosion of steel substrate. At a given concentration of the corrodent, the AE activity is a function of porosity. It is possible to vary the surface roughness of the as-sprayed coating by the laser surface fusion technique.

REFERENCES

1. Clark, P.R., Packer, C.M. and Perkins, R.A., "Development of Coatings for Corrosion/Erosion Protection of Internal Components of Coal Gasification Vessels", Final Report, prepared for United States Dept. of Energy under Contract No. DEAC01-77ET 10245, published on March 15, 1981.

2. Bhat. H., Zatorski, R.A. and Herman, H. and Coyle, R.J., "Laser Treatment of Plasma-Sprayed Coatings", paper to be presented at the 10th International Thermal Spraying Conference, Essen, Federal Republic of Germany, May 2-6, 1983.

3. Dallaire, S. and Ceilo, P., "Pulsed Laser Treatment of Plasma Sprayed Coatings", Metallurgical Transactions B, v13B, Sept. 1982, p. 479-483.

4. Mansfeld, F. and Stocker, P.J., "Acoustic Emission from Corroding Electrodes", Corrosion, v35, n12, Dec. 1979, p. 541-544.

5. Zakharov, Yu.V., Reznikov, Yu.A., Gorbachev, V.I., and Vasilev, V.N., "Use of the Acoustic Method to Investigate General Corrosion", Translation from Zashchita Metallov, v14, n4, July-Aug. 1978, p. 459.

6. Culpan, E.A., Foley, A.G., "The Detection of Selective Phase Corrosion in Cast Nickel Aluminum Bronze by Acoustic Emission Technique", Journal of Materials Science, v17, 1982, p. 953-964.

7. Hartbower, C.E., Gerberich, W.W., Liebowitz, H., "Investigation of Crack-Growth Stress Wave Relationships", Engineering Fracture Mechanics, v1, 1968, p. 291.

8. Lord, Jr., Arthur E., "Acoustic Emission", Physical Acoustics, Principles & Methods, vol. XI, Ed. by W.P. Mason and R.N. Thurston, Academic Press, 1975.

9. Green, G., "Sources of Acoustic Emission During Crack Growth in Ferritic Steels", Materials Science, v15, Nov.-Dec. 1981, p. 505.

10. Egle, D.M., Tatro, C.A., and Brown, A.E., "Frequency Spectra of Acoustic Emission from Nodular Cast Iron", Materials Evaluation, v39, Oct. 1981, p. 1037.

11. Chretien, J.F. and Chretien, N., "A Bibliographical Survey of Acoustic Emission", Non-Destructive Testing, v5, Aug. 1972, p. 220.

12. Draper, C.W. and Benko, J.W., "Simple Determination of Work-Piece Focal Point Window. Tolerances in Laser-Material Processing", Applied Optics, v18, 1979, p. 3205.

13. Erdey, L., in International Series of Monographs on Analytical Chemistry, "Gravimetric Analysis - Part II", Translated by G. Svehla, Ed. by I. Buzas, Pergamon Press, 1965, p. 337, 384.

14. Fischer, I.A., "Variables Influencing the Characteristics of Plasma Sprayed Coatings", International Metallurgical Reviews, Review 164, v17, 1972, p. 117-129.

15. Heymann, F.J., "On the Time Dependence of the Rate of Erosion Due to Impingement or Cavitation", in the Proceedings of the Symposium, "Erosion by Cavitation or Impingement", presented at 69th Annual Meeting of ASTM, Atlantic City, N.J., June 26-July 1, 1966, Published by ASTM, p. 70.

16. Ono, K., Stern, R., Long, M., "Application of Correlation Analysis to Acoustic Emission", ASTM Special Technical Publication, 505, 1972, p.152.

17. Curtis, G., "Acoustic Emission-4, Spectral Analysis of Acoustic Emission", Non-Destructive Testing, v7, April 1974, p. 82.

18. Wadley, H.N.G., Scruby, C.B., and Speake, J.H., "AE for Physical Examination of Metals", International Metals Review, n2, 1980, p. 41.

A MICROSTRUCTURAL STUDY OF PULSED AND CONTINUOUS LASER WELDED STAINLESS STEEL

R. J. Coyle, Jr.
Western Electric Co., ERC
Princeton, N.J.

INTRODUCTION

It is recognized generally that a tremendous potential exists for the application of lasers in welding of metals (1,2). Much of this interest is due to the development of multikilowatt, CO_2 lasers which are capable of deep penetration welding (3). There is less attention accorded to welding by smaller, continuous wave (cw) lasers and pulsed solid state lasers. These lasers can offer distinct advantages, despite their limited depth of penetration.

Lower power, cw CO_2 laser welding usually does not involve the phenomenon of keyhole formation, as in the case of deep penetration welding (4). Low power CO_2 welding is characterized by smaller beads and minimum heat affected zones due to the lower energy input. Pulsed welding as performed by solid state lasers involves high energy inputs but very short interaction times, of the order of millisecond pulses. This relatively short dwell time also creates shallow weld penetrations and narrow heat affected zones (5,6).

The operating characteristics of the aforementioned lasers can be ideal for applications which require a minimum net heat input. Furthermore, the rapid cooling rates and steeper temperature gradients encountered in laser welding may yield unique or unusual weld structures (7).

The objective of this investigation is to compare and evaluate the microstructure of welds made with both lower power cw and pulsed solid state lasers. The material selected for the study is type 316L stainless steel, which is considered to be readily weldable by conventional or laser techniques (8-9). Moreover, the welding and solidification behavior of stainless steels such as 316L has been documented thoroughly in the literature (10-14).

TABLE I

Chemical Composition of 316L Stainless Steel

	Cr	Ni	Mo	Mn	C	P	S	Si	Fe
Weight Percent	17.01	11.02	2.16	1.49	0.02	0.03	0.007	0.35	Bal

EXPERIMENTAL

Material

The chemical analysis of the as-received type 316L alloy is given in Table 1. Weld coupons 12.7mm x 63.5mm x 0.75mm (0.5 x 2.5 x 0.03 in.) thick were sheared with the long axis parallel to the sheet rolling direction. Grain size was varied in two sets of samples by means of annealing at 1060°C for 15 and 30 minutes respectively. The annealed samples were water quenched to room temperature in order to minimize sensitization. Representative microstructures of each of the three material conditions are shown in Figure 1.

Laser Welding

Both continuous and pulsed bead-on-plate welds were made with the weld traverse direction perpendicular to the original rolling direction. Focused spot diameters were adjusted to produce heat affected zone widths of 0.25 to 0.75mm (0.01 to 0.03 in.). The sample surface was flooded with argon through a low pressure shroud mounted coaxially with the beam optical column.

Continuous welding was performed with a CO_2 laser at a power output of approximately 300w. Welds were made at traverse speeds of 12.7, 19.1, and 25.4 mm/s (0.5, 0.75, and 1.0 in/s).

Pulsed welding was performed with a Nd:YAG laser welder at a pulse duration of 3ms and a power density of approximately 3.8 KW/mm^2. A pulse repetition rate of 30 Hz and a traverse speed of 12.7 mm/s (0.5 in/s) was used to produce an approximately fifty percent overlap of target spots.

Representative samples of each type of weld were sectioned and examined by conventional metallographic techniques. In some cases a magnetic particle etchant (15) was employed to delineate regions which contained the magnetic delta ferrite phase.

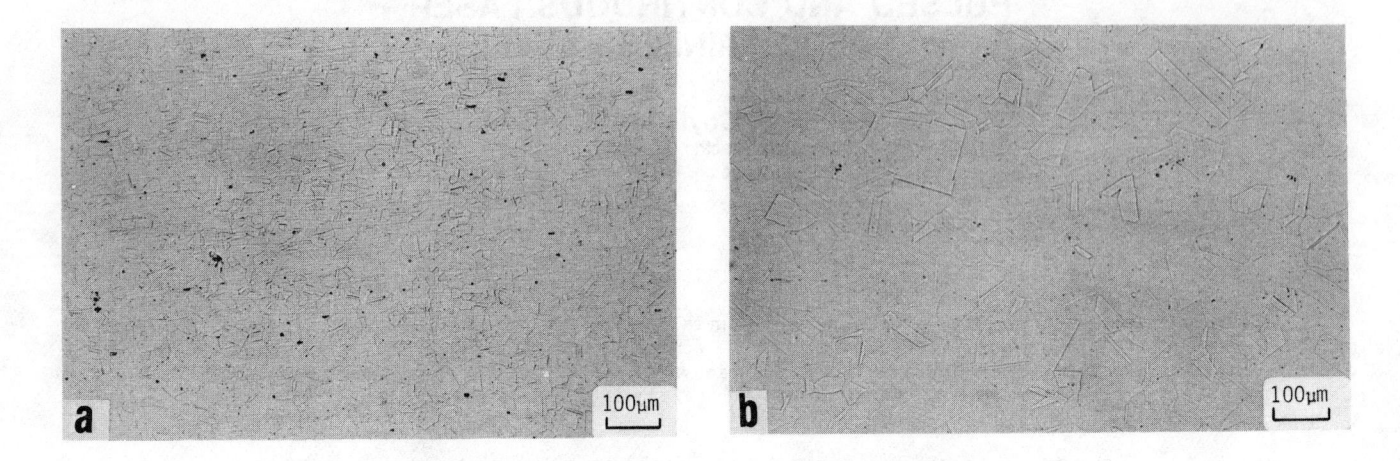

Fig. 1. Typical microstructures of type 316L stainless steel (a) as-received, fine-grained, (b) annealed 15 minutes at 1060 °C, and (c) annealed 30 minutes at 1060 °C.

RESULTS AND DISCUSSION

Pulsed Overlap Welds

The optical micrographs presented in Figure 2 show typical structures and spot overlap patterns found in longitudinal sections of pulsed laser welds of Type 316L stainless steel. A second series of micrographs, taken transverse to the weld travel direction, and at higher magnification is shown in Figure 3. The transverse sections were polished to avoid the overlap region and were treated with the ferromagnetic particle etchant (15) to decorate regions which contain the magnetic ferrite phase.

The fusion zone in the small grain, as-received material, shown in Figure 4 contains adjacent columnar and cellular dendrites. Such solidification structures often are characteristic of rapidly solidified microstructures (16). Some of the grains are fully austenitic (light phase) while others possess a duplex ferrite-austenitic microstructure. The delta ferrite (dark etching, magnetic phase) is located at the interstices of dendrites and cells (Figures 3a and 4) which indicates that the primary mode of solidification is austenitic (13).

If the base metal grain size is increased, the fusion zone grain size becomes greater due to the epitaxial regrowth of the large base metal grains (17-19). Figure 3b shows that a fusion zone which solidifies from a base metal annealed for 15 minutes consists of large grains of primary austenite along with large grains which contain the duplex substructure. The weld solidifies as wide columnar or large equiaxed grains as opposed to the more narrow columnar structure observed in the as-received case. Furthermore, the substructure tends to be more cellular as compared to cellular-dendritic in the fine-grained case.

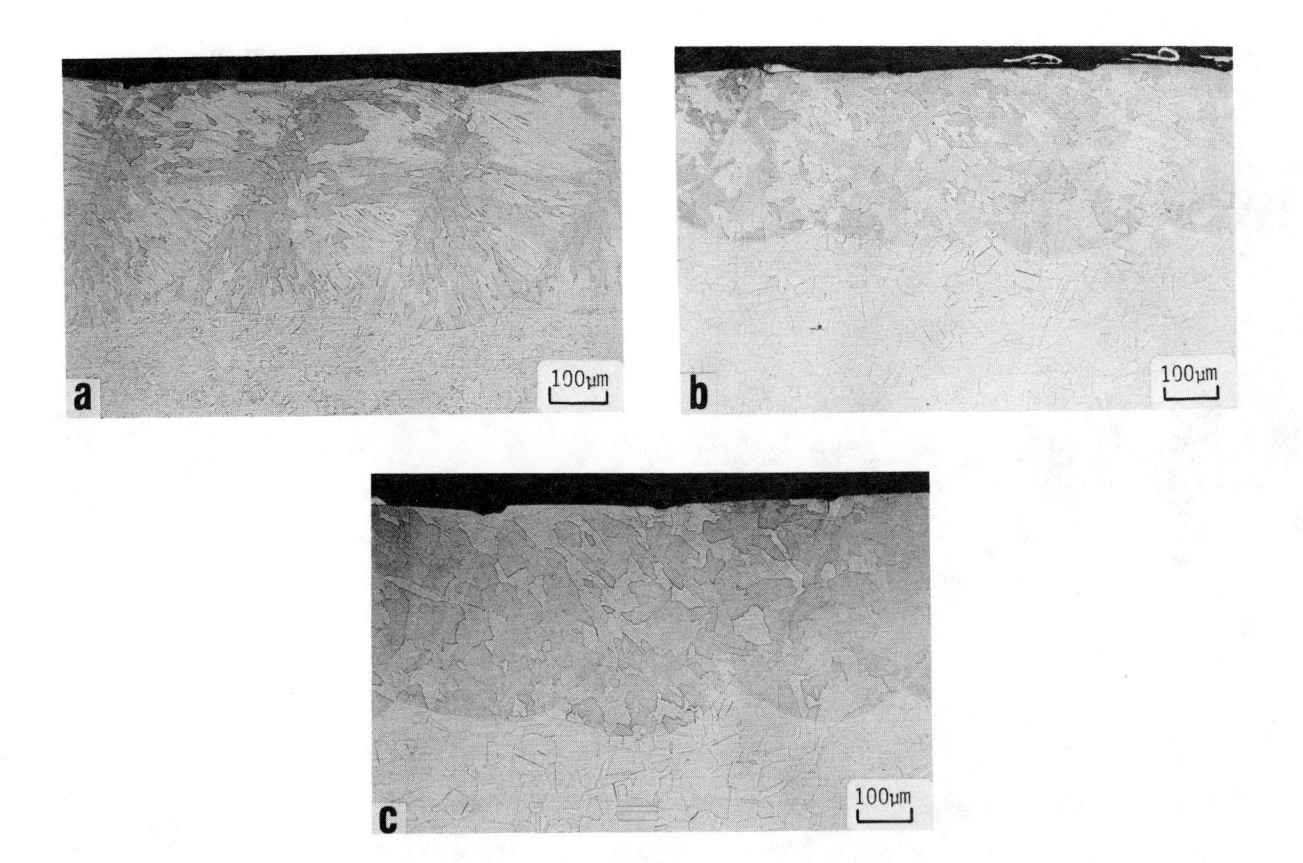

Fig. 2. Longitudinal crossections of pulsed laser welds in 316L base metal (a) as-received, (b) annealed 15 minutes, and (c) annealed 30 minutes. Etchant: oxalic acid, electrolytic.

Figure 3c shows a fusion zone which solidified from a base material annealed for 30 minutes. This weld contains a duplex microstructure with no evidence of totally austenitic grains as in the two previous cases. Examination at higher magnification (Figure 5) reveals a fine substructure of cellular austenite and intercellular ferrite.

The intercellular spacings in the pulsed welds (e.g., Figure 5) appear to be smaller than those observed in conventional (20), high power laser (7), or electron beam (21) welds. Ferrite particle size is also small, in fact, it often can not be resolved by optical microscopy.

The steep temperature gradients attributed to pulsed welding (5) may account for the predominately cellular or cellular-dendritic substructure observed in these welds. The extreme values of temperature gradient and cooling rate found in pulsed welding could also account for the smaller intercellular spacings and ferrite size (22).

No explanation is offered, however, as to why the largest grained base metal yields a completely duplex microstructure. Further investigations are underway of austenite and ferrite morphologies and base metal structure by SEM and STEM to gain further insight into this behavior

As a final observation on the pulsed welds, it is interesting to note the effect of pulse overlap on microstructure. Several transverse sections of pulse overlaps are shown in Figure 6. The microstructures present in the remelted regions have the same solidification mode as the previous pulse. Solidification proceeds in an epitaxial fashion across the new boundary though not necessarily in the previous growth direction (Figure 2). Thus, while pulse overlap may alter the substructure it does not seem to change the solidification mode.

Continuous Welding

Figure 7 shows representative transverse sections of cw bead-on-plate welds made at three different traverse speeds in fine-grained (as-received) 316L. In general, the welds consist of both primary austenitic regions and duplex austenitic-ferritic regions. A large austenitic region is found at the root of the welds produced at faster traverse speeds. As weld speed decreases the volume per cent ferrite increases, the ferrite distribution becomes more uniform, and there is less evidence of single phase regions of austenite.

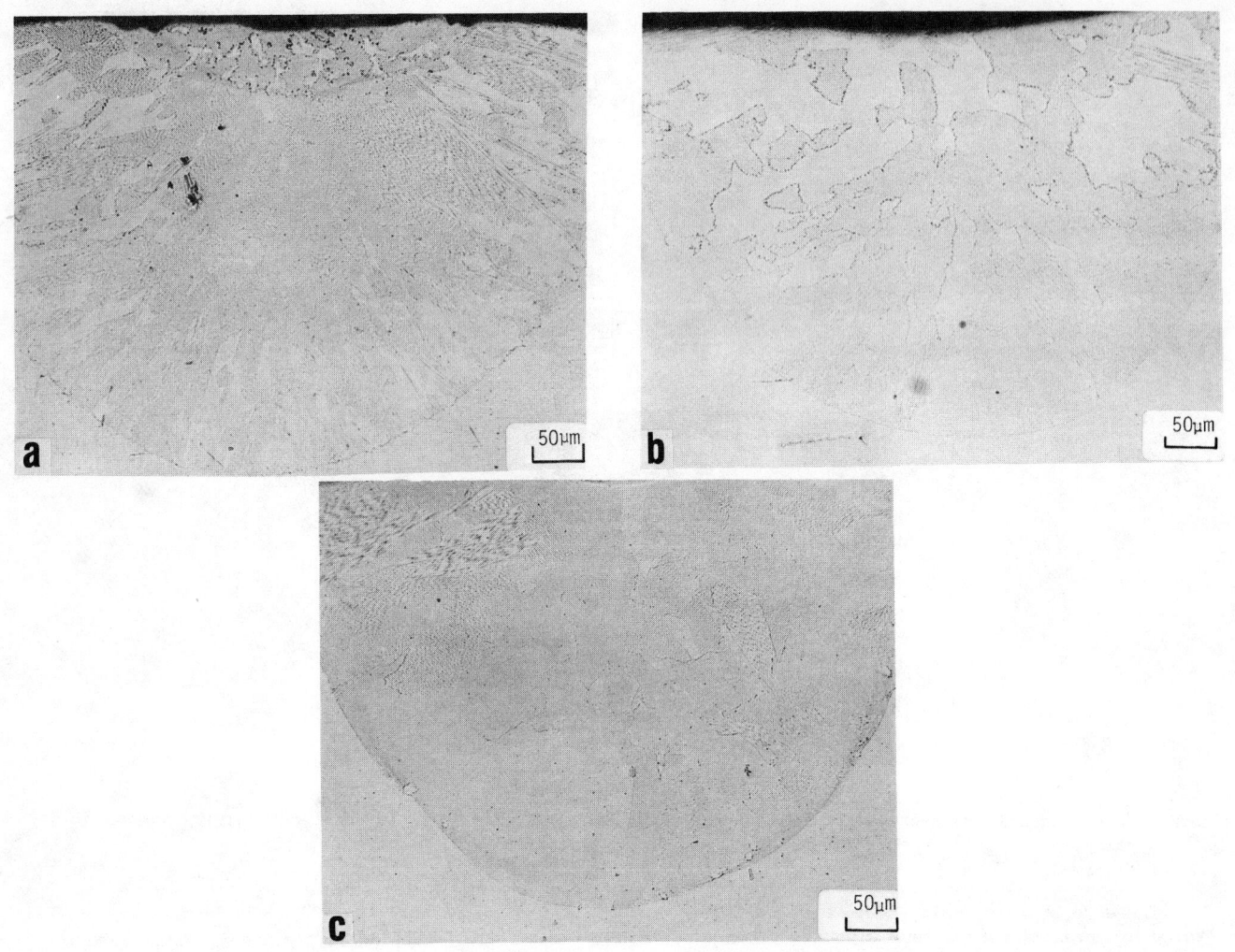

Fig. 3. Transverse sections of welds shown in Fig. 2, at higher magnification, (a) as-received, (b) annealed 15 minutes, (c) annealed 30 minutes. Etchant: KOH electrolytic plus epitaxial magnetic particle (15).

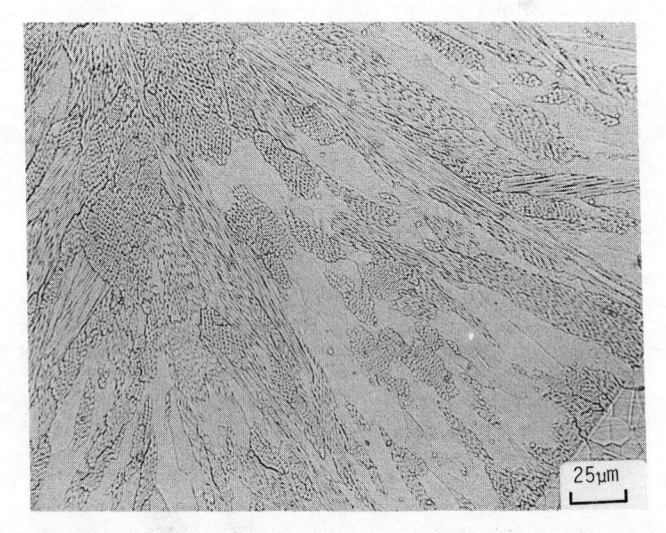

Fig. 4. Fusion zone of pulsed weld in fine-grained 316L. Note columnar grains of austenite and duplex cellular substructure. Etchant: oxalic acid, electrolytic.

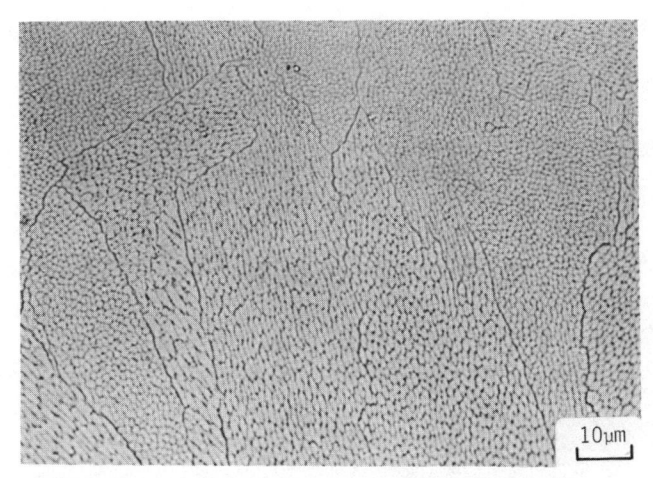

Fig. 5. Cellular substructure of pulsed weld in large-grained 316L (annealed 30 minutes). Weld solidified as primary austenite with ferrite at all boundaries. Etchant: KOH, electrolytic.

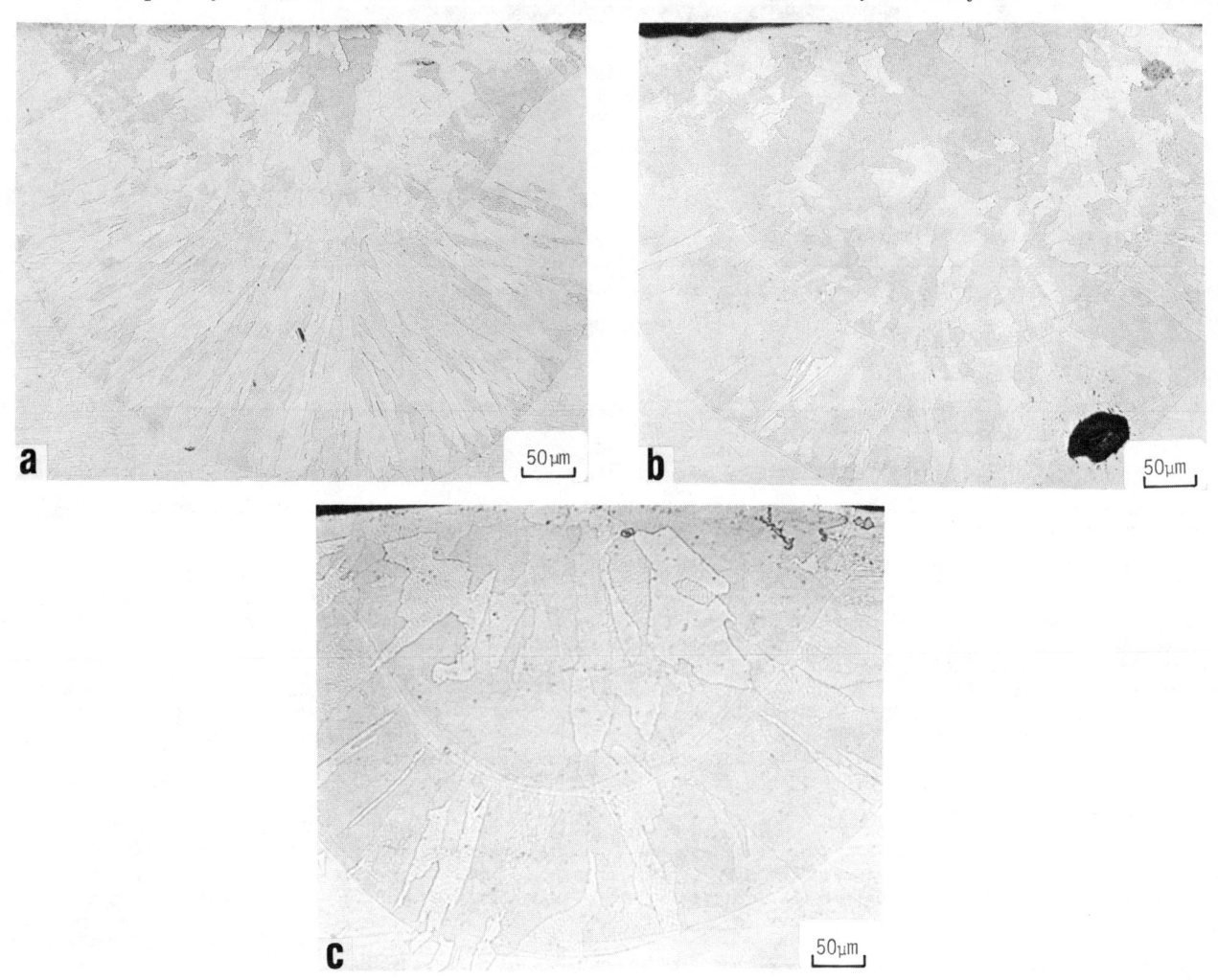

Fig. 6. Transverse sections through pulse overlap regions in 316L (a) as-received, (b) annealed 15 minutes, and (c) annealed 30 minutes. Note epitaxial growth and primary austenitic solidification mode in overlap regions in each welds. Etchant: oxalic acid, electrolytic.

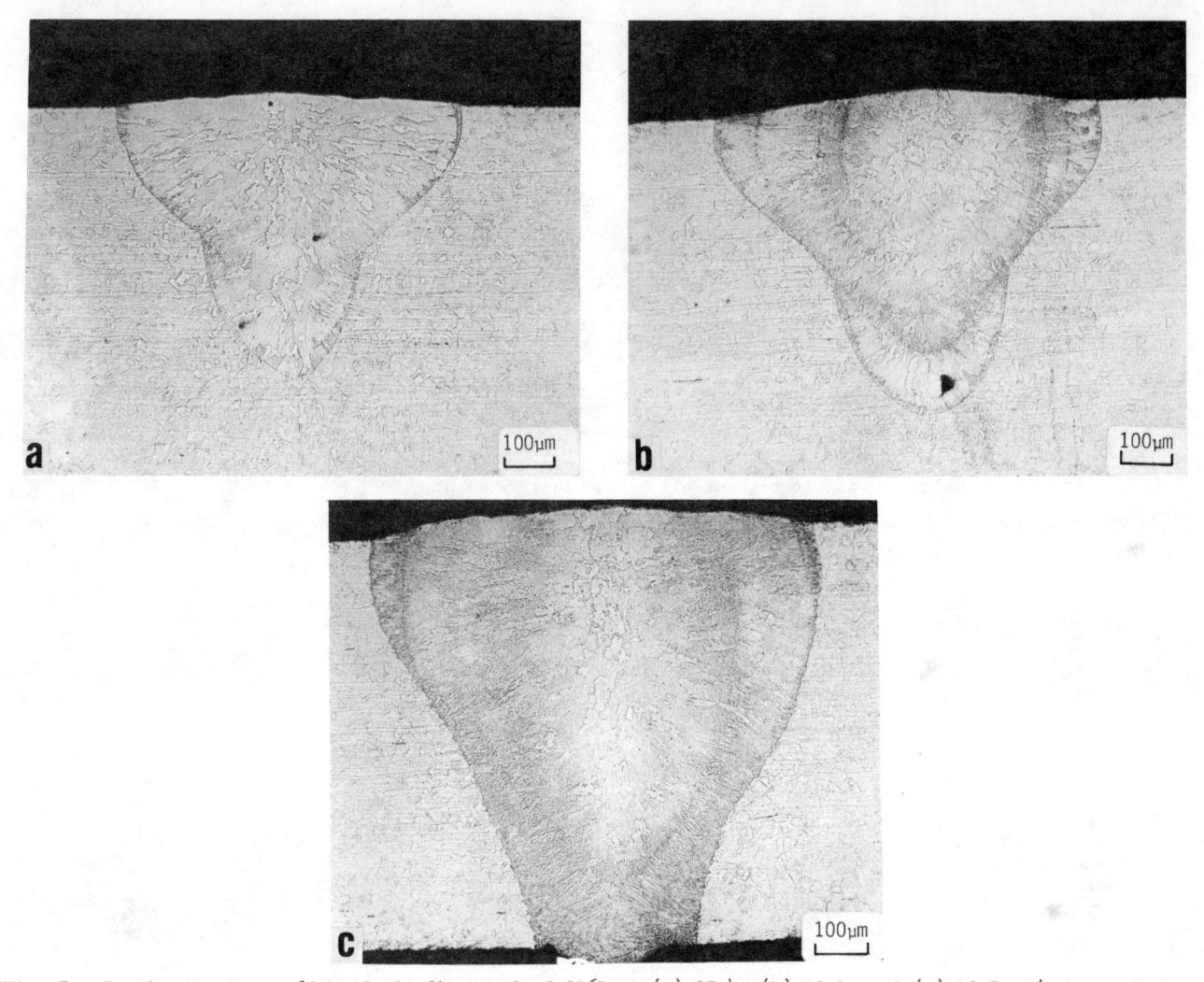

Fig. 7. Continuous wave weld beads in fine-grained 316L at (a) 25.4, (b) 19.1, and (c) 12.7 mm/s traverse speeds. Etchant: oxalic acid, electrolytic.

A considerable variation is found in ferrite morphology within each weld. The photomicrographs in Figure 8 show ferrite which could be classified as vermicular, lacy, lathy, globular or acicular (10-14). Furthermore, ferrite can be found at the cores or boundaries of cells and dendrites in these welds. These observations indicate that solidification can proceed by either a primary ferrite or austenite mode, dependent on local fluctuations in composition and cooling or solidification rates (12). Figure 9 shows a typical transverse section of a cw weld bead in annealed 316L. This weld, which was made at 25.4 mm/s (1.0 in/s), has a different solidification mode than the comparable weld made in the fine-grained material. The primary mode of solidification can be identified as austenitic in the annealed material as compared to mixed austenitic-ferritic in the fine-grained case. Although there is occasional evidence of primary ferrite, it occurs to a lesser extent than the more prevalent austenitic mode.

The weld structures obtained in the large-grained annealed materials are ostensibly more cellular than those obtained in the fine-grained material (Compare Figures 7a and 9a). This appearance may be a manifestation of solidification mode not base metal grain size. The primary austenitic solidification in the annealed case produces cell and dendrite boundaries which are decorated clearly with secondary ferrite, Figure 10a. The presence of austenitic-ferritic solidification in the fine-grained case produces retained ferrite at the cores of the cells along with secondary ferrite at the boundaries, Figure 10b. Hence, the cellular substructure may be more apparent in the annealed case due to the semi-continuous network of ferrite which outlines the boundaries.

When the annealed materials are welded at slower traverse speeds the solidification mode changes from primary austenitic to mixed austenitic-ferritic. This is evident in Figure 11 which shows the variety of ferrite morphologies present in the welds in both annealed base metals. The ferrite content appears to increase with decreasing weld speed as observed in the previous case for the fine-grained material. The ferrite content however, does not seem to vary significantly with grain size at these travel speeds.

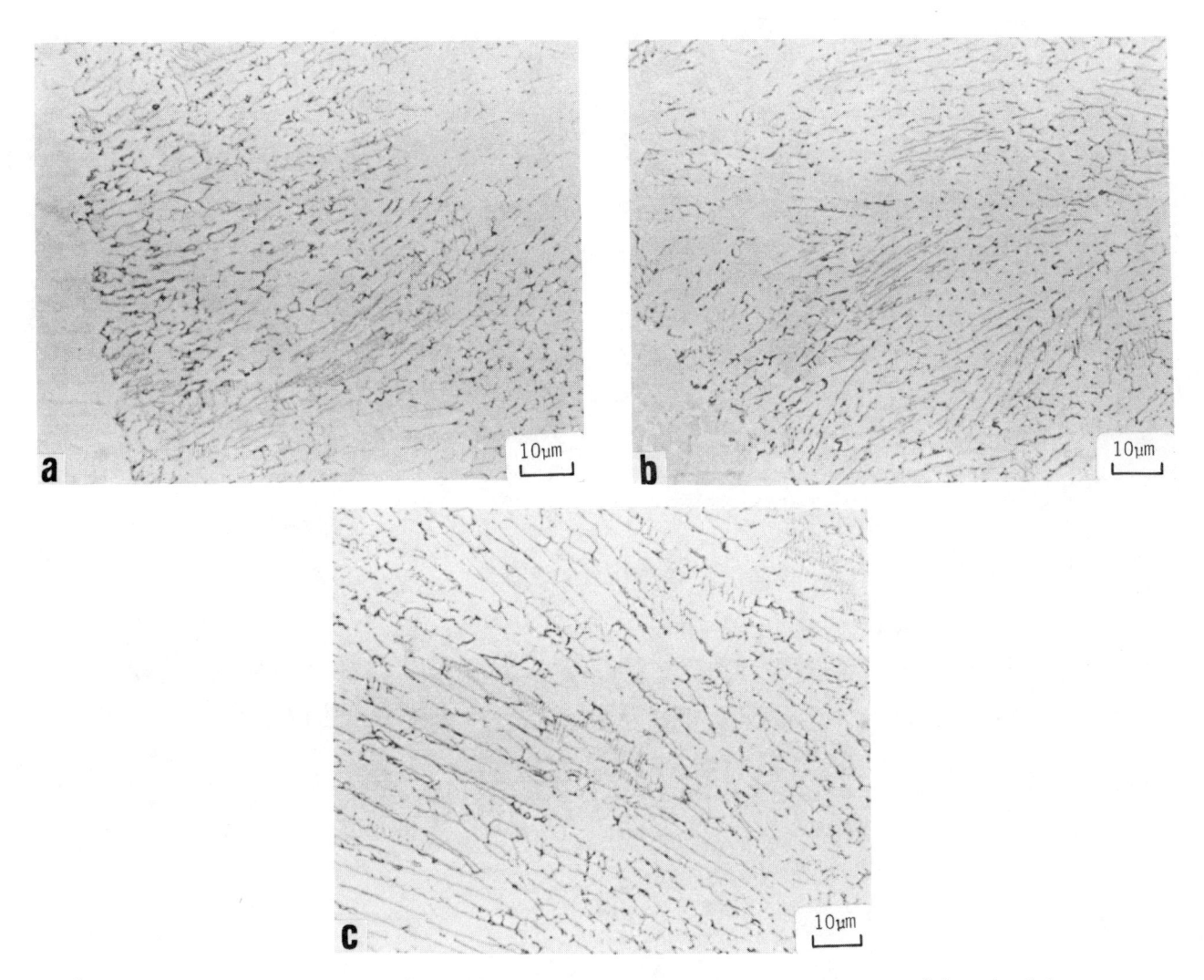

Fig. 8. Examples of ferrite morphologies near fusion line in welds shown in Fig. 7, (a) 25.4, (b) 19.1 and (c) 12.7 mm/s traverse speeds. Etchant: KOH, electrolytic.

Fig. 9. Transverse section of cw weld in annealed (15 minutes) 316L. Traverse speed, 25.4 mm/s. Etchant: oxalic acid, electrolytic.

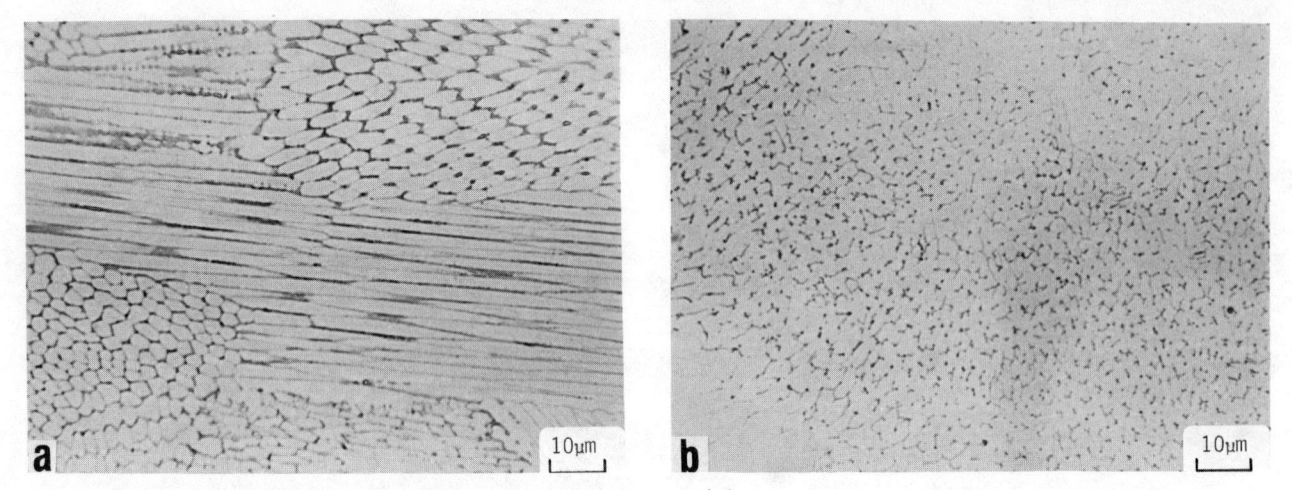

Fig. 10. Substructures in cw welds at high magnification in (a) annealed 316L with primary austenite and intercellular or interdendritic ferrite, and (b) as-received 316L with either primary austenite or primary ferrite present in vermicular morphology. Etchant: KOH, electrolytic.

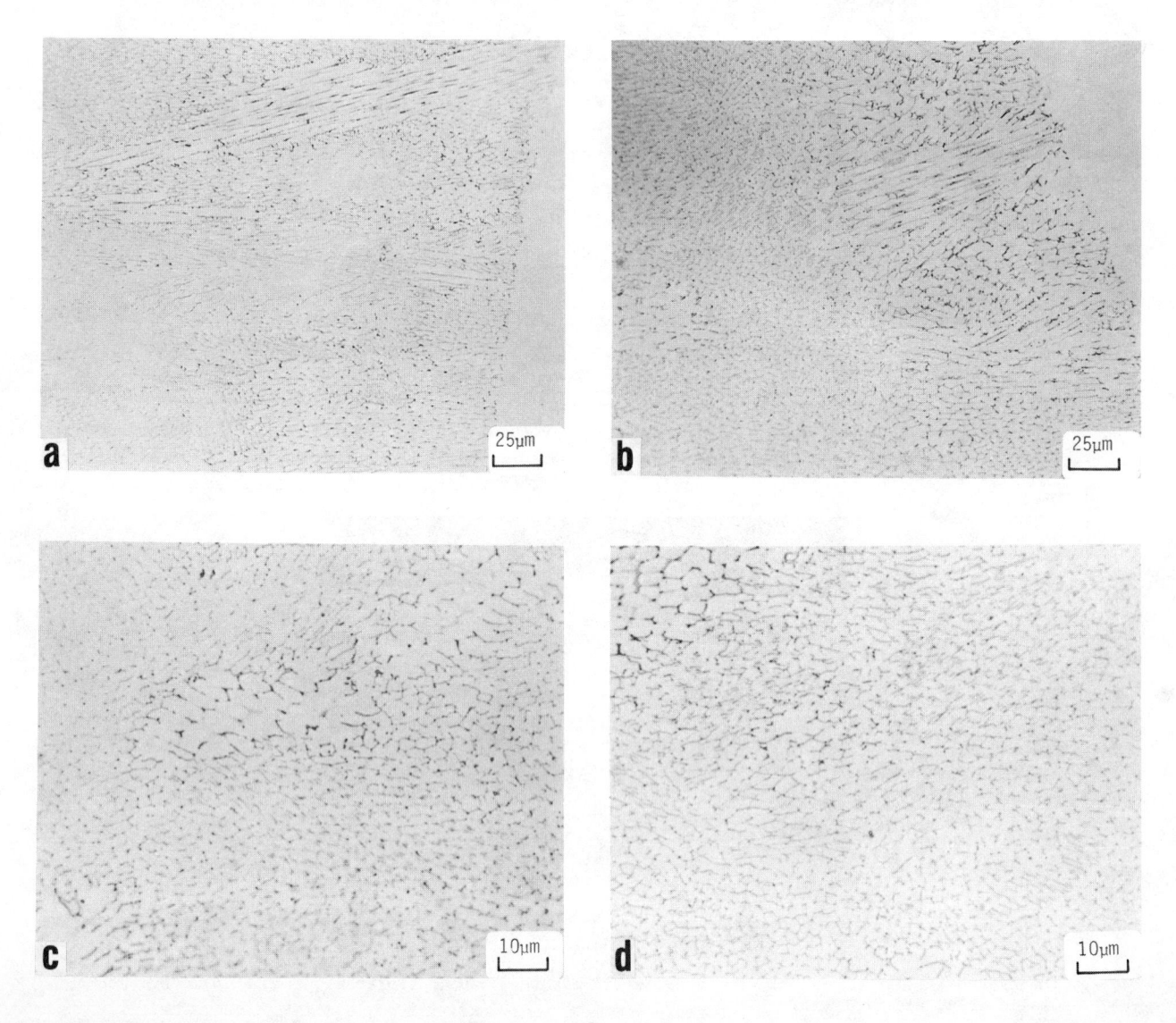

Fig. 11. Ferrite morphologies in cw welded sections of 316L annealed base metals, (a) annealed 30 minutes, 19.1 mm/s weld speed, (b) annealed 15 minutes, 21.7 mm/s. Structures near fusion line, original magn 400x, (c) annealed 15 minutes, 19.1 mm/s, (d) annealed 30 minutes, 12.7 mm/s. Structure away from fusion line, original magn. 1000x. Etchant: KOH, electrolytic.

Thus, for each base metal grain size the fastest traverse speed and hence the fastest cooling rate favored primary solidification of austenite. This trend is consistent with the results of Vitek and David for multi-kilowatt laser welds in Type 308 stainless steel (23). This trend is consistent also with the results presented earlier for pulsed welding, where the primary mode of solidification is austenitic, and the cooling rates are assumed to be faster.

Average interdendritic or intercellular spacings and ferrite particle size appear to be slightly greater in the cw welds (compare Figure 5 and 10a). This observation agrees with the hypothesis that cooling rates are slower in the case of cw welding. Additional quantitative measurements are required to assess the effect of traverse speed (cooling rate) on the substructure and morphology of cw welds.

The ferrite morphologies are varied more in the cw versus the pulsed welds because of the mixed austenitic-ferritic solidification. It is interesting to note that mixed mode solidification occurs at the fastest traverse speed in the fine-grained material whereas primary austenite solidification occurs at the same speed in the annealed material (compare Figures 8a and 10a). This result seems anomalous since the solidification mode in the pulsed welds is independent of grain size. Furthermore, the thermal properties which dictate heat transfer (cooling rate) are structure insensitive. Therefore, further analysis of both solidification and base metal structures is required to resolve this anomalous behavior.

SUMMARY

A comparison has been made of the microstructural characteristics of pulsed Nd:YAG and low power cw CO_2 laser welds in 316L stainless steels. The conculsions arising from this investigation are listed below.

1) Welding conditions which promote faster cooling rates favor primary austenitic solidification in 316L stainless steel. In pulsed welds the primary solidification mode is austenitic. In cw welds the mode is austenitic-ferritic with a greater tendency for primary austenitic solidification at the highest traverse speed.

2) There is an apparent effect of grain size on solidification mode in cw welds made at the fastest traverse speed. The welds in the annealed base metals generally solidify as primary austenite, whereas the comparable weld in the fine-grained base metal solidifies as austenitic-ferritic. However, grain size does not seem to affect solidification made in the rapidly cooled pulsed welds. This anomalous behavior is unresolved in the current study.

3) As a the base metal grain size is increased there is tendency to form more cellular substructures. This effect is not as evident in the cw welds because the mixed mode solidification complicates the microstructural anlaysis.

4) Spacings between cells and dendrites in these laser welds appear to be somewhat smaller than those reported previously. This result is consistent with an explanation based on rapid solidification, particularly in the case of pulsed welding. Further work is necessary to quantify these observations and to compare the results to predicted cooling rates.

5) Several ferrite morphologies are observed in the laser welds. Though the ferrite generally has a very fine structure these morphologies compare to those documented in the literature.

REFERENCES

1. "Trends in Welding and Joining Technology," Metal Progress, Vol. 119, 1, 1981, p. 74.

2. J. Mazumder, "Laser Welding: A State of the Art Review," J. Metals, Vol. 34, 7, 1982, p. 16.

3. C. M. Banas, "High Power Laser Welding - 1978," Opt. Eng., Vol. 17, 3, 1978, p. 210.

4. D. T. Swift-Hook and A.E. F. Gick, "Penetration Welding with Lasers," Weld, J., Vol. 52, 11, 1973, p. 492s.

5. J. E. Anderson and J. E. Jackson, "Theory and Application of Pulsed Laser Welding," Weld, J., Vol. 44, 1965, p. 1018.

6. W. N. Platte, and J. F. Smith, "Laser Techniques for Metals Joining," Weld. J., Vol. 42, 1963, p. 481s.

7. S. A. David and J. M. Vitek, "Solidification Behavior and Microstructural Analysis of Austenitic Stainless Steel Laser Welds," in Lasers in Metallurgy, K. Mukherjee and J. Mazumder, eds., AIME, New York, 1981, p. 247.

8. W. E. Hensley, Handbook of Stainless Steels, D. Peckner and I. M. Bernstein, eds., McGraw-Hill, 1977, p. 26-1.

9. R. A. Willgoss, J. H. P. C. Megaw, and J. N. Clark, "Assessing the Laser for Power Plant Welding," Weld. Met. Fabr., Vol. 47, 1979, p. 117.

10. N. Suutala, T. Takalo, and T. Moisio, "The Relationship Between Solidification and Microstructure in Austenitic-Ferritic Stainless Steel Welds," Met. Trans. A., Vol. 10A, 1979, p. 512.

11. N. Suutala, T. Takalo, and T. Moisio, "Ferritic-Austenitic Solidification Mode in Austenitic Stainless Steel Welds," Vol. 11A, 1980, p. 717.

12. S. A. David, "Ferrite Morphology and Variations in Ferrite Content in Austenitic Stainless Steel Welds," Weld J., Vol. 60, 1981, p. 63s.

13. J. C. Lippold and W. F. Savage, "Solidification of Austenitic Stainless Steel Weldments: Part 1-A Proposed Mechanism," Weld. J., Vol. 58, 1979, p. 362s.

14. J. C. Lippold and W. F. Savage, "Solidification of Austenitic Stainless Steel Weldments: Part 2 - The Effect of Alloy Composition on Ferrite Morphology, Weld. J., Vol. 59, 1980, p. 48s.

15. R. J. Gray, "The Detection of Strain-Induced Martensite in Types 301 and 304 Stainless Steels by Epitaxial Ferro-magnetic Etching," in Microstructural Science, Vol. 1, R. J. Gray and J. L. McCall, eds., Elsevier, New York, 1974, p. 159.

16. T. Z. Kattamis, "Solidification Microstructure of Laser Processed Alloys and Its Impact on Some Properties," in Laser in Metallurgy, K. Mukherjee and J. Mazumder, eds., AIME, New York, 1981, p. 1.

17. W. F. Savage, C. D. Lundin, and A. H. Aaronson, "Weld Metal Solidification Mechanics," Weld. J., Vol. 44, 1965, p. 175s.

18. C. R. Loper Jr., L. A. Shideler, and J. H. Devletian, "The Effect of Heat-Affected Zone Structure on the Structure of the Weld Fusion Zone," Weld. J., Vol. 48, 1969, p. 171s.

19. B. L. Baikie and D. Yapp, "Oriented Structures and Properties in type 316 Stainless Steel Weld Metal," in Solidification and Casting of Metals, The Metal Society, London, 1979, p. 438.

20. J. A. Brooks, J. C. Williams, and A. W. Thompson, "Solidification and Solid State Transformations of Austenitic Stainless Steel Welds," in Trends in Welding Research, ed. S. A. David, ASM, Metals Park, Ohio, 1982. p. 331.

21. P. R. Sahm and F. Schubert, "Solidification Phenomena and Properties of Cast and Welded Microstructures," in Solidification and Casting of Metals, The Metal Society, London, 1979, p. 389.

22. G. J. Davies and J. G. Garland, "Solidification Structures and Properties of Fusion Welds," Int. Met. Rev., Vol. 20, 1975, p. 83.

23. J. M. Vitek and S. A. David, "Microstructural Analysis of Austenitic Stainless Steel Laser Welds," in Trends in Welding Research, S. A. David ed., ASM, Metals Park, Ohio, 1982, p. 243.

This paper is subject to revision. Statements and opinions advanced in papers or discussion are the author's and are his responsibility, not ASM's; however, the paper has been edited by ASM for uniform styling and format.

Printed and Bound by Publishers Choice Book Mfg. Co.
Mars, Pennsylvania 16046

SOLIDIFICATION STRUCTURE AND FATIGUE CRACK PROPAGATION IN LB WELDS

F. W. Fraser and E. A. Metzbower

INTRODUCTION

In an investigation of HY-130 steel laser welds [1], a correlation was found between solidification structure and the path of a stress-corrosion crack propagated in 3.5 percent sodium chloride solution. The specimens, welded with a high power, continuous wave CO_2 laser, were single pass, autogenous butt welds fabricated in 12 mm (0.5 in) thick plate. The fracture mode and solidification structure varied through the thickness of the fusion zone. A cellular structure and fracture by microvoid coalescence were dominant in the upper part of the fusion zone, whereas, a cellular-dendritic solidification structure and intergranular fracture characterized the lower portion of the weld.

The purpose of the research that led to this paper was to investigate, further, the variations in solidification structure through the depth of the fusion zone, and to determine the influence of solidification structure, macro and microstructure on fatigue crack growth in LB welds of structural alloys.

PROCEDURE

Fatigue cracks were propagated in notched and side grooved cantilever beam specimens of HY-130 steel, HY-80 steel and Ti-6Al-2Cb-1Ta-0.8Mo laser beam weldments fabricated from 12 mm (0.5 in) thick plate. The nominal composition of the alloys, the welding parameters and the mechanical properties of the weld metal and base plate, as published by Metzbower and Moon [2], are given in Tables 1, 2, 3 and 4. A description of the welding process is given elsewhere [3]. The notch was placed in the fusion zone, parallel to the welding direction [Fig. 1]. The fatigue crack, propagated in air at a constant displacement and monitored optically with a traveling microscope, was arrested by unloading the specimen prior to failure. The average fatigue crack length was 12 mm (0.5 in). Specimens were prepared in duplicate. A horizontal sectioning technique, illustrated in Figure 2, permitted optical examination of the solidification structure, microstructure and crack path in the top, center and bottom portions of the fusion zone. Figure 3 is a macrograph of a transverse section of the fusion zone, of each alloy, indicating where the horizontal sections were taken. The HY-130 and the HY-80 horizontal sections were etched with a ten percent ammonium persulfate solution to reveal solidification structure and then were repolished and etched with one percent nital for microstructural analysis. The Ti-6211 sections were etched with Knoll's reagent for microstructural analysis only. We were unable to delineate solidification structure in this alloy. In an investigation of solidification structure in Ti alloys, Gould and Williams [4] report a technique for revealing solidification structure in Ti-6Al-6V-2Sn weld metal, but state they were unsuccessful with alloys containing lesser amounts of β stabilizer. They attribute this to the primarily martensitic microstructure of near α alloys, as opposed to the α + β microstructure where a compositional difference between the α and β phases results in preferential etching. The transverse metallographic sections (Fig. 3) were examined for weld contour and direction of grain growth. One fatigued specimen of each alloy was fractured in liquid nitrogen and the fatigue fracture surface was examined in a scanning electron microscope (SEM).

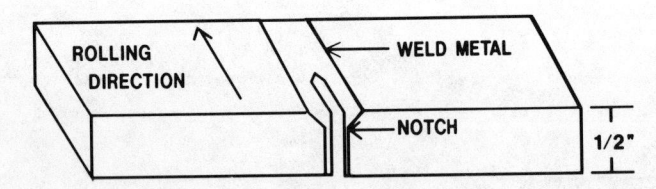

Figure 1. Fatigue specimen schematic showing notch placement in the weld metal.

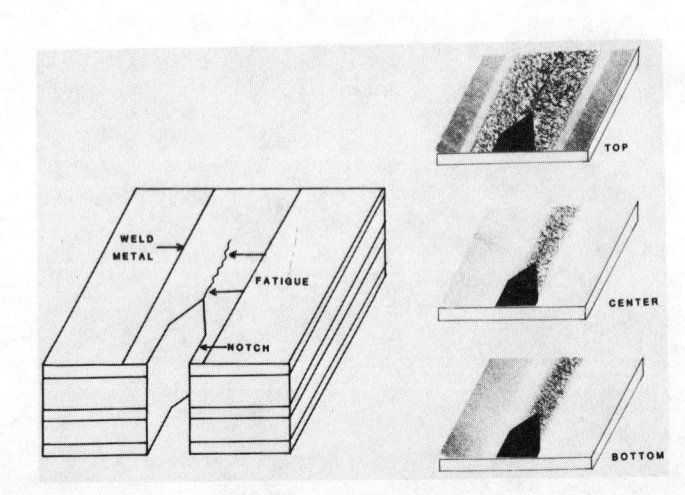

Figure 2. Diagram of the horizontal sectioning technique showing the top, center and bottom metallographic sections, and the notch and fatigue crack in the weld metal.

TABLE 1 - Nominal composition of base plates.

Alloy	Fe	C	Cr	Mo	Ta	Ni	Al	Ti	Cb	Mn
Hy-80	balance	0.18	1.68	0.45	...	3.0	...	...	...	0.30
Hy-130	balance	0.11	0.55	0.55	...	5.0	...	...	...	0.8
Ti-6211	...	....	....	0.8	1.0	...	6.0	balance	2.0	0.8

TABLE 2 - Laser-beam welding parameters.

Material	Power, kW	Speed, mm/s(in./min)	Heat Input, kJ/mm(kJ/in.)
HY-80	10.6	12.7 (30)	0.83 (21.2)
HY-130	11.0	12.7 (30)	0.87 (22.0)
Ti-6211	8.0	27.5 (65)	0.29 (7.4)

TABLE 3 - Mechanical properties of laser welds.

Alloy	σ_{ys} MPa (ksi)	σ_{ult} MPa (ksi)	Elongation percent	Reduction in Area, percent	Fracture Toughness N·m (ft·lb)
HY-80	627 (91)	752 (109)	23.5	75.8	76 (56)
HY-130	952 (138)	987 (143)	18.5	68.4	707 (530)
Ti-6211	758 (110)	855 (124)	11.5	31.7	249 (184)

TABLE 4 - Nominal mechanical properties of base plates.

Alloy	σ_{ys} MPa (ksi)	σ_{ys} MPa (ksi)	Elongation, percent	Fracture Toughness N·m (ft·lb)
HY-80	580 (85)	725 (105)	14	470 (350)
HY-130	890 (130)	... (...)	14	675 (500)
Ti-6211	790 (115)	860 (125)	15	200 (150)

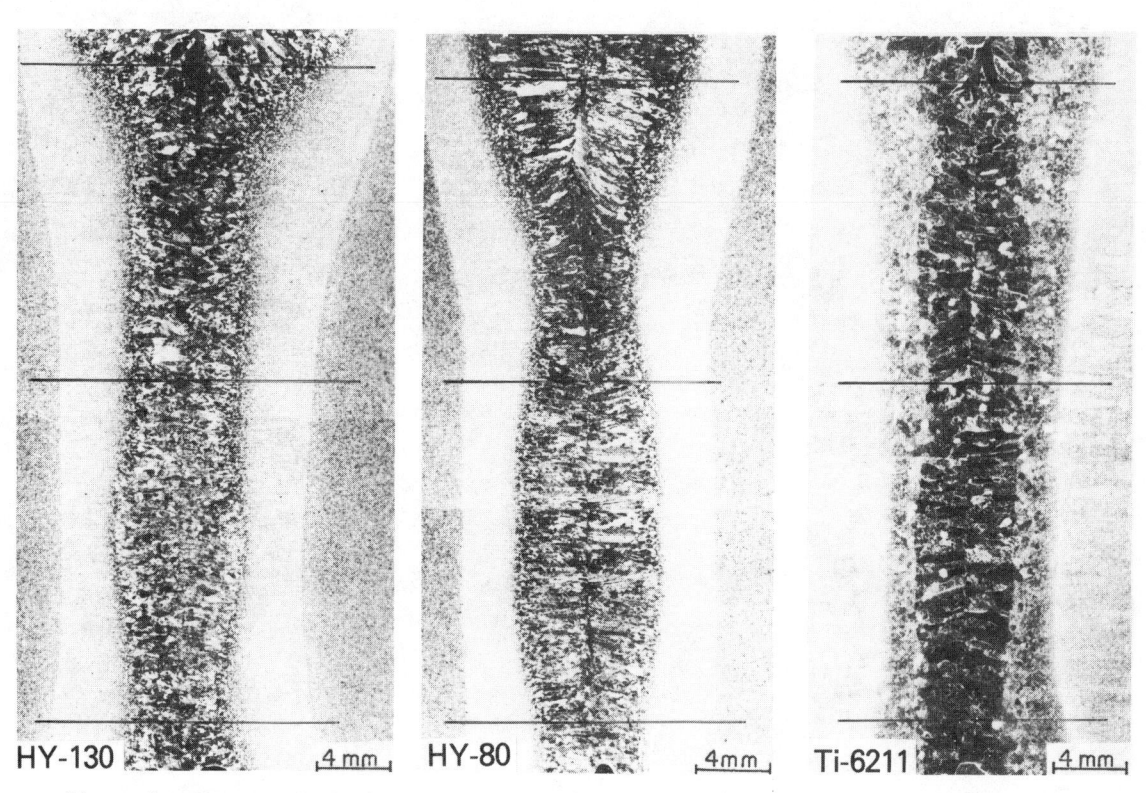

Figure 3. Macrograph showing transverse sections of each alloy. The lines indicate the approximate locations of the longitudinal metallographic sections.

RESULTS AND DISCUSSIONS

HY-130

Figure 4 is a macrograph of the fracture surface showing the fatigue zone and fast fracture area. The top and bottom of the weld and the welding direction are also indicated. The fatigue zone changed from coarse, irregular fracture at the top of the weld to relatively flat fracture at the bottom. SEM micrographs (Fig. 5) show both coarse, intergranular (IG) fracture and flat, transgranular fracture at the top of the specimen (A) and (B), and rough, transgranular fracture in the center (C) and at the bottom of the weld (D) indicative of crack branching.

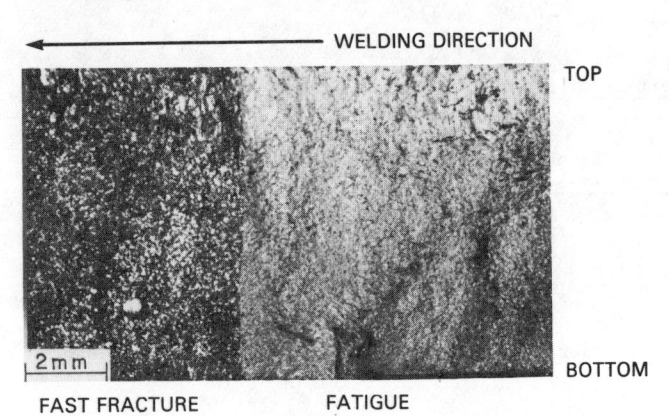

Figure 4. Macrograph of the HY-130 fracture surface showing the top and bottom of the weld, the welding direction, the fatigue area and the fast fracture area.

The path of the crack through the macrostructure in the top, center and bottom of the fusion zone is shown in Figure 6. Due to the very narrow fusion zone it is difficult to ensure notch placement along the weld center line. Prior to notching, each specimen was polished and etched. The center line was then scribed under a microscope. Nevertheless, as is evident from Figure 6, the notch fell off center in this specimen. However, in each horizontal section, the crack migrated toward the center of the weld and then continued to propagate in the vicinity of the weld center line. The crack path in the metallographic sections correlates with the fracture modes found in the top, center and bottom areas of the fusion zone on the fracture surface (Fig. 5). In the top section, the grains appear equiaxed and the crack path is both intergranular and transgranular. In the center and bottom metallographic sections, the grains are columnar and the crack path is transgranular. The apparent difference in grain shape, from equiaxed in the top section to columnar in the center and bottom sections, is the result of a change in the direction of grain growth in the fusion zone. Grains in the molten weld pool grow epitaxially from the unmelted base metal. The direction of maximum grain growth is normal to the fusion boundary and parallel to the direction of heat flow. As the contour of the fusion boundary changes, so does the direction of grain growth. In the broad, shallow pool at the top of the fusion zone (Fig. 3), the columnar grains extend upward as they grow from the base plate toward the weld center line. Therefore, the grains in a section taken normal to this upward direction of columnar grain growth appear equiaxed. In the remainder of the fusion zone, the fusion boundary is relatively straight and, for the most part, the

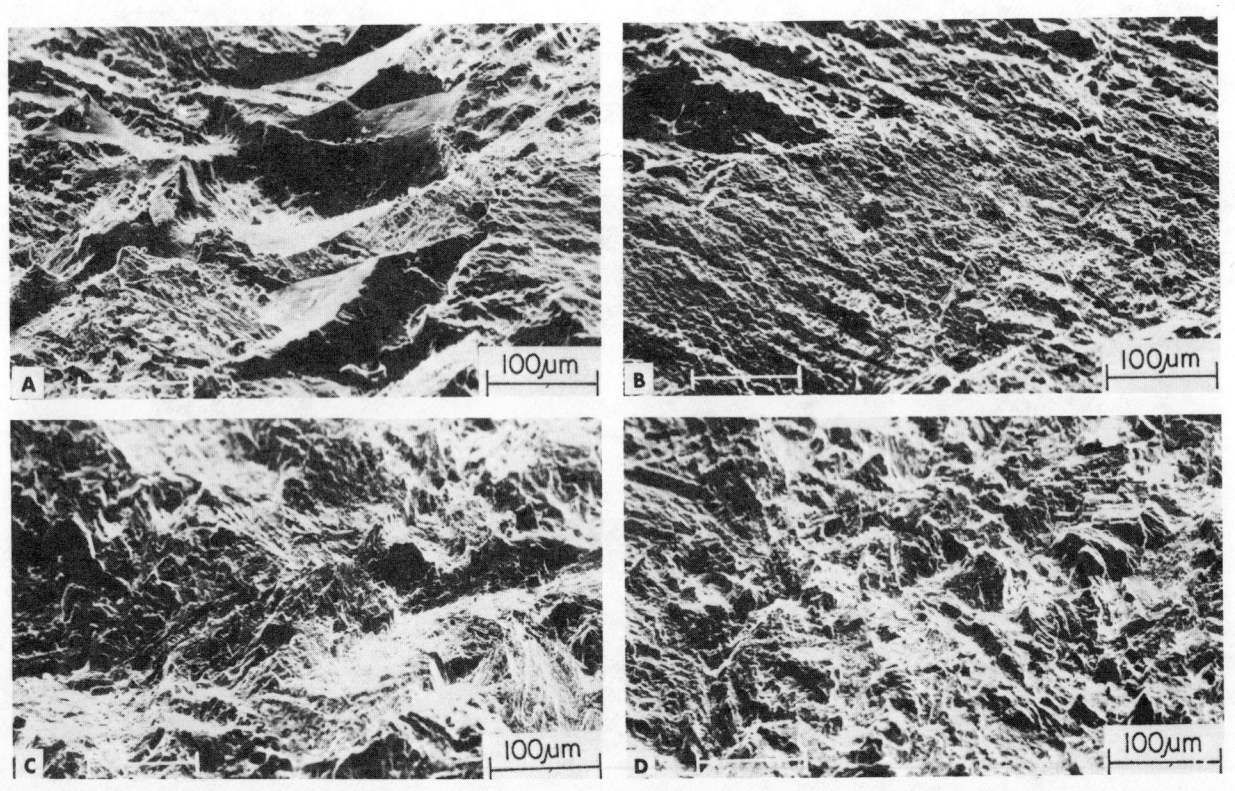

Figure 5. HY-130 fatigue fracture surface. Top of fusion zone: (A) intergranular, (B) flat, transgranular; center and bottom of fusion zone: (C) and (D) rough, transgranular indicative of crack branching.

                TOP   CENTER   BOTTOM

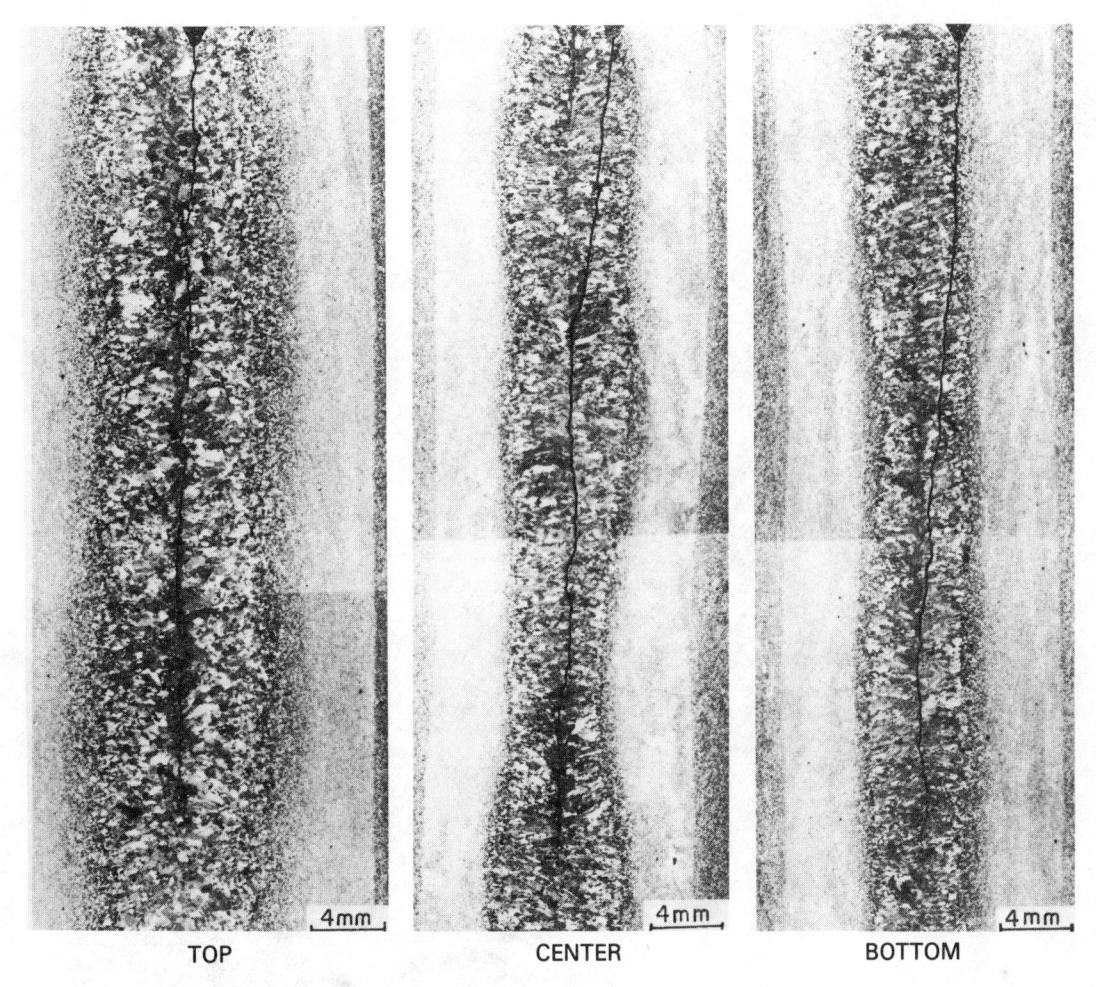

Figure 6. HY-130 fatigue crack path through the top, center and bottom horizontal metallographic sections.

columnar grains grow parallel to each other as the solidification front advances toward the weld center line. A horizontal section through this region of the fusion zone shows a columnar grain morphology.

The microstructure from the top to the bottom of the fusion zone is illustrated in Figure 7. A martensitic structure developed throughout the depth of the weld. The only apparent difference in the microstructure was the high degree of directionality of the elongated platelets, within the grains, in the top section of the fusion zone. The solidification structure, on the other hand, changed from predominately cellular in the broad weld pool at the top of the fusion zone to cellular-dendritic throughout the remainder of the weld. As determined previously, and illustrated by Figure 8, the path of a stress-corrosion crack is influenced by the solidification structure. On the other hand, a fatigue crack path is dictated by the microstructure (Fig. 9). In Figure 9A, a change can be seen in the direction of the crack path with a change in the orientation of the martensite platelets. There appears to be no correlation between the crack path and the solidification structure shown in Figure 9B.

In general, whether the solidification mode is cellular or cellular-dendritic, the structure changes from fine at the fusion boundary, where the temperature gradient is steepest and solidification is relatively slow, to coarse at the center of the weld, where the temperature gradient is flatter and solidification is more rapid. Figure 10 is typical of this change in solidification structure. However, variations in this typical pattern can be found throughout the fusion zone, as illustrated in Figures 11, 12, 13 and 14. Figure 11 shows the solidification structure along the fusion boundary at opposite sides of the weld. In Figure 11A, a fine, cellular structure is shown at the fusion boundary. Figure 11B, on the other hand, shows a coarse, cellular-dendritic structure typical of the rapidly cooled weld metal found, usually, at the center of the fusion zone. Rapid cooling of the molten metal at the fusion boundary can lead to incomplete mixing in the fusion zone. In a butt weld, since there is no difference in the chemical composition of the weld metal and base plate, such an unmixed area would escape detection. In a laser weld of HY-80 plate fabricated with an Inconel insert, however, an unmixed area along the fusion boundary was, indeed, found. The lack of mixing was verified by microprobe analysis. Figure 12 shows an unmixed area at the fusion boundary. The line indicates the approximate location of the microprobe trace shown below the micrograph. The trace shows the relative change in Ni concentration from the base metal through the unmixed zone and into the weld metal. This melted but unmixed zone is comparable to that found by Savage, et. al, in conventional

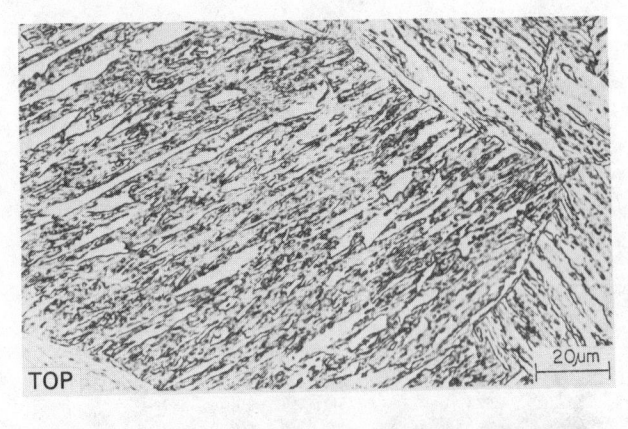

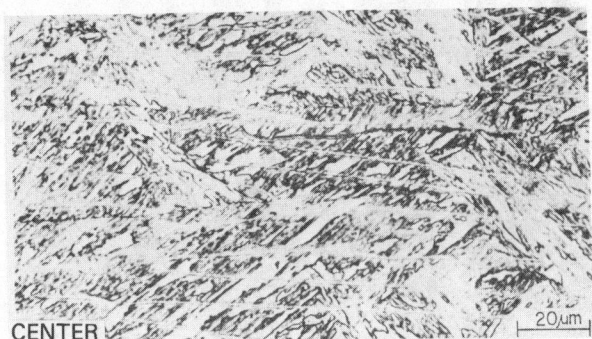

Figure 7. Martensitic microstructure in the top, center and bottom sections of the HY-130 fusion zone.

Figure 8. HY-130 solidification structure and stress-corrosion cracks. Stress-corrosion cracks coincide with solidification structure.

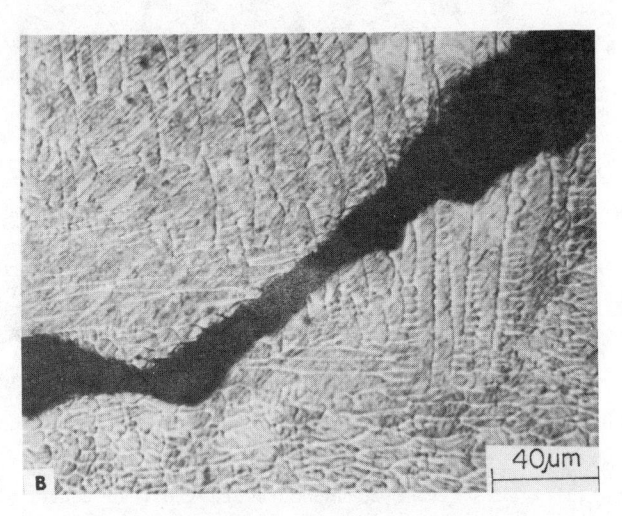

Figure 9. HY-130 fatigue crack path showing a correlation between crack path and microstructure in A (nital etch), but no correlation with solidification structure in B (ammonium persulfate etch).

welds [5]. Evidence of grain boundary liquation, a potential source of hot cracking or microfissuring [6], can also be seen in the heat affected zone (HAZ) in Figure 11A. A second variation in solidification morphology is seen in Figure 13 where the direction of solidification at the center of the weld is parallel to the welding direction. David and Liu [7] associate a change in the direction of fusion zone grain growth with a change in the shape of the weld pool. In addition to an elliptical and a teardrop pool, they identify a transition shape in which the growth of grains inward toward the weld center line is interrupted by grains growing along the center line in the welding direction, as seen in Figure 13. Lastly, Figure 14 shows the apparent formation of additional solidification fronts within the fusion zone, accompanied by a change in the cell size and the direction of solidification, with the formation of each front.

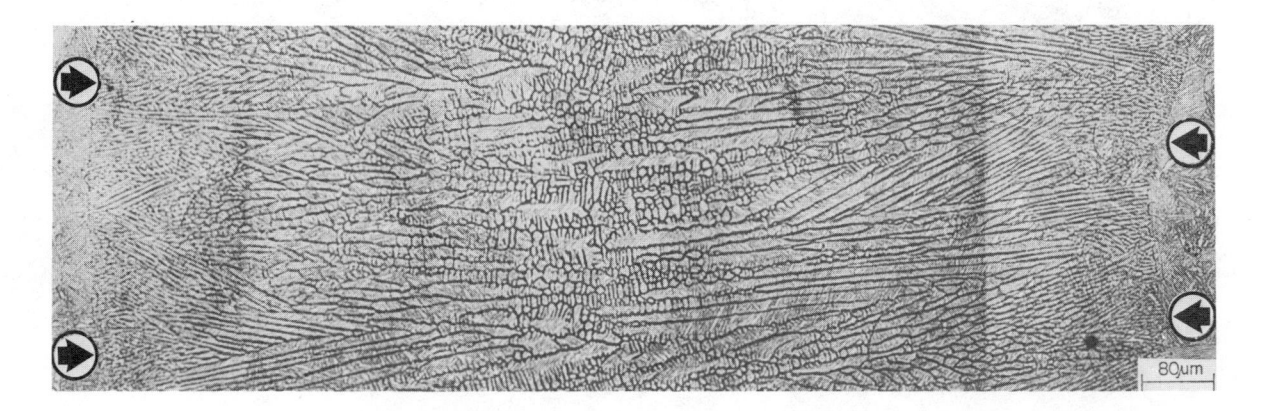

Figure 10. Typical change in solidification structure from fine cellular at the fusion boundary (arrows) to coarse cellular-dendritic at the weld center.

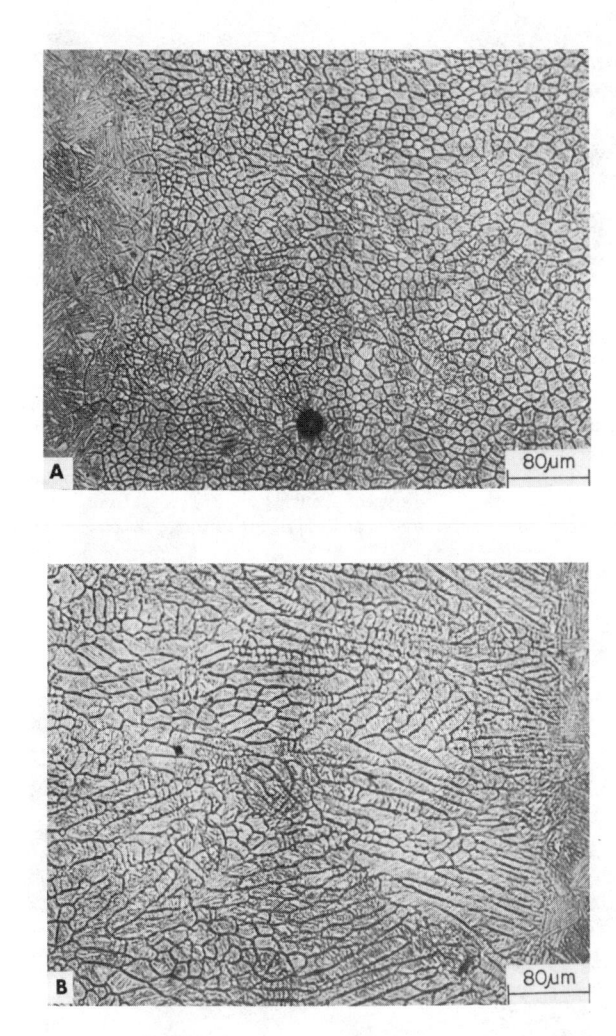

Figure 11. Variations in solidification structure along the fusion boundary: (A) fine cellular; (B) coarse cellular-dendritic. Grain boundary liquation can be seen in HAZ in (A).

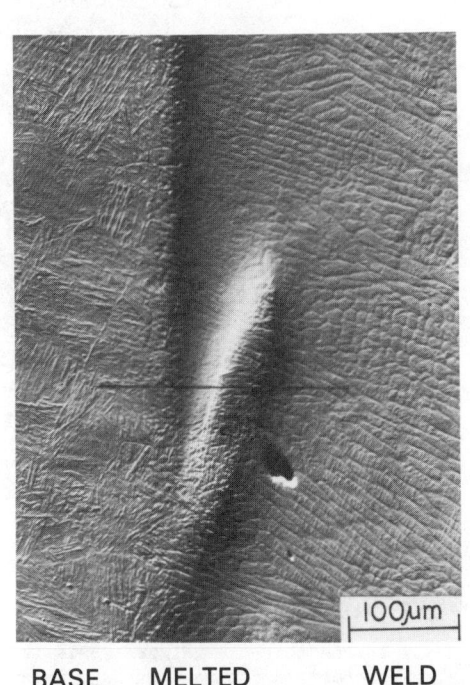

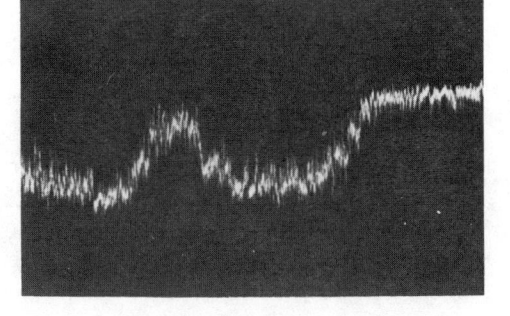

Figure 12. HY-80 weld with Inconel insert showing a melted, unmixed zone at the fusion boundary and variation in Ni concentration, in counts/sec, across the unmixed zone. The line, in the upper micrograph, indicates path of the electron beam during microprobe analysis.

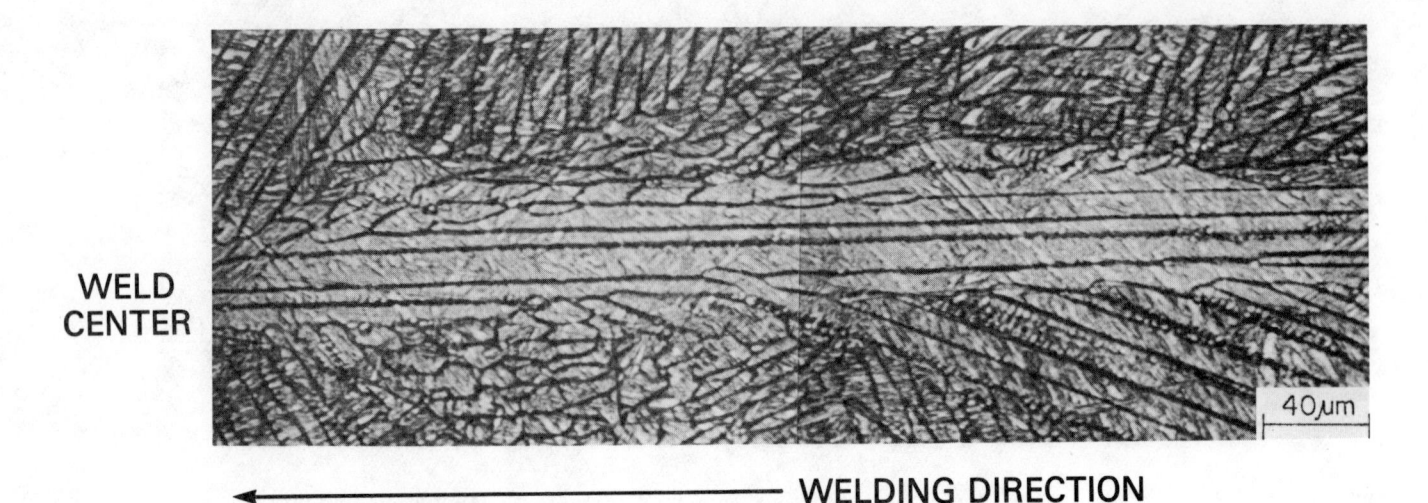

Figure 13. Solidification structure showing the direction of solidification parallel to the welding direction.

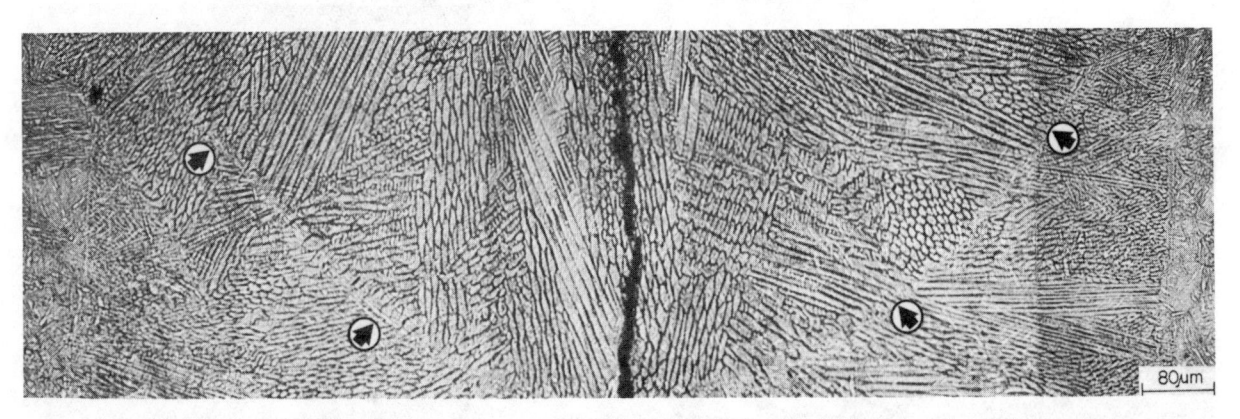

Figure 14. Solidification structure showing additional solidification front (arrows) with changes in cell size and solidification direction.

HY-80

In contrast to the HY-130 specimen (Fig. 4), the fracture surface of the HY-80 specimen (Fig. 15) shows no discernible change from the top of the fusion zone to the bottom. Figure 16 is typical of the entire fracture surface and shows IG separation and solidification structure in addition to fatigue.

The transverse metallographic section of the HY-80 weld (Fig. 3) shows parallel grain growth throughout the fusion zone. This pattern of grain growth is seen in all of the horizontal metallographic sections (Fig. 17), along with extensive, transverse solidification cracking. The

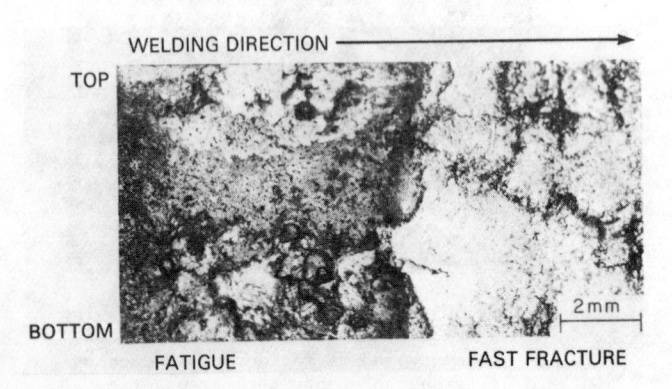

Figure 15. Macrograph of the HY-80 fracture surface showing welding direction, top and bottom of the fusion zone, and the fatigue and fast fracture areas.

Figure 16. HY-80 fracture surface showing IG separation and solidification structure in addition to fatigue.

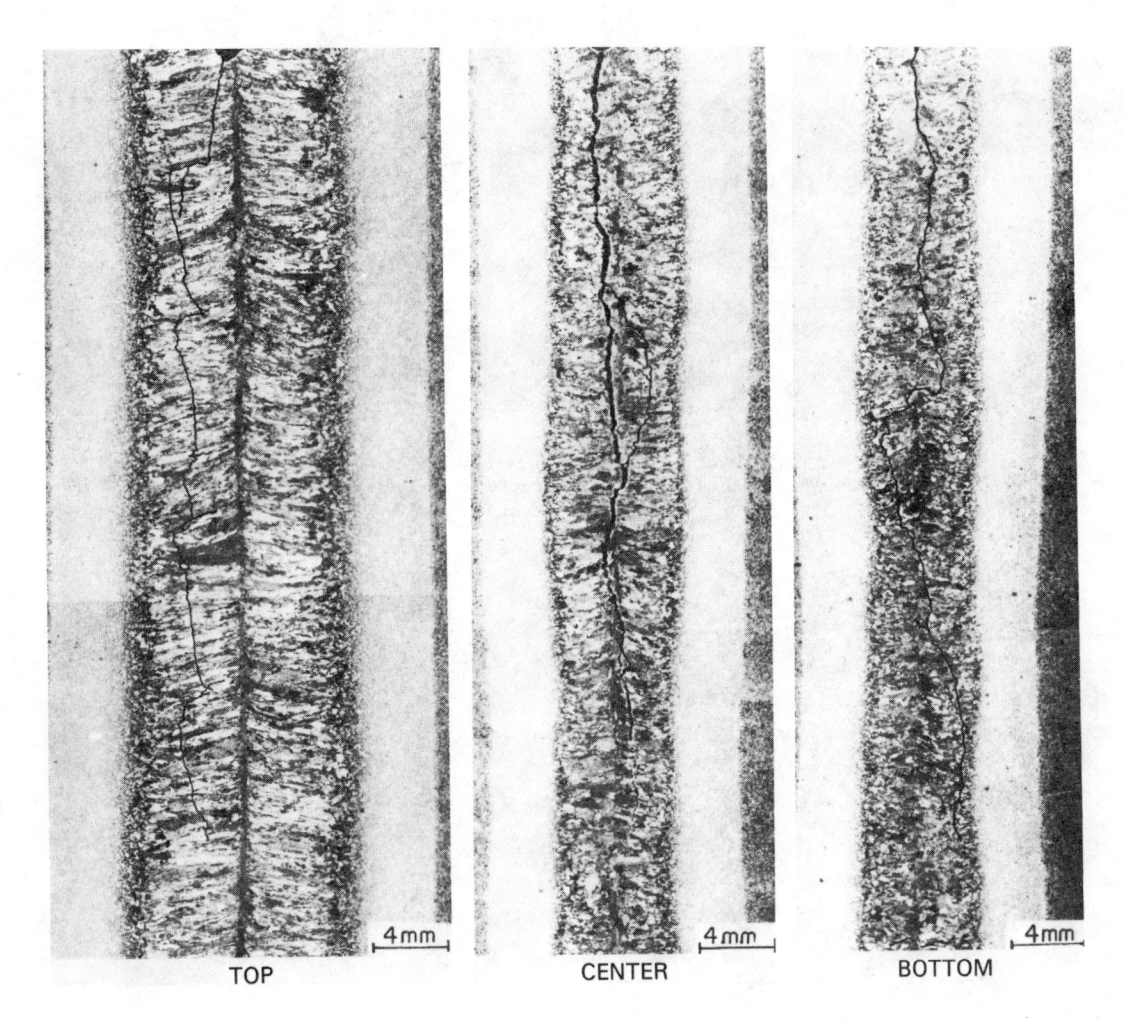

Figure 17. HY-80 LB weld showing fatigue crack path in top, center and bottom horizontal sections.

solidification cracks were most prominent in the top section. The fatigue crack propagated between the solidification cracks giving a terraced, or stepped, appearance to the crack profile in the top section of the weld. Numerous short transverse cracks, found in the center and bottom metallographic sections, led to discontinuities in the path of the propagating fatigue crack and altered the crack growth direction, macroscopically. A portion of the crack path from the center section is shown in Figure 18. Figure 18A shows the solidification structure. The fatigue crack propagated between a series of short, transverse solidification cracks. Figure 18B shows the same area etched to reveal the microstructure. A comparison of the two micrographs indicates that the path of the fatigue crack was dictated, primarily, by the existing solidification cracks. Similar transverse solidification cracks, in GTA autogenous welds of AISI type 330 tubing, were reported by Koch and Hill [8], and attributed to silicon segregated to the grain boundaries during welding. Metzbower and Moon reported solidification cracking along the center line of HY-80 LB welds [9], and found increased amounts of silicon in conjunction with solidification structure on the fracture surface of DT specimens. Energy dispersive X-ray analysis of the fatigue fracture surface, in this investigation, showed a few scattered silicon inclusions, however, microsegregation to the extent reported by Koch and Hill, and Metzbower and Moon was not detected. Microprobe analysis of the metallographic sections, likewise, showed no increase in silicon content, above matrix level, along the edge of the solidification cracks. Neither were sulfur nor phosphorus, two impurity elements associated with hot cracking in steels [10], detected. However, it is possible that either, or both, of these elements could be associated with the solidification cracks in concentrations below the detectable limits for X-ray analysis. In an investigation of grain boundary segregation, by auger electron spectroscopy, Parks, et al. [11] found S, P, C and Si segregated to a depth of several atomic monolayers at grain boundary surfaces in 304 SS weldments. Although the concentrations of the impurity elements were at least an order of magnitude higher than the matrix values, it is doubtful that these thin films would be detected by electron beam microprobe analysis.

In general, a martensitic microstructure, along with some bainite, developed throughout the fusion zone of the HY-80 weldment (Fig. 19). The solidification structure was mainly cellular-dendritic, similar to that shown in Figure 10. Also, as in the HY-130 specimen, center line solidification parallel to the welding direction was found in different areas through the depth of the HY-80 fusion zone. In these regions, the microstructure was composed of highly directional, elongated platelets (Fig. 20).

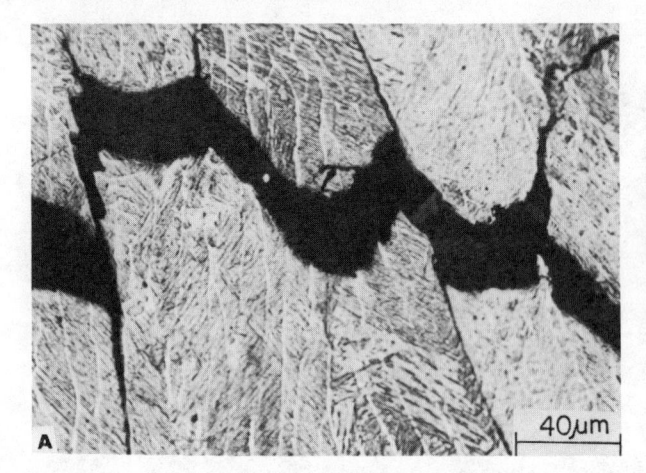

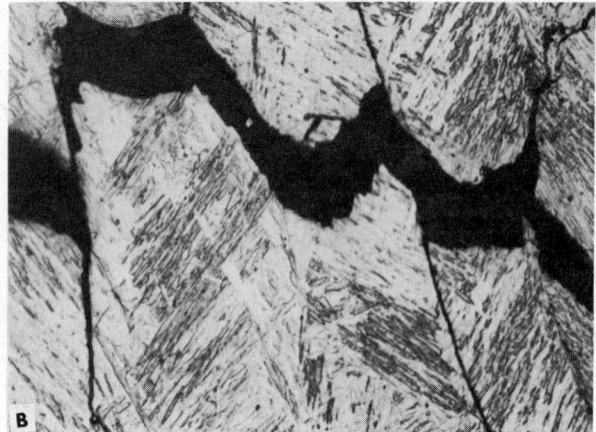

Figure 18. HY-80 fatigue crack path. (A) solidification structure (ammonium persulfate etch); (B) microstructure (nital etch).

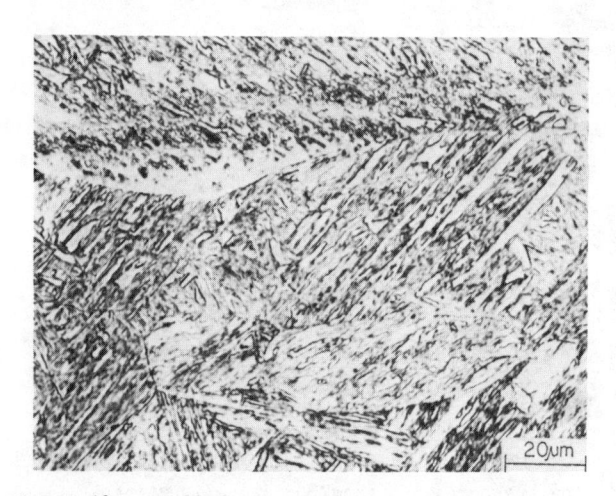

Figure 19. HY-80 fusion zone microstructure composed of martensite with some bainite.

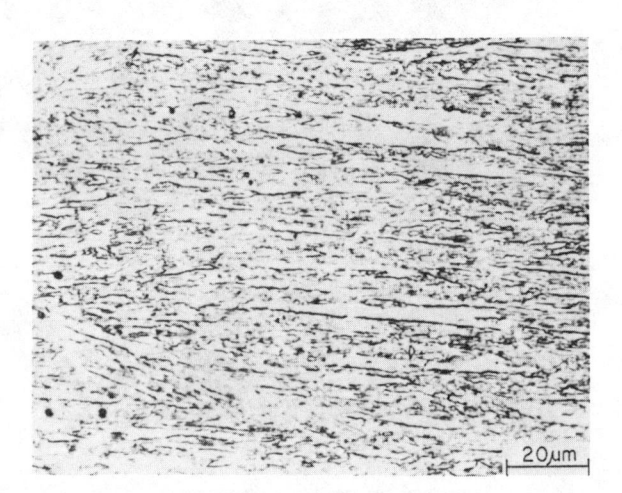

Figure 20. HY-80 fusion zone microstructure, along the center line, showing elongated martensitic platelets.

Ti-6211

The macrograph of the Ti alloy (Fig. 21) shows that the specimen fractured obliquely. Only a fraction of the fatigue zone fell in the weld metal.

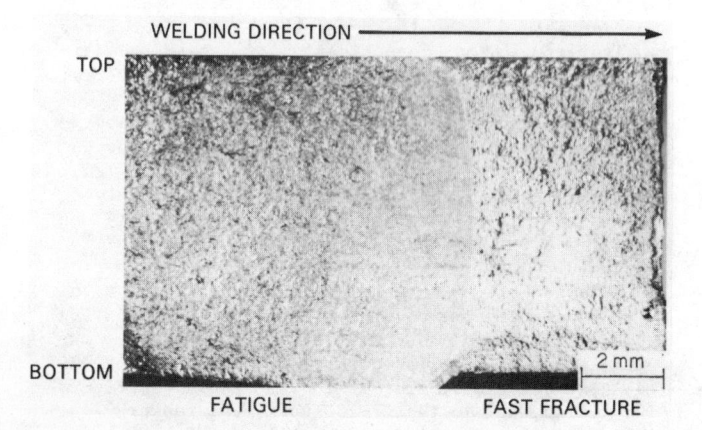

Figure 21. Macrograph of the Ti-6211 fracture surface showing welding direction, top and bottom of the fusion zone, and the fatigue and fast fracture areas.

Figure 22, the top horizontal metallographic section, also shows the crack propagating out of the weld metal and into the heat affected zone. The center and bottom sections showed a similar crack path. The contour of the weld (Fig. 3) shows a broad pool at the top that is similar to, but shallower than, that seen in the HY-130 specimen. Due to the shallowness of the top pool and the depth of the side groove, the top horizontal metallographic section fell just below, rather than within, this area. Consequently, unlike the HY-130 specimen, the grain orientation was the same in all three horizontal sections.

The fracture surface (Fig. 23) showed elongated ridges near the top of the weld, relatively flat fracture in the weld center, and a slightly rougher texture at the bottom of the weld. Figure 24 is representative of the microstructure found in all 3 sections of the fusion zone. The martensitic alpha structure includes a Widmanstätten formation along with both elongated aligned and branched platelets. In an investigation of microstructural effects on fatigue crack propagation in Ti-6Al-4V, Yoder, et al. [12] found that grain size enlargement was beneficial to fatigue crack propagation resistance. Propagation of the fatigue crack out of the coarse columnar grain microstructure in the Ti-6211 weld

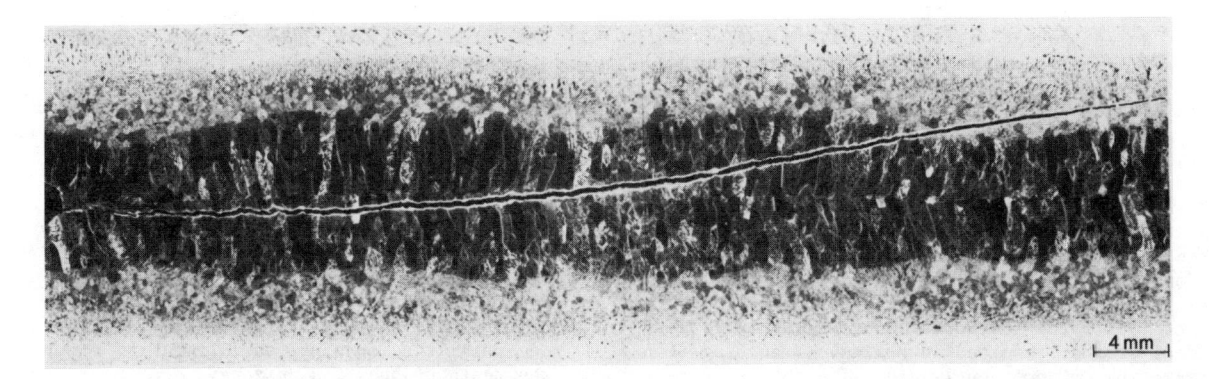

Figure 22. Ti-6211 Fatigue crack path in the top horizontal metallographic section.

Figure 23. Ti-6211 fracture surface showing fatigue fracture in the top, center and bottom of the fusion zone.

Figure 24. Martensitic alpha microstructure typical of the Ti-6211 fusion zone.

metal and into the finer grained structure in the HAZ might be interpreted as indicating a greater resistance of the weld metal to fatigue crack growth. However, from this prelimary investigation, we are unable to say whether the microstructure, the grain size or residual welding stresses played the dominant role in directing the path of the fatigue crack.

The interaction of the laser beam and the plasma during welding is believed to determine the contour, or profile, of the fusion zone. How varying the welding parameters and controlling the plasma affects the contour of the weld is the subject of an investigation currently in progress. In the broad, shallow pool at the top of the HY-130 and Ti-6211 weldments (the nail head contour) the columnar grains are inclined upward due to the changing direction of the maximum thermal gradient in the molten pool, as the normal to the fusion boundary changes. On the other hand, the more uniform direction of thermal gradient in the broad Vee shaped pool at the top of the HY-80 fusion zone, where the fusion boundary is relatively straight and the normal to the fusion boundary is relatively constant, results in the growth of columnar grains primarily inward, toward the weld center line. In the narrow, straighter portion of the fusion zone below the upper weld pool, the pattern of parallel columnar grain growth, normal to the fusion boundary and the weld center line, is the same in all three weldments. The bulge seen near the center of the profile in each weld, but most pronounced in the HY-80 specimen, is presumably due to internal side wall reflections during welding. The wall-focusing effect of a laser beam was investigated by Arata and Miyomoto [13] using a transparent acrylic resin. The shape of the resulting cavities showed a succession of narrow and wide regions, through their depth, similar to the LB weld profiles.

The microstructure and grain orientation dictated the path of the fatigue crack in the HY-130, HY-80 and Ti-6211 LB welds investigated. However, in the event of severe solidification or hot cracking, as found in the HY-80 weldment, the location and frequency of the solidification cracks determined the path of the fatigue crack, macroscopically. Between solidification cracks, however, the microstructure, not the solidification structure, determined the path of the crack. In the HY-130 specimen, due to the orientation of the grains in the upper weld metal, the direction of the propagating fatigue crack was nominally parallel to the longitudinal axis of the grains. The orientation of the grain boundaries, along with the alignment of the microstructure within the individual grains, resulted in minimal branching of the crack tip. Consequently, the fracture surface showed broad, flat areas of transgranular fatigue along with scattered IG separation at the top of the weld. In the center and bottom regions of the weld, the crack propagated along the transverse axis of the columnar grains. This orientation, coupled with the unaligned microstructure, resulted in extensive crack tip branching and a rough, irregular surface on the fractured specimen. The fusion zone of the Ti-6211 specimen, which was similar in contour to the HY-130 specimen, showed a similar change in fracture mode from the top to the bottom of the weld zone. However, IG separation was not detected. In the HY-80 specimen, since the grain orientation and microstructure were basically the same throughout the depth of the fusion zone, there was no difference in fracture mode from the top to the bottom of the weld. In addition to fatigue, solidification cracks and grain boundary separation were found across the entire fracture surface.

Typically, a fine cellular or cellular-dendritic solidification structure forms at the fusion boundary and becomes increasingly coarse as the weld center line is approached. Variations in solidification structure through the depth of the fusion zone, in both the HY-130 and the HY-80 specimens, illustrate the degree of complexity of heat flow patterns in LB welds. The predominantly cellular structure, found throughout the broad weld pool at the top of the HY-130 specimen, indicates a slower rate of cooling in that area than in the remainder of the weld. In the HY-80 fusion zone the solidification mode was generally cellular-dendritic. However, in sections along the center of the weld the structure was entirely cellular and solidification took place parallel to the welding direction. This same phenomenon was noted in the HY-130 specimen as illustrated by Figure 13. The change in the direction of grain growth, from that typical of a teardrop pool with grains growing normal to the welding direction, to center line solidification with grains growing in the welding direction, again indicates a change in the heat flow pattern. Another variation was seen in the coarse cellular structure found at the fusion boundary (Fig. 11), indicating a faster cooling rate than that generally found in this region. When a filler metal is incorporated in an LB weld, we believe such rapid cooling at the fusion boundary can result in a lack of mixing of the melted base metal with the filler metal, as demonstrated by the HY-80 weld fabricated with an Inconel insert (Fig. 12). Finally, complex heat flow patterns in LB welds can result in additional solidification fronts (Fig. 14) and can give the illusion of additional weld passes.

CONCLUSIONS

1. The grain orientation and microstructure dictate the path of a fatigue crack in LB welds of structural alloys.

2. As the weld contour changes, the direction of maximum thermal gradient changes and, consequently, the direction of columnar grain growth.

3. Typically, a fine cellular or cellular-dendritic structure forms at the fusion boundary and becomes increasingly coarse as the weld center line is approached. However, due to complex heat flow patterns in LB welds, changes can occur in the mode of solidification throughout the depth of the fusion zone.

4. Extremely rapid solidification along the fusion boundary can result in a melted but unmixed zone of inhomogeneous composition.

REFERENCES

1. E. A. Metzbower, D. W. Moon and F. W. Fraser "Laser Welding of Structural Alloys," Proceedings of the International Conference on Welding Technology for Energy Applications, Gatlinburg, Tenn., May 16-19, 1982 pp. 313-329. Also, F. W. Fraser and E. A. Metzbower, "Stress-Corrosion Crack Propagation in HY-130 Laser Welds," presentation, AIME Annual Meeting, Dallas Texas, Feb. 14-18, 1982.

2. E. A. Metzbower and D. W. Moon, "Fractography of Laser Welds," ASTM STP 733, Fractography and Materials Science, Ed. L. N. Gilbertson and R. D. Zipp, Nov. 1979, pp. 131-149.

3. E. M. Breinan and C. M. Banas, "Welding with High Power Lasers," Advances in Metal Processing, Sagamore Army Materials Research Conference Proceedings, 25, Ed. S. Burke, R. Mehrabian, V. Weiss, July 1978, pp. 111-131.

4. J. E. Gould and J. C. Williams, "Solidification Structures and Phase Transformations in Welded Ti Alloys," Titanium 1980, Ed. H. Kimura and O. Izumi, TMS-AIME, Warrendale, Pa., 1980, Vol. 4, pp. 2337-2346.

5. W. F. Savage, E. F. Nippes and E. F. Szekeres, "A Study of Weld Interface Phenomena in a Low Alloy Steel," Welding Journal, 55(9), Sept. 1976, 260-S to 268-S.

6. W. F. Savage, "Solidification, Segregation and Weld Defects," Weldments: Physical Metallurgy and Failure Phenomena, Proceedings of the Fifth Bolton Landing Conference, Ed. R. J. Christoffel, E. F. Nippes, H. D. Solomon, August 1978, pp. 1-18.

7. S. A. David and C. T. Lin, "High-Power Laser and Arc Welding of Thorium-Doped Iridium Alloys," Welding Journal, 61(5), May 1982, 157-S to 163-S.

8. J. B. Koch and K. A. Hill, "Failure Analysis of Type 330 Welds by Scanning Electron Microscopy," Welding Journal 59(8), August 1980, 242-S to 244-S.

9. E. A. Metzbower and D. W. Moon, Fractography of Laser Welds," ASTM STP 733, Fractography and Materials Science, Ed. L. N. Gilbertson and R. D. Zipp, Nov. 1979, pp. 131-149.

10. J. C. Lippold and W. F. Savage, "Solidification of Austenitic Stainless Steel Weldments: Part III - The Effect of Solidification Behavior on Hot Cracking Susceptibility," Welding Journal 61(12), Dec. 1982, 388-S to 396-S.

11. J. Y. Parks, S. Danyluk and D. E. Busch, "Scanning Auger Electron Spectroscopy Studies of Grain-boundary Segregation in Type 304 Stainless Steel," Proceedings of the Fifth Bolton Landing Conference, Ed. R. J. Christoffel, E. F. Nippes, H. D. Solomon, August 1978, pp. 165-179.

12. G. R. Yoder, L. A. Cooley and T. W. Crooker, "A Comparison of Microstructural Effects on Fatigue-Crack Initiation and Propagation in Ti-6Al-4V," NRL Memorandum Report 4758, March 5, 1982.

13. Y. Arata and I. Miyomoto, "Wall-Focusing Effect of Laser Beam," Advanced Welding Technology, the Second International Symposium of the Japan Welding Society, August 1975, pp. 125-130.

LASER WELDING OF STEELS AND NICKEL ALLOYS

R. F. Duhamel and C. M. Banas
United Technologies Research Center

ABSTRACT

Multikilowatt laser welding experiments were conducted on low-carbon steels, 300 and 400 series stainless steels and several nickel-base alloys, including Hastelloy-X and Inconel 718. Tests were conducted with industrial laser systems rated at continuous duty power levels from 3 to 12 kW.

Beam focusing for welding was provided by two-element, metal mirror focusing heads having nominal f/numbers from 6 to 12. These included flat-spherical mirror pairs in on-axis configurations, flat-spherical mirror combinations with slight off-axis alignment and other mirror combinations specifically designed to reduce aberrations. Plasma suppression and workpiece shielding were provided by inert gas flow over the weld zone. Generally, welds were formed with the part traversing under a fixed beam, although a few welds were formed with a stationary workpiece and a moving beam. Salient features of the test equipment are discussed.

Single-pass weld penetrations to 2.5 cm and thin material welding speeds to 25.4 cm/sec were demonstrated. Welding tests were performed at power levels from 2-19 kW. Results are presented in terms of representative metallographic weld penetration cross sections as well as in terms of penetration as a function of specific welding energy. The welding response of the different materials is compared.

INTRODUCTION

High-power laser processing is currently undergoing intense development. Reliable, industrially-suited, multikilowatt laser systems are being supplied to industry by a number of manufacturers. It is anticipated that production applications of lasers will increase dramatically within the next several years and that the laser, now only two years beyond its 21st birthday, will establish a significant niche in production materials processing.

The reasons for rapid acceptance of the laser in manufacturing applications include process economy and unique thermal processing capabilities.[1-5] In welding applications, single-pass weld penetrations up to 5 cm have been generated, welding speeds of centimeters per second have been attained in thin materials[6] and unique weld characteristics have been demonstrated.[7] In local thermal surface-treating tasks, the laser has been used for transformation hardening of ferrous alloys, for surface alloying, for hardfacing and for rapid melting and solidification of surfaces for attainment of specific properties.[4,8] Cutting, one of the earliest production applications of lasers, continues to receive significant attention with fully-automated laser cutting systems available from more than a half dozen manufacturers.

The rapid development of laser processing has resulted in the generation of a large quantity of information relative to processing capabilities. Because of the significant differences in laser systems and processing techniques, prediction of processing parameters for a specific task, based on available information, is not always convenient. In the following, an attempt is made to simplify prediction of laser welding performance by representing it solely in terms of the commonly-used parameter of specific welding energy.

EXPERIMENTAL APPARATUS AND PROCEDURE

Laser Equipment

The welds described in this paper were formed using multikilowatt, cross-beam, convectively-cooled, cw, CO_2 lasers in which the optical beam, convective cooling gas circulation and electric discharge are orthogonal. The 10.6μ wavelength optical beam is extracted from a high-voltage, low-current glow discharge established in a mixture of helium, nitrogen and carbon dioxide. This mixture is circulated through the lasing region at a pressure of approximately one-tenth of an atmosphere. Unstable oscillator (annular beam profile) cavity optics are normally utilized to effect single-mode operation with high efficiency and good focusability for deep-penetration laser welding. Stable oscillator (Gaussian mode) or top hat (multimode) optics can also be provided for the modular systems if specific applications warrant their use.

Most of the tests described were conducted with industrially-rated laser systems employing a modular construction concept as shown in Fig. 1. The individual modules contain separate gas-circulating, electrode and heat exchanger components and share common cavity optics. At the present state of development, the output of a single module is rated at 3 kW for continuous duty. Consequently, the three-module system shown in Fig. 1 is rated at 9 kW. To date, up

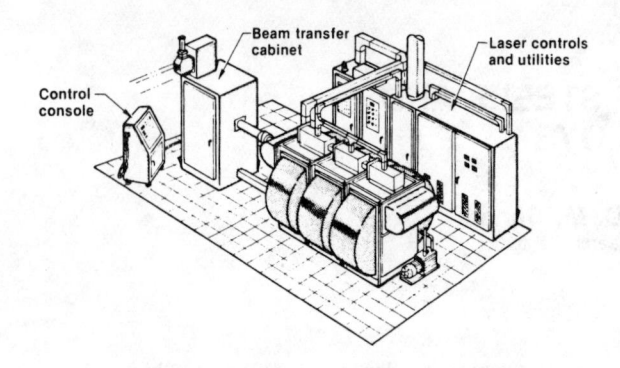

Fig. 1. 3 Module - 9 kW Laser System

to four modules have been incorporated in a single unit, and larger units appear technically and economically feasible. It also has been observed that output efficiency increases with the number of modules and that multi-unit modular systems have demonstrated short-term operation at power levels of at least 150% of their continuous-duty rating.

Ancillary Equipment

As has been noted, the laser units operate at approximately 0.1 atmosphere. Since the output beam is in the infrared portion of the spectrum, normally transparent materials cannot be utilized for beam transmission to the atmosphere. For one- and two-module units (3-6 kW), this requirement is met with antireflection (AR) coated, zinc-selenide optics. At higher power levels, however, the durability of solid output windows becomes increasingly less satisfactory. To obviate the problems associated with infrared windows, an aerowindow has been developed for the higher-power systems. The aerowindow, shown in Fig. 2, contains a precisely-shaped aerodynamic flow passage through which compressed air flows at high speed. The unit is designed so that the inner isobar of the flow corresponds to the laser operating pressure while the outer corresponds to atmospheric pressure. Thus, the 10:1 pressure differential between the atmosphere and the laser chamber is maintained without the requirement for a solid window.

In practice, aerowindow flow conditions are maintained such that there is a slight outflow of laser gases to the atmosphere. This is done to ensure that no detrimental atmospheric air leakage occurs into the laser. A focusing mirror is utilized in the last pass of the optical cavity to reduce the beam size at the aerowindow aperture. If the beam is to be transmitted for appreciable distances before utilization for processing, recollimating optics are utilized beyond the aerowindow to provide a parallel output beam of the desired diameter.

In typical materials processing tests, the beam is transmitted from the output window to a retractable mirror/shutter into the focusing optics and then onto the workpiece, as shown in Fig. 2.

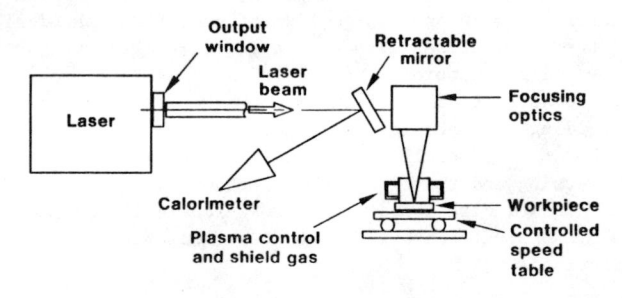

Fig. 3. Experimental Welding Arrangement

When closed, the retractable mirror/shutter directs the beam into a calorimeter for power measurements. If power monitoring is required during processing, a calibrated infrared sensor is utilized to monitor power scattered from one of the beam-directing mirrors in the beam transmission system.

Focusing Optics

Power densities in the range from 10^6 to 3×10^7 W/cm^2 are required for deep-penetration laser welding. Keyhole welding does not occur at lower power densities, while higher power densities lead to excessive vaporization, material expulsion and poor weld characteristics. At the multikilowatt power levels involved, these high power densities are achieved with incident spot diameters of 0.5 to 1 mm diameters, which are readily obtained with dual-element, metal-substrate, reflective focusing optics. Focusing systems in the range of f/6 to f/12 normally satisfy the above conditions and provide satisfactory depth of focus for laser welding of both thin and thick materials.

Two representative focusing systems are shown in Fig. 4. The off-axis system is the least expensive to construct and incurs only minimal aberrations if the

Fig. 2. Industrial Laser Aerowindow

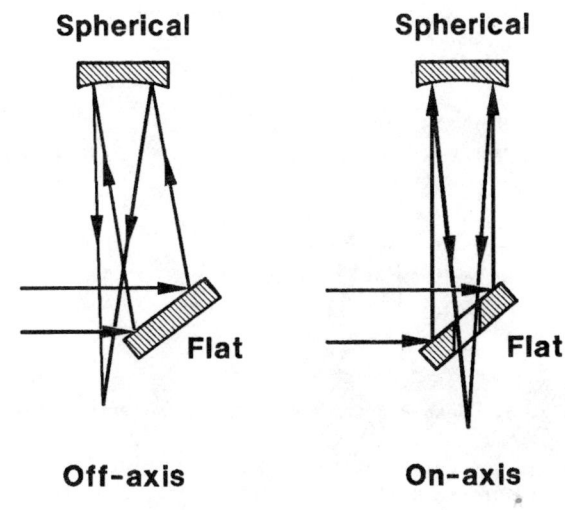

Fig. 4. Reflective Focusing Optics

off-axis angle is maintained below 6 degrees. In fact, the inherent slight focal spot ellipticity has been shown, in some instances, to be of advantage in high-speed welding applications involving motion along the major axis of the ellipse.

The annular output characteristics of the unstable oscillator beam are employed to advantage in the on-axis system shown in Fig. 4. The flat, beam-directing mirror initially encountered by the beam contains a central hole which permits the focusing beam to pass through the center. This technique provides minimal aberration with inexpensive spherical optics, but must be used in the near-field to avoid central region power loss.

Other focusing systems, including single-unit, off-axis parabolic elements and corrected aspheric mirror pairs, also are practical. Forced convection cooling is generally utilized to minimize thermal distortions. Reflective surfaces have included polished aluminum, polished copper, vapor-deposited gold on electroless nickel on copper and polished molybdenum. The latter material has been found to be particularly durable since its melting temperature far exceeds the temperature of molten droplets of steels and nickel-base alloys. Thus, the mirror surface retains its integrity under severe production operating conditions and requires only careful cleaning for restoration.

Plasma Suppression and Shielding

Plasma suppression and protection of the weldment from atmospheric contamination is generally accomplished with a flow of inert gas. Inadequate suppression leads to the generation of a beam-absorbing plasma above the workpiece[9] which inhibits laser energy delivery to the fusion zone. Helium is the preferred gas for plasma suppression due to its high ionization potential and resistance to breakdown but other gases may also be used.

The degree of sophistication of workpiece shielding is strongly dependent upon the sensitivity of the material to atmospheric contamination as well as upon the end utilization of the weldment. For many carbon steel, stainless steel and nickel alloy applications, plasma suppression provisions (which also shield the interaction zone) are adequate for atmospheric protection. In more stringent applications, trailer shielding similar to that used in more conventional welding processes can be employed. Root bead shielding can also be provided when necessary. For critical applications and reactive materials, welding may be performed in an enclosed chamber purged with inert gas. Argon, or mixtures of argon and helium, may be effectively used for trailer, root bead and enclosed chamber shielding provided that helium is utilized locally at the interaction zone for plasma suppression.

Welding

At the workstation, the workpiece is normally placed under the focusing optics in the downhand or flat welding position. Downhand welding is the most convenient method, although out-of-position laser welding also has been demonstrated. In most instances, the surface of the material is located at or near the point of minimum spot diameter. Relative motion between the focused beam and the workpiece is provided by CNC-controlled motion of the part or the beam. Bead-on-plate penetrations are often utilized to generate a broad range of parametric welding information at least expense. Experience has shown that the bead-on-plate is equivalent to a tight butt weld and can be readily generated without the encumbrance of joint preparation and seam-tracking requirements.

Joint preparation normally requires machined or ground mating surfaces with unbroken corners. In butt joint configurations, a fitup to within 3% of the nominal material thickness is required for generation of a sound weldment. Further, the faying surfaces must be free from contaminants. In many materials, simple degreasing is adequate; in more critical applications, acid cleaning may be essential and further cleanliness may be obtained by final rinse with ethyl alcohol just prior to welding.

When all of the weld prep requirements have been met, the workpiece may be laser-tack-welded to prevent joint separation during welding caused by welding forces approaching material yield stress. Extremely robust fixturing would be required to inhibit part separation due to these forces. Taper in and/or taper out of the weld bead can be effected either by variation in processing speed or by controlled ramping of the laser power. Standard welding fume exhaust equipment is utilized for removal of undesirable weld byproducts.

DISCUSSION OF WELDING RESULTS

Background

As noted earlier, the high-power laser has demonstrated unique materials processing capabilities. A principal reason for this uniqueness stems from the fact that laser beams can be focused to extremely high power densities, power densities unattainable by any other means except with high-voltage electron beam equipment.

Since power density is not an easily visualized physical entity, it is useful to interpret the intense energy concentration afforded by the laser in terms of an "equivalent temperature." This approach leads to a "feel" for the intensity and provides a simple yardstick for comparison of laser processing with other welding processes.

It is recalled that the energy radiated from a perfect thermal radiator is proportional to the fourth power of the radiator temperature in accordance with the Stefan-Boltzmann relationship

$$W = \sigma T^4$$

in which W is the radiated power per unit of area, σ is the Stefan-Boltzmann radiation constant and T is the absolute temperature of the source. Further, it is a consequence of the laws of thermodynamics that no thermal source may radiate more energy than a perfect or "blackbody" source at the same temperature and that a perfect radiator is also a perfect absorber.

Solving the Stefan-Boltzmann relationship for temperature yields

$$T \equiv [W/\sigma]^{0.25}$$

An "equivalent" laser spot temperature based on the above expression defines an incident beam power density. It is noted that a perfect thermal source would have to operate at this temperature in order to provide comparable power density.

For reference, it is noted that the approximate lower intensity limit for laser keyhole welding of 10^6 W/cm^2 corresponds to a thermal source temperature of 20,500°K. At the upper end of the intensity range of 3×10^7 W/cm^2, the equivalent thermal source temperature is 48,000°K! These values compare to much lower temperatures (e.g., approximately 3,300°K for oxyacetylene, 5,200°K for shielded metal arc and 6,400°K for atomic hydrogen) for more conventional welding processes. It can be readily seen that the intense energy concentration afforded by the laser provides the capability for vaporization of any known material.

Vaporization is, in fact, the key to the deep-penetration, or keyhole, welding capability of the laser. The rate of radiant energy delivery to a localized region of the workpiece surface is so rapid that a metal vapor column is established in the material (Fig. 5). The metal vapor pressure maintains the keyhole against the fluid dynamic forces of the liquid metal surrounding it. Beam energy is thus permitted to flow directly into the material and is not just deposited at the material surface. With appropriate motion of the part relative to the beam, the deep-penetration cavity is translated through the material with melting taking place at the leading edge and solidification at the trailing edge. The keyhole

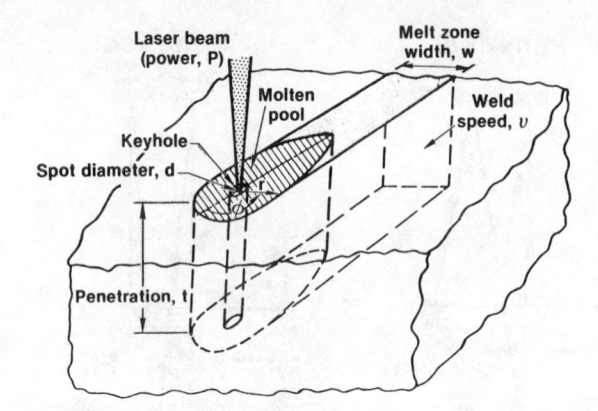

Fig. 5. Deep-Penetration Weld Characteristics

region is generally depressed slightly below the material surface while the solidified top bead, with its characteristic chevron solidification pattern, presents a slight reinforcement. The general shape of the processing zone surface can be likened to the bowl of a spoon. The deep-penetration beam welding process has been thermally modeled in terms of an unsteady, cylindrical heat source.[10]

Bead Profiles

Representative bead penetration profiles are shown in Figs. 6-9 for the three subject materials. The characteristics of laser welds in carbon steel of three different thicknesses are compared in Fig. 6.

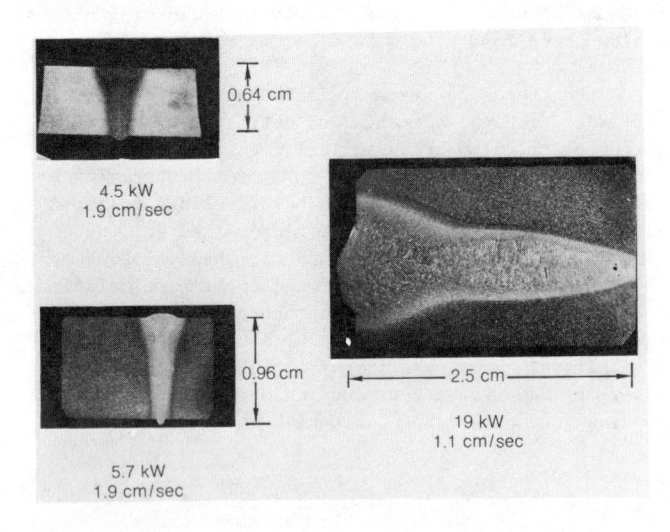

Fig. 6. Typical Carbon Steel Bead Profiles

Included is a 2.5 cm penetration which is currently the approximate limit for single-pass weld penetration for industrially-rated laser equipment. This penetration is shown in the horizontal position in which it was formed.

All three bead profiles exhibit a depth-to-width ratio of approximately 4:1. Some broadening of the top bead is noted in all samples, particularly the thinnest one. This behavior is generally attributed to plasma effects at the interaction point but may also be due, in part, to power density factors. It is noted that similar characteristics have been observed in some electron beam welds in which plasma effects were not a factor.

A slight waist is noted at about 1/3 of the penetration in the 2.5 cm sample. This bead shape is not desirable since solidification occurs first at the narrowest point and tends to trap evolved gases in the fusion zone as well as to isolate liquid metal which is then prone to shrinkage cracking on solidification. Some defects are, in fact, to be noted in the root bead area of the profile. The existence of these defects underscores the fact that the penetration is near the limit for the 19 kW power delivered to the workpiece.

Stainless steel bead profiles shown in Fig. 7 are similar to those obtained in carbon steel except top

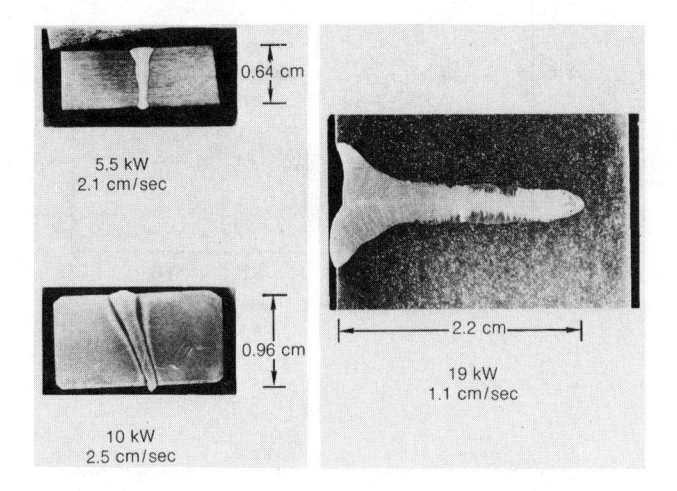

Fig. 7. Typical Stainless Steel Bead Profiles

bead broadening is slightly more pronounced. Apart from the top bead, the remainder of the penetrations exhibit very narrow fusion zones with depth-to-width ratios of approximately 6:1. This is probably due to the lower thermal diffusivity of stainless steel which tends to minimize lateral thermal energy transfer. A further consequence of the low thermal diffusivity is shift in grain orientation in top and bottom portions of the weld, indicating the influence of surface heat transfer.

As with the carbon steel maximum penetration sample, the 2.2 cm penetration in stainless steel is shown in the horizontal welding position in which it was formed. The influence of gravity on the face bead characteristics is quite evident in the cross section shown. Pronounced face bead broadening exhibited by this sample indicates plasma effects. Such effects may be more pronounced in stainless than in carbon steels due to the presence of alloying elements which can act to reduce plasma breakdown potential.

The angled penetration in 0.96-in.-thick material is shown to illustrate that incidence angle does not markedly influence penetration characteristics. In fact, other tests have shown that effective coupling occurs for beam incidence angles up to 75 deg from the normal. This behavior can provide an advantage over electron beam welding in the joining of ferromagnetic materials.

The nickel alloy bead profiles shown in Fig. 8 demonstrate an even more pronounced tendency for face

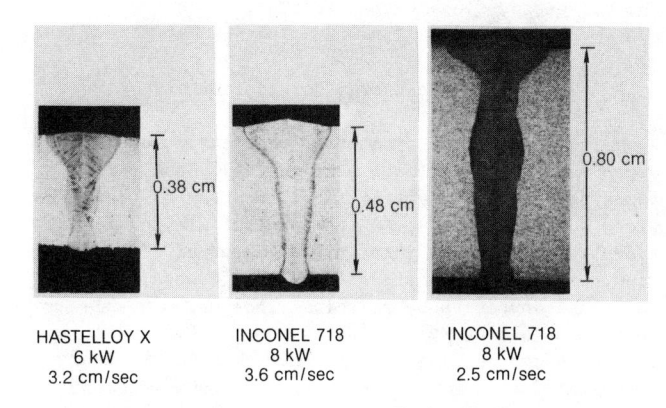

Fig. 8. Typical Nickel Alloy Bead Profiles

bead broadening than do the stainless steel profiles. Again, as in stainless steel, this broadening may stem from the influence of alloying elements. Aluminum, for example, is generally present in both stainless steel and nickel alloys but to a greater degree in the latter. Aluminum has a lower vaporization temperature than the base material as well as a significantly lower ionization potential. These characteristics could lead to enhanced thermal ionization in the metal vapor at the interaction point and a resultant decrease in plasma breakdown potential. In Ref. 11, it is noted that even relatively small quantities of aluminum can have marked effects on penetration and bead characteristics in arc welds; similar behavior for laser welds can be inferred.

A side-by-side comparison of bead penetration profiles in 0.64-cm-thick samples of the three subject materials is shown in Fig. 9. The penetrations have

Fig. 9. Bead Profile Comparisons

been selected to emphasize some of the differences in welding response of the three materials discussed in the preceding paragraphs. Of particular note are the extremely large "champagne glass" face beads in the nickel alloy weld and the three distinct grain orientation zones in the stainless steel. As noted previously, the latter are due to the influence of surface cooling effects on the weld zone. These effects are heightened in stainless steel because of its low thermal diffusivity, which reduces the rate of lateral thermal energy flow in the material.

Specific Energy Correlation

The response of a material to laser welding is dependent, in general, on beam quality, beam power, focusing optics characteristics, location of the workpiece surface relative to the point of minimum focus, travel speed, plasma joint fitup, material composition, weld prep and other factors. Of these, power and speed appear to be principal independent variables controlling welding performance. For this reason, laser welding results have often[2,5,6] been represented in terms of penetration versus speed with power as a parameter -- or as penetration versus power with speed as a parameter.

If it is recalled that dividing the power by the speed yields the specific welding energy delivered to the workpiece,

$$SWE = P/V,$$

penetration can be plotted directly as a function of this single parameter. Specific welding energy is, of course, a commonly used quantity for more conventional welding processes. In such use, the quantity usually noted is total energy consumed (i.e., energy from the plug) without correction for losses in the process itself. With respect to laser welding, a similar procedure is followed. The laser beam energy used in the computations below is the total beam energy without correction for reflection from the material surface. For reference, it is noted that reflection losses of 5-20% may occur in laser welding of the subject materials.[6]

Penetration is shown as a function of specific welding energy for carbon steel, stainless steel and nickel alloys in Figs. 10, 11 and 12, respectively. In each case, the data appear randomly scattered. Although there is a general increase in penetration with specific welding energy, energy values for constant penetration differ by factors as large as two or three. This behavior is not surprising in view of the fact that all of the other process controlling variables have been ignored in the attempted correlation. Some of these, e.g., plasma suppression effectiveness and incident beam spot size, exert first order influence on welding performance. At first glance, therefore, representation of laser welding penetration data solely in terms of specific welding energy does not appear fruitful.

As speed is increased for given welding conditions, however, fusion zone width decreases and a

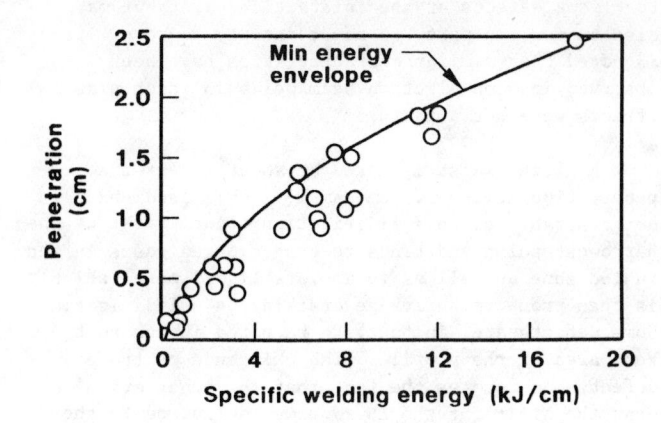

Fig. 10. Carbon Steel Welding Performance

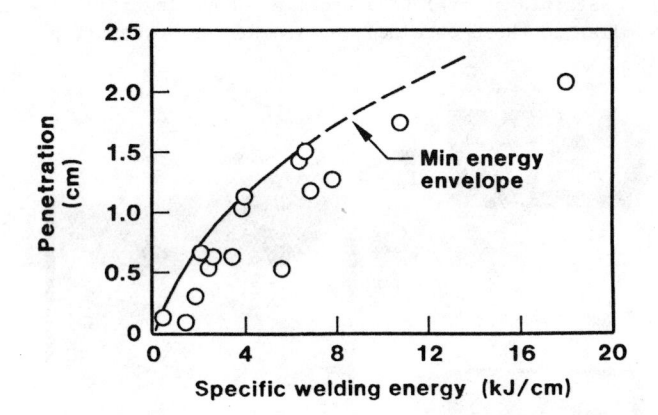

Fig. 11. Stainless Steel Welding Performance

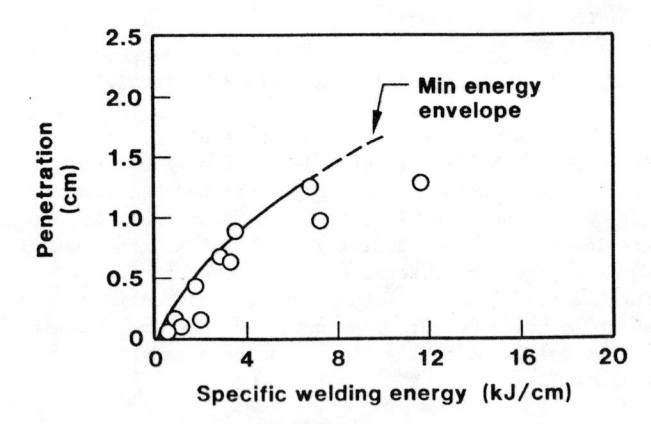

Fig. 12. Nickel Alloy Welding Performance

fusion zone profile is obtained which can be associated with "optimized" welding conditions. It is often found that such welds are formed with the least expenditure of energy, i.e., with minimum specific welding energy. Minimum specific energy is, of course, represented by the left-hand boundary of the experimental points shown in Figs. 10-12.

If a curve is drawn to bound the least specific energy required to form a weld joint, the relatively smooth curve shown in Fig. 10 for carbon steel is obtained. Similar behavior pertains in Figs. 11 and

12 for stainless steels and nickel alloys, respectively. It can be assumed that minimum specific energy represents an "optimum" performance level for laser welding and appears to form a reasonable basis for a priori selection of welding parameters. The general characteristic of the bounding curves is roughly parabolic in conformity with the expectation that energy should increase with the square of the penetration provided that depth-to-width remains constant.

The variation of penetration with minimum specific welding energy is similar for the three classes of materials investigated, as shown in Fig. 13.

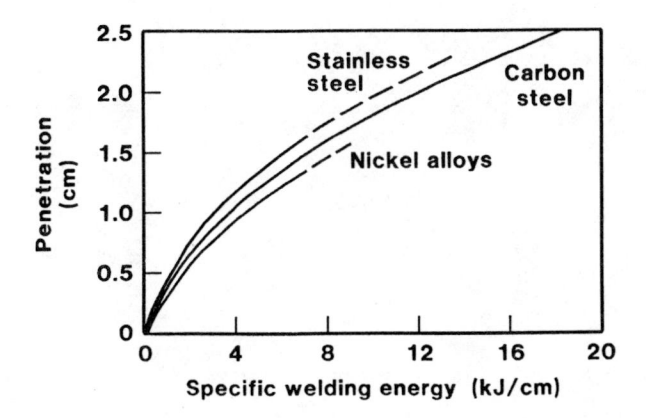

Fig. 13. Comparison of Minimum Welding Energy

Direct comparison of the limiting energy curves for the three materials shows that slightly higher energy is required for the nickel alloys. This general trend is in conformity with material thermal diffusivity as noted in Table I. As one might argue, there is a greater conduction loss in material with higher diffusivity -- leading to increased weld energy requirements.

TABLE I. THERMOPHYSICAL PROPERTIES

	T_M*°C	$\bar{C}_p$, cal/gm °C	$\bar{\alpha}$, cm^2/sec
Steel	1925	.17	.092
Stainless Steel	1925	.13	.027
Nickel Alloys	1991	.17	.140

*$T_M = T_{melt} + \dfrac{H_f}{\bar{C}_p}$

Despite the generally well-behaved trends, the differences noted are small and it should be cautioned that there may be other factors influencing the results. These include the previously-noted differences in absorption efficiency, plasma screening and other laser processing factors.

In spite of the above-noted factors and the effects of other processing parameters such as focusing optics, plasma suppression provisions, surface cleaning procedures, etc., the specific welding energy minimum appears to be a convenient correlation parameter for high-power laser welding results. Additional welding tests will serve to evaluate the general validity of this concept.

CONCLUDING REMARKS

A possible framework for simple representation of high-power laser welding data has been advanced. This framework, which utilizes the criterion of minimum specific welding energy as a critical parameter, appears to provide a convenient basis for selection of appropriate laser welding parameters for carbon steels, stainless steels and nickel-base alloys. The differences in performance in these three materials appear (within the framework of minimum specific welding energy) to correlate with material thermophysical properties.

ACKNOWLEDGMENT

The assistance of Mrs. B. True and Ms. P. Talbot in preparation of the manuscript is gratefully acknowledged.

REFERENCES

1. Brown, C. O. and C. M. Banas, Deep-Penetration Laser Welding. Paper presented at the AWS 52nd Annual Meeting, San Francisco, California, April 26-29, 1971.

2. Banas, C. M., High Power Laser Welding - 1978. Optical Engineering 17:3 (May-June 1978).

3. Engel, S. L., Laser Cutting of Thin Materials. SME Technical Paper MR74-960 (1974).

4. Belforte, D. A., High Power Laser Surface Treatment. SME Technical Paper 1077-373 (1977).

5. Locke, E. V., E. D. Hoag and R. A. Hella, Deep-Penetration Welding with a High-Power CO$_2$ Laser. IEEE Journal of Quantum Electronics, Vol. QE-8, No. 2, February 1972.

6. Banas, C. M. and R. Webb, Macro-Materials Processing. Proceedings of the IEEE, Vol. 70, No. 6, June 1982.

7. Breinan, E. M. and C. M. Banas, Fusion Zone Purification During Welding with High Power CO$_2$ Lasers. Presented at the Second International Symposium of the Japan Welding Society, Osaka, Japan (August 25-29, 1975).

8. Breinan, E. M., B. H. Kear and C. M. Banas, Processing Materials with Lasers. Physics Today (November 1976).

9. Fowler, M. C. and D. C. Smith, Ignition and Maintenance of Subsonic Plasma Waves in Atmospheric Pressure Air by cw CO$_2$ Laser Radiation and Their Effect on Laser Beam Propagation. Journal of Applied Physics, Vol. 46, No. 1, pp 138-150, January 1975.

10. Tong, H. and W. H. Giedt, Depth of Penetration During Electron Beam Welding. ASME Paper No. 70-WA/HT-2.

11. Block-Bolten, A. and T. W. Eagar, Selective Evaporation from Weld Pools. MIT Report 5-55-82.

8301-021

CRITICAL THERMAL RADIUS IN LASER SOLDERING

U. I. Chang
Engineering and Research Staff
Ford Motor Company

ABSTRACT

This paper introduces the concept of "critical thermal radius" as a means of predicting thermal results in laser material processing. Critical thermal radius (c-radius) is defined as the radius of an area on the workpiece where the desired thermal effect is achieved. The c-radius concept is expected to be instrumental in studying the effects of process parameter variations.

An example of the formulation, calculation and experimental validation of a c-radius is presented in this paper. Laser soldering of an electronic module was selected as the process to investigate. Thus, the c-radius was defined to be the radius of the molten spot on the solder pads of the electronic module.

An analytical expression for the c-radius was obtained as the result of thermal analysis of the system (workpiece) and intensity distribution analysis of the laser beam. The c-radius was expressed in terms of the process parameters and the materials' thermal properties.

The process parameters include power, beam on-time, beam spot diameter and specimen temperature. The materials' thermal properties include those properties related to thermal resistances, heat capacities and surface absorptivity.

Experiments were performed to measure the melt radii on the solder pads. The measured melt radii were compared with the calculated c-radii and were found to be in good agreement. The c-radius expression was also proven to be useful in predicting other process characteristics such as the threshold power of melting.

INTRODUCTION

Beam spot diameter is a common measure in laser thermal processing to define the beam size on the workpiece. However, this diameter by itself is seldom useful in predicting the amount of the thermal work accomplished. For example, the drilled hole diameter is different from the beam spot diameter[1], and the weld width varies with the power and speed even though the beam spot diameter remains the same[2,3]. The purpose of this paper is to introduce the concept of critical thermal radius as a means of predicting thermal performance.

The critical thermal radius is the radius of an area on the workpiece within which the desired thermal effect is achieved. It may represent the radius of the melt area in laser soldering, the nugget diameter or weld width in laser welding, the width of the hardened case in laser heat treating, or the diameter of a drilled hole in laser drilling, as appropriate.

One example of the determination and use of c-radius is presented in this paper, namely the solder melt radius in laser soldering. A simple thermal model of one-dimensional heating led to an expression for the required heat input to melt solder. This heat input was matched with the local intensity of the incoming beam to obtain the c-radius expression. Several types of beam profiles were analyzed for corresponding c-radius equations.

Experiments were performed with a 375W CO_2 laser whose output beam was expected to be near (95%) Gaussian. The experimental results were found to be in reasonable agreement with the analytical predictions of melt radius calculated from the c-radius expression for a Gaussian beam.

ANALYTICAL MODEL

Specimen Configuration

The specimen is an electronic module with approximately 0.006" (0.015cm) thick solder pad (see Figure 1) on a very thin layer of Pd-Ag conductor which was printed on a 0.022" (0.056cm) thick alumina substrate. The alumina substrate is cemented to a 0.060" (0.152cm) thick aluminum base plate. Figure 2 shows a cross-section of the specimen through a solder pad.

Thermal Analysis

When a laser beam impinges on a metallic absorbing surface, the laser light is absorbed by interaction with electrons. A quantum of optical energy is absorbed by an electron which, in turn, dissipates its energy by colliding with lattice phonons and other electrons. Since the mean free time between the collisions of electrons is of the order of 10^{-12} or 10^{-13} sec.[4,5], one may consider that the optical energy is instantaneously turned into heat in a typical laser thermal processing operation, where the process time is orders of magnitude longer than piconseconds.

The penetration depth of the light wave in a good conductor lies in the range of 10^{-6}cm [5,6] or 10 to 100 atomic layers. Therefore, the laser absorption is considered to occur at the surface in cases where the processing depth is over 10^{-5}cm range.

Figure 1. (Photograph showing solder pad, Pd-Ag conductor, alumina substrate, and aluminum base plate)

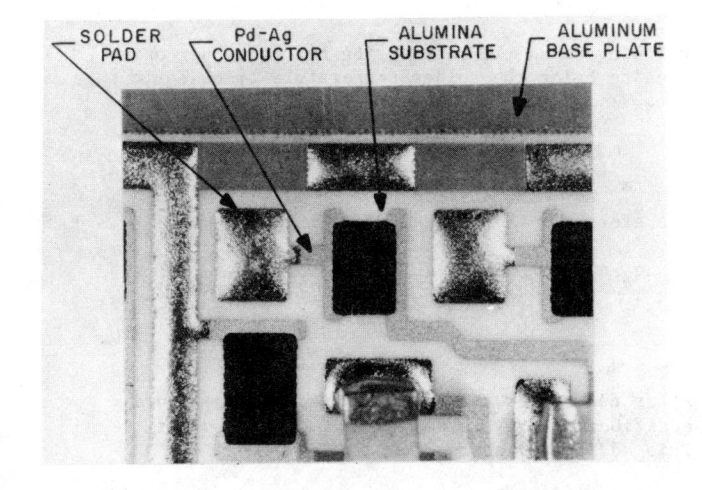

With the above assumptions, the system in Figure 2 may be represented as shown in Figure 3.

The system shown in Figure 3 yields a second order differential equation for T_1. The analysis can be further simplified by proportioning the heat flow into two components, accumulated heat and conducted heat. The input heat flux q is partially accumulated in the solder pad and the alumina substrate and the rest is conducted away to a heat sink. (See Figure 4.)

In Figure 4, T_2 is the average temperature of the alumina substrate through the thickness. If one assumes a parabolic temperature profile within the alumina substrate (Figure 5), the average temperature of the substrate, T_2, is given by,

$$T_2 = \frac{T_1}{3} \quad \text{(See Appendix 1)} \tag{1}$$

(For symbol definitions, see Nomenclature)

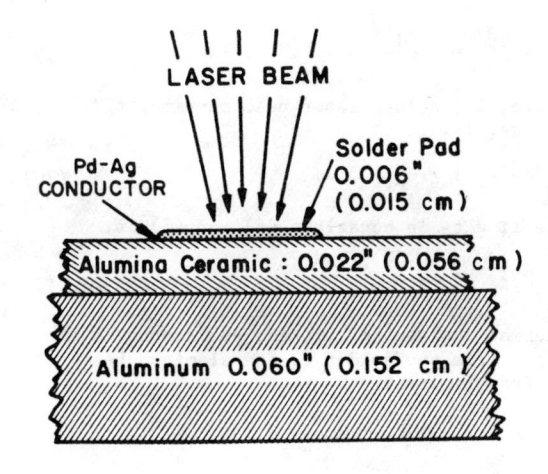

Figure 2. Cross-section of the Specimen for Thermal Analysis

The power densities required for laser soldering is relatively low compared to those for deep penetration welding (requiring the keyhole effect). For soldering the beam is typically defocused to a relatively large spot area. It is, therefore, assumed that the primary mode of heat transfer in laser soldering is conduction.

Additional assumptions adopted to simplify the heat flow analysis of the system (Figure 2) are:

- Thermal properties of the materials and the thermal resistances do not change with temperature;
- The convection and the radiation losses are negligible;
- The heat flow is one-dimensional;
- The solder pad is at a uniform temperature throughout the thickness; and
- The aluminum base is the heat sink.

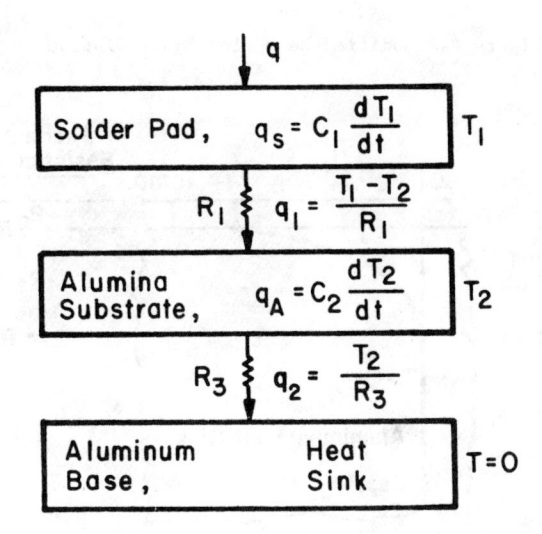

Figure 3. Heat Flow Block Diagram

The energy balance of the system in Figure 4 yields a first order differential equation (2).

$$q = q_s + q_A + q_3$$

$$= C_1 \frac{dT_1}{dt} + C_2 \frac{dT_2}{dt} + \frac{T_1}{R}$$

$$= \left(C_1 + \frac{C_2}{3}\right) \frac{dT_1}{dt} + \frac{T_1}{R} \tag{2}$$

where $C_1 = c_1 \rho_1 V_1$, $C_2 = c_2 \rho_2 V_2$

(For symbol definitions see Nomenclature and Figure 4.)

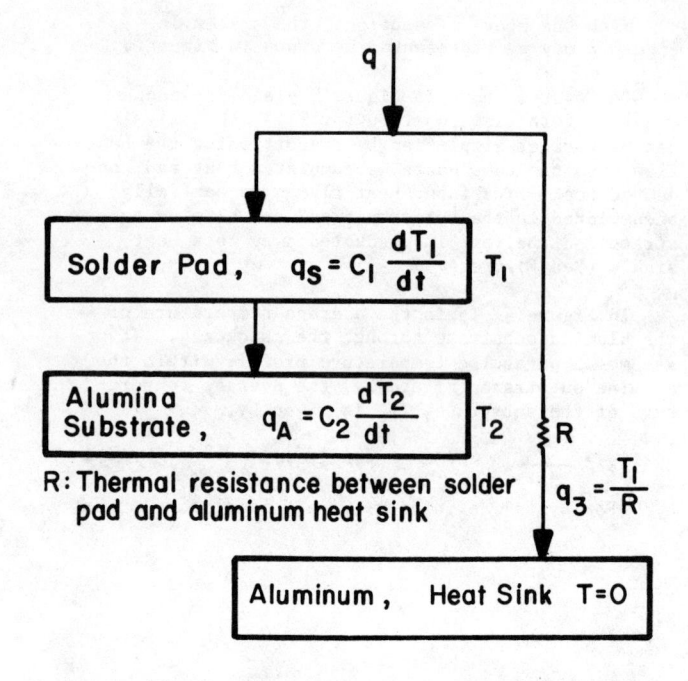

Figure 4. Modified Heat Flow Block Diagram

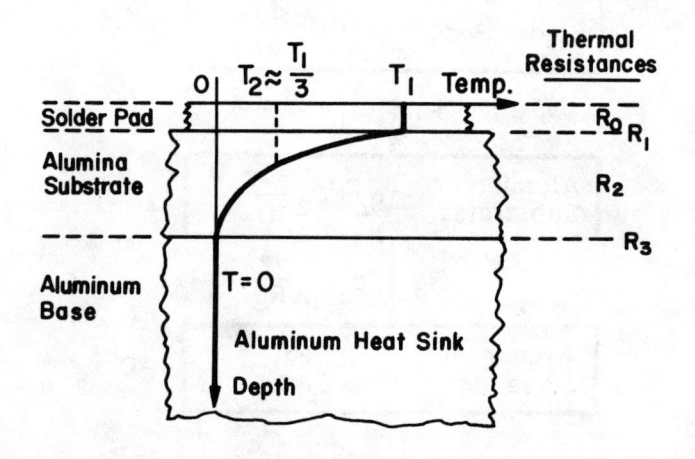

Figure 5. Simplified Temperature Profile Along a Cross-section of the Alumina Specimen

The solution of the differential equation (2) with the initial conditions ($T_1 = 0$ when $t = 0$) is,

$$T_1 = R \times q \left[1 - \exp\left(-\frac{t}{RC}\right) \right] \quad (3)$$

where
$$C = C_1 + \frac{C_2}{3} \quad (4)$$

The time constant, τ, for the solder temperature to reach 63.2% ($= 1 - e^{-1}$) of the steady-state temperature is,

$$\tau = RC \quad (5)$$

Defining q_c as the rate of the heat flow per unit area to bring the solder temperature T_1 to T_m in t_c sec,

$$T_m = R \times q_c \left[1 - \exp\left(-\frac{t_c}{RC}\right) \right] \quad (6)$$

Therefore,

$$q_c = \frac{T_m}{R \left[1 - \exp\left(-\frac{t_c}{RC}\right) \right]} \quad (7)$$

After the solder reaches the melting temperature, additional energy is needed to melt the solder. This additional energy, ΔH, is

$$\Delta H = L (\rho_1 \Delta V) \quad (8)$$

where L = latent heat of fusion for solder
ρ_1 = density of solder
ΔV = small volume of melted solder

If we define Δt to be the time to supply ΔH at a rate of q_c

$$\Delta t = \frac{\Delta H}{q_c} = \frac{L\rho_1 \Delta V}{q_c} \quad (9)$$

Therefore, the actual laser beam on-time, t_c', to melt solder is

$$t_c' = t_c + \Delta t \quad (10)$$

and the time t_c in equations (6) and (7) is,

$$t_c = t_c' - \frac{L\rho_1 \Delta V}{q_c} \quad (11)$$

The thermal resistance R is the resistance to the heat flow from the solder to the aluminum base heat sink. From Figure 5,

$$R = R_0 + R_1 + R_2 + R_3$$

$$= \frac{d_1}{k_1} + R_1 + \frac{d_2}{k_2} + R_3 \quad (12)$$

The time, t_o, for the applied heat flux to reach the alumina-aluminum interface may be approximated by[7]

$$t_o = \frac{d_1^2}{6\alpha_1} + \frac{d_2^2}{6\alpha_2} \quad (13)$$

Footnote:
If the impinging laser beam intensity is not uniform, the heat flux q will not be uniform. In this case, the foregoing analysis can be applied to a small segment of the solder pad where q is considered uniform. The thermal behavior of the entire solder pad, then, is analyzed by considering the solder pad as a bundle of small segments packed in parallel. This argument is valid based on the one-dimensional heat flow assumption.

where d_1 = thickness of the solder pad

d_2 = thickness of the alumina layer

α_1 = thermal diffusivity of the solder ($= \frac{k_1}{c_1 \rho_1}$)

α_2 = thermal diffusivity of the alumina ($= \frac{k_2}{c_2 \rho_2}$)

Therefore, the foregoing temperature equations, including equation (7), are expected to be valid for times greater than t_0.

Spatial Beam Intensity Distribution Analysis of a Gaussian Beam

Many of material processing lasers, especially many CO_2 lasers, are claimed to have a Gaussian (TEM$_{00}$ mode) beam profile[8]. It appears appropriate here to review some of the characteristics of a Gaussian beam.

The radius of a Gaussian beam is typically defined as the radius at which the intensity falls to $1/e^2$ of its peak value[9]. (86.5% of the total beam energy is contained within the area of this radius... see Appendix 2.) The intensity profile of a Gaussian beam (TEM$_{00}$ mode) is given by (see Figure 6)

$$p(r) = p_p \exp\left[-2\left(\frac{r}{a}\right)^2\right] \quad (14)$$

where $p(r)$ = beam intensity at the radius r
  p_p = peak beam intensity
  a = Gaussian radius at which the intensity falls to $(1/e^2)$ of the peak intensity

The peak intensity is given by (see Appendix 2),

$$p_p = \frac{2P}{\pi a^2} \quad (15)$$

where P is the total power of the Gaussian beam.

Therefore,

$$p(r) = \frac{2P}{\pi a^2} \exp\left[-2\left(\frac{r}{a}\right)^2\right] \quad (16)$$

Some of the characteristics of a Gaussian beam are illustrated in Figure 6. Characteristics of other types of beams are shown in Appendix 3.

Critical Thermal Radius (c-radius) for a Gaussian Beam

When a Gaussian beam melts a spot of radius c on the solder pad (as shown in Figure 7), the beam intensity p_c at the edge of the melt spot (r=c) is, (from equation (16))

$$p_c = \frac{2P}{\pi a^2} \exp\left[-2\left(\frac{c}{a}\right)^2\right] \quad (17)$$

This radius c is selected as the "critical thermal radius" since it defines the area where the desired thermal effect (i.e., melting) is achieved. This radius is then,

$$c = \frac{a}{\sqrt{2}} \left[\ln \frac{2P}{\pi a^2} - \ln p_c\right]^{1/2} \quad (18)$$

The heat flux q_c in equation (7) is related to the beam intensity p_c by the surface absorptivity* of the solder, A.

$$q_c = A \times p_c \quad (19)$$

The c-radius (critical thermal radius) is now, (from equations (7), (18) and (19))

$$c = \frac{a}{\sqrt{2}} \left[\ln \frac{2P}{\pi a^2} - \ln \frac{T_m}{A \times R \left\{1 - \exp\left(-\frac{t_c}{RC}\right)\right\}}\right]^{1/2}$$

---
* Surface absorptivity as defined in this paper is the ratio of the absorbed flux to the incident flux. Alternative terminology is "absorbed fraction." The value is (1 - reflectance).

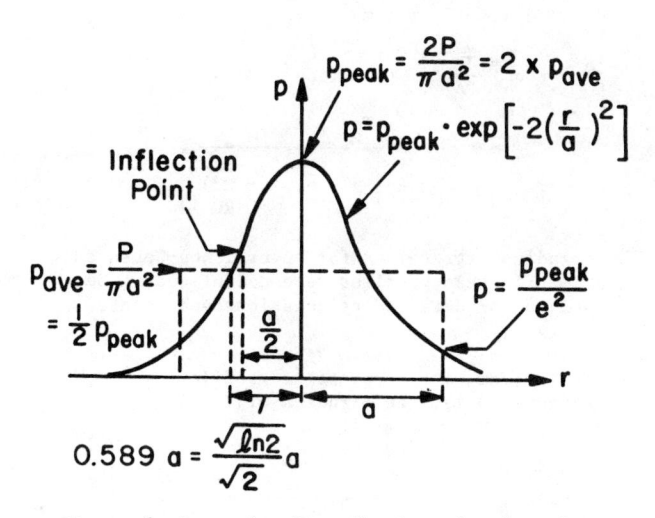

Figure 6. Intensity Distribution of a Gaussian (TEM$_{00}$) Beam (Re. Appendix 2)

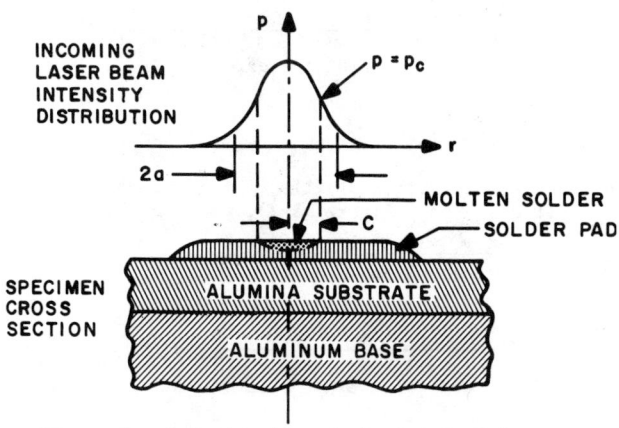

Figure 7. Relationship of the Critical Beam Intensity P_c and the Critical Thermal Radius c.

$$= \frac{a}{\sqrt{2}} \left[\ln \frac{2P \times A \times R \{1 - \exp(-t_c/RC)\}}{\pi a^2 \times T_m} \right]^{1/2} \quad (20)$$

$$\text{if } P > \frac{\pi a^2}{2} \times \frac{T_m}{A R \{1 - \exp(-\frac{t_c}{RC})\}}$$

$c = 0$, otherwise.

c-Radius for Other Axisymmetric Beams

If the intensity, p, of an axisymmetric beam is expressed as a function of radius, r, power, P, and some characteristic radius, m, then the radius may be expressed in terms of P, p and m.

$$r = f(P, p, m) \quad (21)$$

At the radius $r = c$, the intensity p is p_c by definition. The c-radius can then be expressed as,

$$c = f(P, p_c, m) \quad (22)$$

where $p_c = \frac{q_c}{A}$ (equation (19))

$$= \frac{T_m}{A \times R \left[1 - \exp(-\frac{t_c}{RC})\right]}$$

The c-radius expressions for several non-Gaussian axisymmetric beam profiles were calculated. (See Appendix 3 for detailed calculations and related figures.)

o For a beam with a parabolic profile (Figure A3-1), the c-radius will be,

$$c = m \left[1 - \frac{\pi m^2}{2P} \times \frac{T_m}{A R \left[1 - \exp(-\frac{t_c}{RC})\right]} \right]^{1/2} \quad (23)$$

$$\text{if } P > \frac{\pi m^2}{2} \frac{T_m}{A R \left[1 - \exp(-\frac{t_c}{RC})\right]}$$

$c = 0$, otherwise

o For a beam with a profile of frustum of right circular cone (Figure A3-2), the c-radius is,

$$c = m - \frac{\pi(m^2 + mn + n^2)(m-n)}{3P} \times \frac{T_m}{A R \left[1 - \exp(-\frac{t_c}{RC})\right]} \quad (24)$$

$$\text{if } P > \frac{\pi}{3} \frac{(m^2 + mn + n^2) T_m}{A R \left[1 - \exp(-\frac{t_c}{RC})\right]}$$

$c = 0$, otherwise

o For a beam with a uniform profile (Figure A3-3), the c-radius is,

$$c = m, \text{ if } P > \frac{\pi m^2 T_m}{A R \left[1 - \exp(-\frac{t_c}{RC})\right]} \quad (25)$$

$c = 0$, otherwise

o For a beam with a triangular profile (Figure A3-4), the c-radius is,

$$c = m - \frac{\pi m^3}{3P} \frac{T_m}{A R \left[1 - \exp(-\frac{t_c}{RC})\right]} \quad (26)$$

$$\text{if } P > \frac{\pi m^2}{3} \frac{T_m}{A R \left[1 - \exp(-\frac{t_c}{RC})\right]}$$

$c = 0$, otherwise

Minimum Power to Melt Solder

The minimum power, P_{min}, to melt solder at a given time, t_1, can be obtained by setting $c = 0$ in the c-radius equations. With a Gaussian beam of $1/e^2$ radius, a, the P_{min} becomes,

$$P_{min}(t = t_1) = \frac{\pi a^2}{2} \times \frac{T_m}{A R \{1 - \exp(-\frac{t_1}{RC})\}} \quad (27)$$

The minimum power to melt solder when $t \to \infty$ is then,

$$P_{min}(t \to \infty) = \frac{\pi a^2}{2} \times \frac{T_m}{AR} \quad (28)$$

EXPERIMENTAL PROCEDURES

A CO_2 laser (Photon Sources Model 300) was used to melt the solder pad on an alumina substrate (see Figure 1 and Figure 2).

The radius of the melt area was measured to compare with the theoretical prediction. According to the manufacturer's specifications, the mode structure of the output laser beam is 95% in TEM_{00}[10] up to the rated power of 375W. No accurate measurement was made to determine the actual mode structure, but a plexiglas burn pattern was made to check the approximate beam shape. It was assumed that the beam used in the experiment was essentially Gaussian based on the manufacturer's specifications and the plexiglas burn pattern.

RESULTS AND DISCUSSION

Comparison of Analytical c-radius with Experimental Results

The c-radii were measured experimentally for comparison with the calculated values. Two different techniques were used to measure the melt radii (c-radii).

One technique was to measure the diameter of the melt area on a solder pad after predetermined laser exposure. The molten solder area showed a finer surface texture than undisturbed surface as shown in Figure 8. Excessive power or extended beam on-time resulted in substrate damage as shown in Figure 9.

The other technique was to measure the diameter of a crater formed by capillary action of the molten solder. By placing a brass lead on the solder pad, the molten solder was pulled into the lead/solder pad interface leaving a well-defined melt boundary (see Figure 10).

The measured solder melt radii and the calculated c-radii were found to be in good agreement. Since the beam used in the experiment was considered essentially Gaussian, the c-radius equation for a Gaussian beam (equation (20)) was used in the calculation.

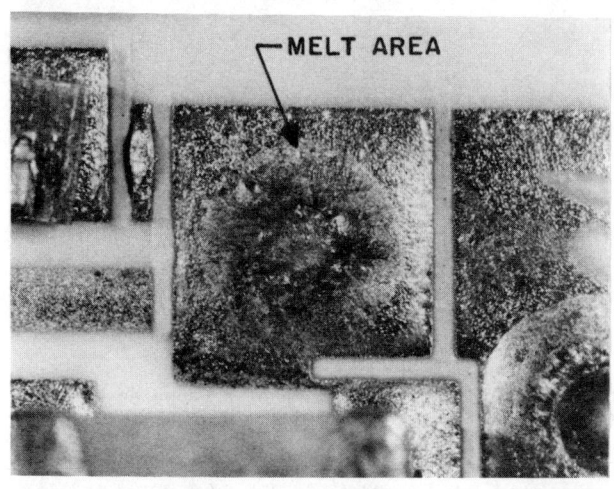

Figure 8. Melt Area on Solder Pad Without Lead (20X)

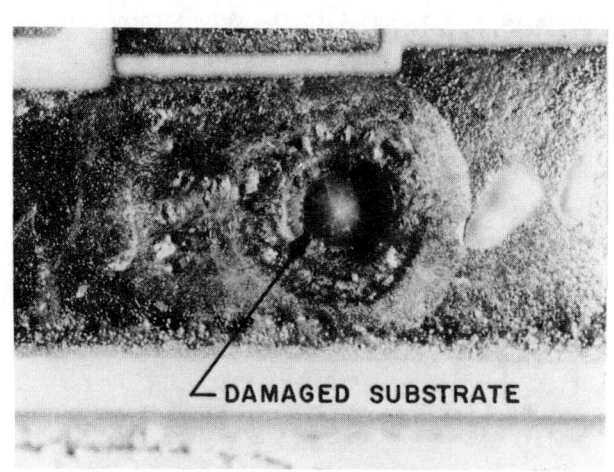

Figure 9. Melted and Damaged Substrate (20X)

Figure 10. Melt Area on Solder Pad With Lead (20X)

The time, t_o, (equation (13)) for the heat flux to reach the heat sink (aluminum base) is calculated to be 0.014 sec. Therefore, all of the beam time in the experiments are selected to be longer than 0.014 sec. The small volume of melted solder ΔV, for practical measurement of melt radii was assumed to be 0.0025 cm^3 (i.e., 0.001" thick layer of melt on a unit area 1cm^2) and the calculated value of $(L\rho_1 \Delta V)/q_c$ was approximately 0.025 t_c'. Therefore,

$$t_c = 0.975 \, t_c' \qquad (29)$$

The time constant, RC in equation (3) gives an idea of the time for the solder pad to reach a steady-state temperature where the rate of heat input is equal to the rate of heat conducted away. The calculated value of τ (= RC) is 0.069 sec ($R = 3.5°C$ sec cm^2/cal, $C = C_1 + C_2/3 = 0.01964$ cal/$°C$ cm^2). At $t_c = \tau$, the solder temperature is expected to be 63.2% of the steady-state value.

The same time constant, $\tau = RC$, is seen in equation (20). Therefore, it is expected that the c-radius will reach approximately 98.2% of the steady-state value at $t_c = 4\tau = 4 \times 0.069 = 0.276$ sec. (The corresponding beam on-time, t_c', is then 0.283 sec.)

Minimum Power to Melt Solder

The minimum power to melt solder for a long beam on-time ($t \to \infty$) is calculated to be 23.3 watts (equation (28)). Since the power level of the particular laser used could not be scaled down to this level, a pulsing technique was used to lower the effective power. Approximately 35 ~ 40W of effective power was needed to melt the solder when the beam on-time was 2.5 sec.

SUMMARY

The concept of the critical thermal radius was introduced in this paper. The critical thermal radius (c-radius) is the radius (or width in case of a moving heat source) of an area where the desired thermal effect is accomplished.

An example of the c-radius was presented. The particular c-radius selected was the melt radius on a solder pad in laser soldering. A thermal analysis of the workpiece and the spatial distribution analysis of the beam led to an expression of the c-radius in terms of characteristic beam spot diameter, beam power, surface absorptivity, thermal resistances, specific heats and densities of materials, laser beam on-time, melting point of the solder and the specimen temperature. A simple thermal model of one-dimensional heating was assumed in the analysis. Several specific c-radius expressions for different beam profiles were obtained.

The thermal analysis was performed on the lumped subsystems rather than on the continua of the multi-layer specimen. The lumped approach is justified because the purpose of the thermal analysis was to determine the thermal outcome of the solder pad, not the temperature profile within the continua of the composite specimen. The lumped approach greatly simplified the analysis and has been proven effective and useful.

Experiments were performed to measure melt radii on solder pads using a 375W CO_2 laser. The beam mode was assumed to be TEM_{oo} as specified by the manufacturer. The measured melt radii were compared with analytical predictions of melt radius calculated from the c-radius expression for a Gaussian beam. They were found to be in reasonable agreement.

Prediction of effective soldering parameters was possible with the use of the c-radius expression. Also the analysis for c-radius was useful in understanding some of the observed thermal behaviors displayed in the particular laser soldering arrangement.

ACKNOWLEDGMENTS

The author is grateful to Drs. W. J. Evans, J. W. Grant, B. W. Schumacher and S. A. Weiner, and Messrs. H. T. Johnson and B. T. Bajorek, all with E&R Staff, Ford Motor Co., for helpful review and suggestions in preparing the manuscript.

Appendix 1: Temperature Distribution in Alumina Substrate

The temperature distribution, T(z,t), along the passage of heat conduction in a semi-infinite solid with a constant heat flux q can be represented by[11],

$$T(z,t) = \frac{q}{k} \int_z^\infty \text{erfc} \frac{z}{2\sqrt{\alpha t}}$$

$$= \frac{2q\sqrt{\alpha t}}{k} \text{ierfc} \frac{z}{2\sqrt{\alpha t}}$$

$$= \frac{2q}{k} \left[\left(\frac{\alpha t}{\pi}\right)^{1/2} \exp\left(-\frac{z^2}{4\alpha t}\right) - \frac{z}{2} \text{erfc} \frac{z}{2\sqrt{\alpha t}} \right]$$

where

q = heat flux
k = thermal conductivity
z = depth
α = thermal diffusivity

Applying the above equation, the average temperature of the alumina substrate in the thickness direction, T_2, may be calculated. (See Figure 5 for the cross-section of the specimen.)

$$T_2 = \frac{1}{d_2} \int_o^{d_2} \frac{2q_1 \sqrt{\alpha_2 t}}{k_2} \text{ierfc} \frac{z}{2\sqrt{\alpha_2 t}} dz$$

(For symbol definitions, see Nomenclature)

To simplify the calculation of the average alumina substrate temperature, T_2, a parabolic temperature profile may be assumed (see Figure 5). This approximation predicts the temperature with an error of approximately +8%[7] from the exact value. In this calculation, it is also assumed that the effect of R_1 and R_3 cancels out each other in the calculation of T_2.

$$T(z) = T_1 \left[\frac{d_2 - z}{d_2} \right]^2 \qquad (A1-1)$$

The average temperature of the alumina substrate, T_2, is then

$$T_2 = \frac{1}{d_2} \int_o^{d_2} T_1 \left[\frac{d_2 - z}{d_2} \right]^2 dz = \frac{T_1}{3} \qquad (A1-2)$$

Appendix 2: Beam Intensity Distribution (General Information)

The intensity distribution of a Gaussian beam with spot diameter (@ $1/e^2$) 2a is given by,

$$p(r) = p_p \exp\left[-2\left(\frac{r}{a}\right)^2\right] \qquad (A2-1)$$

The beam power P is equal to the volume formed by rotating this normal curve (see Fig. A-1).

$$P = \int_o^\infty p(r) \cdot 2\pi r \, dr \qquad (A2-2)$$

Let $x = -2\left(\frac{r}{a}\right)^2$, then $dx = -\frac{4}{a^2} r \, dr$.

Substituting the above relationships into equation (A2-2) yields,

$$P = \int_o^\infty p_p \exp\left[-2\left(\frac{r}{a}\right)^2\right] 2\pi r \, dr$$

$$= \int_0^{-\infty} p_p \exp[x] \, 2\pi \left(-\frac{a^2}{4}\right) dx$$

$$= \frac{\pi a^2}{2} p_p \left[e^x\right]_{-\infty}^{0} = \frac{\pi a^2}{2} p_p$$

Therefore,

$$p_{peak} = p_p = \frac{2P}{\pi a^2} \quad (A2\text{-}3)$$

Therefore,

$$p(r) = \frac{2P}{\pi a^2} \exp\left[-2\left(\frac{r}{a}\right)^2\right] \quad (A2\text{-}4)$$

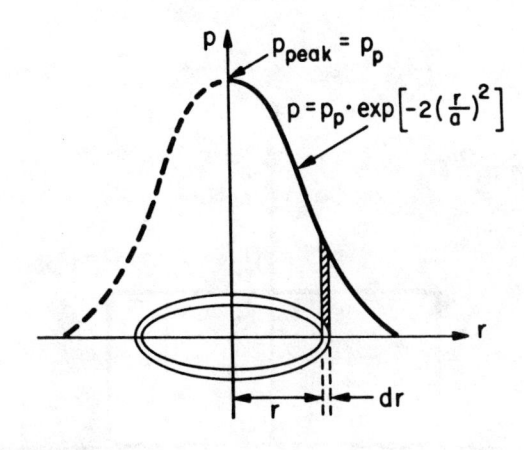

Figure A-1. Integration of a Gaussian Beam Profile

From equation (A2-3), one can see that the peak intensity (or peak power density) is twice the arithmetic average intensdity ($P/\pi a^2$), which is commonly used in the laser material processing community.

$$P_{peak} = 2 \times \frac{P}{\pi a^2} = 2 \times \text{arithmetic average power density}$$

By substituting $p = P/\pi a^2$ into equation (A2-4), it can be seen that the average intensity line intersects the Gaussian profile at

$$r = \sqrt{\frac{\ln 2}{2}}\, a = 0.589\, a. \text{ (see Figure 6)}$$

The power contained in the area of Gaussian radius, a, is

$$\int_0^a p(r) \, 2\pi r \, dr = \frac{\pi a^2}{2} p_p \left[e^x\right]_{-2}^{0}$$

$$= \frac{\pi a^2}{2} \cdot \frac{2P}{\pi a^2} \left[1 - e^{-2}\right]$$

$$= P\left[1 - e^{-2}\right]$$

$$= 0.8647\, P. \text{ (approximately 86\% of the total power)}$$

From the characteristics of a normal curve, the inflection point occurs at $r = a/2$. Some of the characteristics of a Gaussian beam are illustrated in Figure 6.

Sometimes, the spot diameter is defined as that of a circle where the intensity falls to $1/e$ of the peak value.

$$p(r) = p_p \frac{1}{e} = p_p \exp\left[-2\left(\frac{r}{a}\right)^2\right]$$

Then, $\quad r = \dfrac{a}{\sqrt{2}}$

Therefore, the $\dfrac{1}{e}$ radius is $\dfrac{1}{\sqrt{2}}$ (70.7%) of the $\dfrac{1}{e^2}$ radius.

Appendix 3: c-Radii for Axisymmetric Beams Other Than Gaussian

Parabolic Profile

If the beam profile is parabolic as shown in Figure A3-1, the intensity p at a radius r is,

$$p = p(r) = p_p \left(1 - \frac{r^2}{m^2}\right) = \frac{2P}{\pi m^2}\left(1 - \frac{r^2}{m^2}\right) \quad (A3\text{-}1)$$

Then

$$r = m\left(1 - \frac{p}{p_p}\right)^{1/2} = m\left(1 - \frac{\pi m^2}{2P} p\right)^{1/2} \quad (A3\text{-}2)$$

By definition, $r = c$ when $p = p_c$.

Since $p_c = \dfrac{q_c}{A} = \dfrac{T_m}{A R \left[1 - \exp\left(-\dfrac{t_c}{RC}\right)\right]}$ (see eq. 19 & 7)

$$c = m \left[1 - \frac{\pi m^2}{2P} \cdot \frac{T_m}{A R \left\{1 - \exp\left(-\dfrac{t_c}{RC}\right)\right\}}\right]^{1/2} \quad (A3\text{-}3)$$

if $P \geqslant \dfrac{\pi m^2}{2} \cdot \dfrac{T_m}{A R \left\{1 - \exp\left(-\dfrac{t_c}{RC}\right)\right\}}$

$c = 0$, otherwise.

$$P = \frac{1}{2}\pi m^2 p_p$$

$$p(r) = p_p\left(1 - \frac{r^2}{m^2}\right)$$

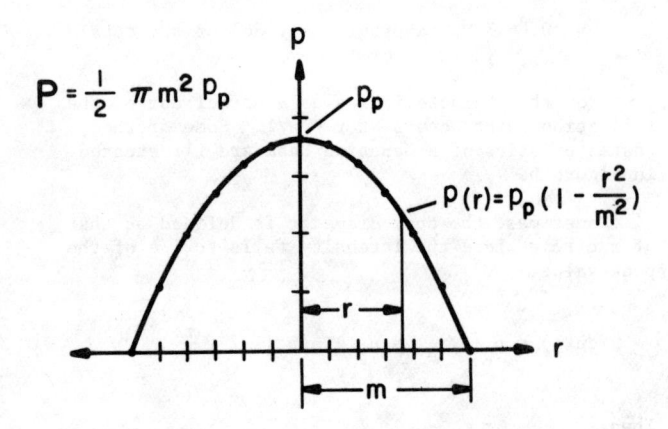

Figure A3-1. A Parabolic Beam Profile

$$P = \frac{\pi p_p}{3}(m^2 + mn + n^2)$$

$$p(r) = p_p \frac{m-r}{m-n}$$

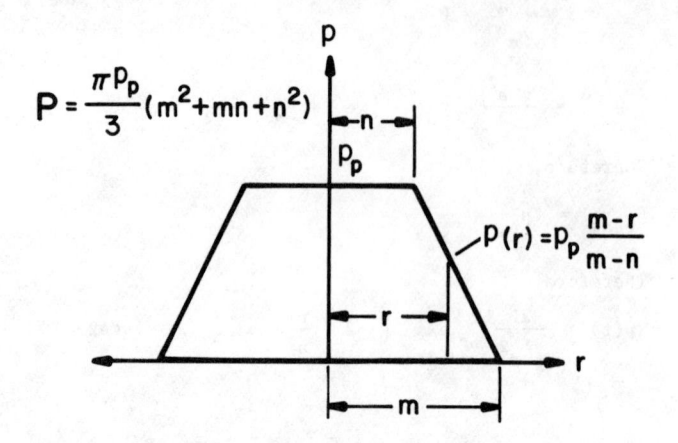

Figure A3-2. Frustum Beam Profile

Beam Profile of Frustum of Right Circular Cone

The intensity, p, at a radius, r, of the beam shown in Figure A3-2 is,

$$p = p(r) = p_p = \frac{3P}{\pi} \frac{m}{(m^2 + mn + n^2)} \quad \text{when } 0 \leq r \leq n$$

(A3-4)

$$p = p(r) = \frac{3P}{\pi} \frac{m}{(m^2 + mn + n^2)} \frac{(m-r)}{(m-n)} \quad \text{when } n \leq r \leq m$$

For $n \leq r \leq m$

$$r = m - \frac{\pi(m^2 + mn + n^2)(m-n)}{3P} p \quad \text{(A3-5)}$$

Therefore,

$$c = m - \frac{\pi(m^2 + mn + n^2)(m-n)}{3P} \times \frac{T_m}{AR\left\{1 - \exp\left(-\frac{t_c}{RC}\right)\right\}} \quad \text{(A3-6)}$$

if $P > \frac{\pi}{3} \frac{(m^2 + mn + n^2)}{AR\left\{1 - \exp\left(-\frac{t_c}{RC}\right)\right\}} T_m$

$c = 0$, otherwise.

A uniform beam (or top hat beam as commonly called) is a special case of the frustum beam (when $m = n$). Therefore for a uniform beam (Figure A3-3)

$$c = m, \text{ if } P > \frac{\pi m^2 T_m}{AR\left\{1 - \exp\left(-\frac{t_c}{RC}\right)\right\}} \quad \text{(A3-7)}$$

$c = 0$ otherwise.

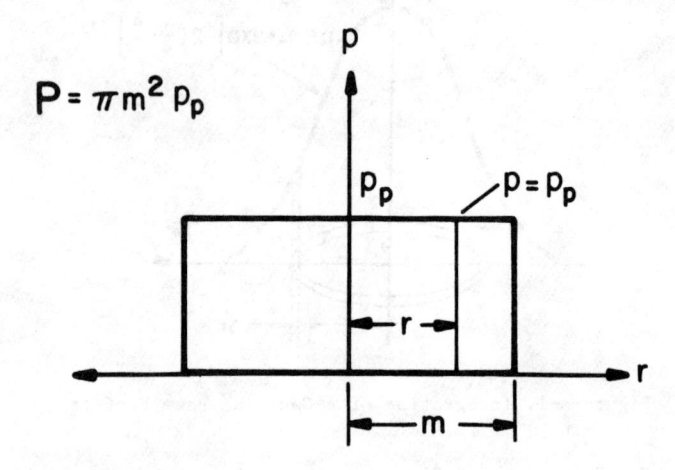

$$P = \pi m^2 p_p$$

$$p = p_p$$

Figure A3-3. Uniform Beam Profile

The frustum beam profile may be used as an approximation of some of higher order multimode beams.

Triangular Beam Profile

For a beam with a triangular profile, (Figure A3-4),

$$p = p(r) = p_p \frac{m-r}{m} = \frac{3P}{\pi m^2} \frac{m-r}{m} \quad \text{(A3-8)}$$

$$r = m\left(1 - \frac{p}{p_p}\right) \quad \text{(A3-9)}$$

Therefore

$$c = m\left[1 - \frac{\pi m^2}{3P} \frac{T_m}{AR\left\{1 - \exp\left(-\frac{t_c}{RC}\right)\right\}}\right] \quad \text{(A3-10)}$$

If $P > \frac{\pi m^2}{3P} \frac{T_m}{A R \left\{ 1 - \exp\left(-\frac{t_c}{RC} \right) \right\}}$

c = 0, otherwise

A triangular beam is a special case (n = o) of a frustum beam.

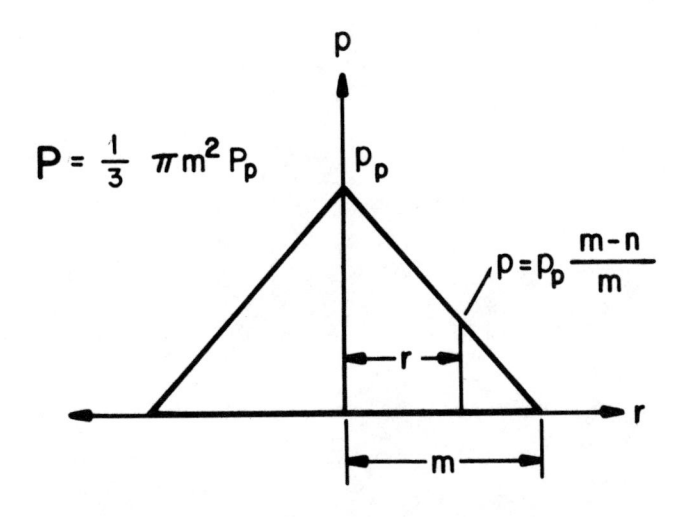

$P = \frac{1}{3} \pi m^2 P_p$

$p = p_p \frac{m-n}{m}$

Figure A3-4. Triangular Beam Profile

REFERENCES

1. U. I. Chang, "Laser Drilling of Pyrex Glass for Pressure Sensor Manufacturing," Proceedings of the 1st International Laser Processing Conference, Nov. 16-17, 1981, Anaheim, CA.

2. S. L. Engel, "A Guide to Laser Metal Welding - To 1500 Watts," Lasers in Modern Industry, Edited by J. F. Ready, Published by SME, 1979.

3. U. I. Chang, K. W. Casey, "Laser Welding of Exhaust Gas Oxygen Sensor," Proceedings of a Conference, "Applications of Lasers in Materials Processing," April 18-20, 1975, Washington D. C., Edited by E. A. Metzbower.

4. J. F. Ready, "Effects of High-Power Laser Radiation," Academic Press, New York, London, 1971.

5. M. vonAllmen, "Coupling of Beam Energy to Solids," the Proceedings of a Symposium on Laser and Electron Beam Processing of Materials, Cambridge, MA, Nov. 27-30, 1979. Sponsored by the Materials Research Society.

6. N. Bloembergen, "Fundamentals of Laser-Solid Interactions," Proceedings of an ASM Conference, Applications of Lasers in Materials Processing April 18-20, 1976, Washington D.C., Edited by E. A. Metzbower.

7. V. S. Arpact, "Conduction Heat Transfer," Addison-Wesley Publishing Co. Menro Park, CA (1966) p.80.

8. Laser Focus Buyers' Guide, January 1982 issue, 17th Edition p.144-148.

9. Laser Institute of America, "Guide for Material Processings by Lasers," Prepared by Laser - Material Processing Committee, 1977, The Paul M. Herrod Co.

10. Photon Sources Technical Summary, "CW CO_2 Laser, 50 to 1000 Watts," No. TS501000/0474.

11. H. S. Carslaw and J. C. Jaeger, "Conduction of Heat in Solids," Oxford Press, 2nd Ed. (1954), p.75.

NOMENCLATURE

A	= Surface absorptivity of solder at 10.6 µm, (1-Reflectance), absorbed fraction.
a	= Gaussian radius at $1/e^2$ point
C	= Heat capacity of the system
C_1, C_2	= Heat capacities of solder and alumina, respectively ($C_1 = c_1\rho_1 V_1$, $C_2 = c_2\rho_2 V_2$)
c	= Critical thermal radius
c_1, c_2	= Specific heats of solder and alumina, respectively
d_1, d_2	= Thicknesses of solder pad and alumina substrate, respectively
k_1, k_2	= Thermal conductivities of solder and alumina, respectively
L	= Latent heat of fusion (for solder)
m, n	= Characteristic beam radius
P	= Laser beam power
P_{min}	= Minimum power to melt solder
p	= Intensity (or power density) of a axisymmetric beam at radius r
p_c	= The critical intensity of a axisymmetric beam
p_p	= The peak intensity of a axisymmetric beam
q	= Rate of heat flow per unit area
q_1	= Rate of heat flow through unit area of solder/alumina interface
q_2	= Rate of heat flow through unit area of alumina/aluminum interface
q_3	= Rate of heat flow per unit area to aluminum heat sink
q_A	= Heat accumulation in unit area of alumina substrate
q_c	= Critical rate of heat flow per unit area to bring the solder temperature to melting point

q_s	=	Heat accumulation in unit area of solder pad
R	=	Thermal resistance per unit area of the system
R_o	=	Thermal resistance per unit area through solder pad
R_1	=	Thermal resistance per unit area of solder/alumina interface
R_2	=	Thermal resistance per unit area through alumina substrate
R_3	=	Thermal resistance per unit area of alumina/aluminum interface
r	=	Radius
T	=	Temperature rise
T_1	=	Average temperature of the solder pad in the thickness direction minus specimen temperature
T_2	=	Average temperature of the alumina substrate in the thickness direction minus specimen temperature
T_m	=	Melting temperature of the solder minus specimen temperature
t	=	Time
t_1	=	A given time
t_c	=	Critical time to bring solder temperature to T_m
t'_c	=	Critical beam on-time to melt solder
t_o	=	Time for the applied heat flux to reach the heat sink
ΔV	=	Small volume of melted solder on unit area (to be visible in the measurement of melt radii)
V_1	=	Volume of solder on unit area
V_2	=	Volume of alumina on unit area
z	=	Depth
α_1, α_2	=	Thermal diffusivities of solder and alumina, respectively
ρ_1, ρ_2	=	Densities of solder and alumina, respectively
τ	=	Time constant

8301-022

LASER HARD-SURFACING OF TURBINE BLADE SHROUD INTERLOCKS

R. M. Macintyre

This paper describes the development, testing and implementation of a novel manufacturing technique for the application of hard-surfacing alloy to a high performance engine component. It demonstrates:-

(i) the effective application of laser technology

(ii) the necessity to adapt design and manufacturing techniques to take advantage of a new process capability

(iii) the integration of improvements in different areas of technology such as materials, process control and electronics to achieve the optimum manufacturing benefit

BACKGROUND

The RB.211 high pressure turbine blade is a shrouded blade manufactured from a cast nickel-based superalloy optimised for creep strength. It operates in the engine at a flame temperature of approximately 1600 K.

The tip shrouds on the blade are designed with interlocking edges to combat the wear that takes place on these edges due to blade vibration. The geometry of the interlock design is intended to place the wear face at the optimum angle and allow its area to be maximised. The wear pad size is nevertheless limited to approximately 4.5mm by 3mm.

Excessive wear of the interlock would lead to blade rejection on overhaul or, if allowed to go undetected, precipitate resonance leading to blade fracture. The wear pad is therefore faced with a cobalt-based hard-surfacing alloy with the objective of extending the wear resistance to at least match the designed creep life of the blade which is targetted at 10,000 hours of service operation.

PRIOR TECHNIQUE

Blades were previously hard-surfaced using a manual tungsten inert gas (T.I.G.) welding process using the alloy in the form of a small diameter wire. The cobalt based alloy is very sensitive to dilution with nickel as this affects the microstructure in the hard-surfaced deposit. This has a deleterious effect on the high temperature wear resistance of the alloy. The optimum level of nickel dilution could not be achieved without a "double-pass" welding technique which involved grinding a weld preparation on the blade shroud, weld depositing the hard-surfacing alloy, partially grinding back this first deposit, applying a second layer of alloy and then finish grinding the two-layer deposit. The sequence is shown diagrammatically in Figure 1. The need to control the thickness of each layer to avoid nickel dilution from the blade material required great dexterity and control by the welder in the manipulation of the hard-surfacing wire and welding torch and great concentration for the several minutes required to produce the welded deposit.

PROBLEMS

Problems were caused in the prior technique by the difficulty of producing hard-surfaced deposits which were fully fused to the base material while keeping to a minimum the amount of the blade material which was melted, in order to achieve the required levels of dilution. Operator fatigue and variations in individual skill led to inconsistency in the finished deposit in terms of both nickel content and deposit hardness. There were variations in the amount of material applied, the deposit sometimes extending beyond the edge of the weld preparation into the inside radius of the shroud interlock which could cause a cracking problem. The blade material being very highly alloyed is susceptible to cracking in the heat-affected zone of any weld. The T.I.G. welding technique with its relatively high total heat input tended to produce such heat-affected zone cracking in the blade material.

It was felt that the laser offered the potential of a higher power density heat source able to fuse hard-surfacing alloy in powder form onto a substrate with a much lower total heat input. Initial work was carried out at Rolls-Royce, Barnoldswick and at Imperial College, London under a C.A.S.E. award using pre-placed powder on flat substrates, with some success. Ref. 1, It was, however, soon recognised that the small size and shape of the interlock would prevent this application technique being used.

THE LASER

The laser used for this work is a fast axial flow carbon dioxide laser with an output power up to a nominal rated 2kW. It produces a parallel beam approximately 20mm in diameter. This is delivered via a 45° mirror to a potassium chloride lens by which means it can be focussed to a diameter as small as 0.3mm giving a power density of the order of $10^4 W/mm^2$, considerably in excess of that available from a T.I.G. welding torch.

LASER HARD-SURFACING TECHNIQUE

As the hard-surfacing powder could not be preplaced on the small pad area it was decided to adopt a blown powder technique. The technique eventually developed is shown diagrammatically in Figure 2. A defocussed beam is used to produce a melt pool between 1mm and 2mm wide, typically 1.3mm for RB.211 blades. The hard-surfacing material, in powder form is blown into the melt pool by a stream of inert gas from a nozzle at one side of the beam. As the powder enters the melt pool it is fused to the base material by the laser beam. The workpiece is traversed under the beam and powder delivery nozzle as shown in the diagram to produce a bead of hard-surfacing alloy. If a rectangular or other shape pad is required then adjacent, overlapping tracks are laid down to cover the area required. A pad of any required thickness can be built up by applying successive layers as shown.

The powder delivery system shown in Figure 3 consists of a small hopper with a metering orifice in its base. Powder feeds from this, by gravity, to a delivery tube where it is carried by a stream of argon gas to the nozzle. The hopper is fitted with a vibrator to ensure an even powder flow. The powder flow rate is controlled primarily by changing the metering orifice and also by varying the gas flow rate through the nozzle.

The various movements required to hard-surface a turbine blade are performed using a five-axis manipulator (X, Y, Z plus two rotational axes), all axes being fitted with stepper motor drives and controlled via a microprocessor. One of the rotational axes serves to turn the blade over so that both interlocks on one blade can be hard-surfaced in a single set-up. Although the control system is open-loop it has an inherent accuracy of $\pm$ 0.02mm.

PROCESS OPTIMISATION

The main process parameters, which were varied during the investigatory work, were beam power, spot size, traverse speed and powder delivery rate. The influence of each of these on the quality of the hard-surfaced deposit was checked. The spot size was kept deliberately small to give a narrow bead width, enabling the required shape of wear pad to be built up very precisely. The beam power and spot size together determine the beam power density on the workpiece and this was kept between the quite close limits required to maintain good fusion of the hard-surfacing alloy to the blade material. Typical power densities used are in the range 10^2 to $10^3 W/mm^2$. The power was therefore limited to less than one kilowatt. Once the optimum power density range had been determined the total heat input for a given deposit using specific parameters was calculated and those parameters selected which gave a minimum heat input without too great a penalty in total processing time. The tolerance band for each of the main parameters was investigated and those parameters selected which enabled the greatest tolerances to be used commensurate with acceptable results.

The finalised technique built up each interlock in four layers, applied alternatively to each side by turning the blade over, thus allowing a few seconds for each layer to cool down while a layer was being applied to the opposite interlock. All the blade movements required to complete both interlocks were contained in a programme in the microprocessor and carried out as a complete sequence in a single hard-surfacing operation, with a cycle time of approximately seventy-five seconds. The operations sequence shown diagrammatically in Figure 1 for comparison, was therefore simplified compared to the prior technique, consisting of the grinding of a weld preparation, hard-surfacing and a finish grind.

EVALUATION AND TESTING

Test pieces in the form of simulations of the interlock geometry or actual blades were subject to extensive laboratory examination. Hard-surfaced deposits were sectioned and examined for excessive porosity, lack of fusion to the base material and for the appropriate metallurgical structure indicating an undiluted deposit. Test pieces and blades were checked for nickel content using energy dispersive analysis on the scanning electron microscope at Rolls-Royce, Barnoldswick. Hammer wear tests were carried out simulating as closely as possible the type of wear which occurred in the engine to produce wear figures for comparison with similar tests utilising the existing technique.

Finally, engine blades were hard-surfaced using the developed technique and subjected to modification approval testing and cyclic endurance testing and directly compared to conventionally hard-surfaced blades in the same engine build.

RESULTS

By careful balancing of parameters and microprocessor programme refinement, deposits were obtained which were free from porosity, fully fused to the blade material and had a nickel content only 1 - 2% higher than the original cobalt-based hard-surfacing alloy, a result only matched by the most rigorous execution of the conventional technique.

The hammer wear tests demonstrated wear rates at the high blade operating temperature equal to the best achieved with this material composition. These results were confirmed by the engine testing in which measured wear on the laser hard-surface blades matched that on the conventionally treated blades. A blade after engine running is illustrated in Figure 4 showing the size and position of the wear surface.

On the basis of this evaluation including pre- and post-engine test laboratory evaluation the process was cleared for civil aero-engine use.

INTRODUCTION TO PRODUCTION

The process development which led to the successful engine testing was carried out using a general purpose workstation which was unnecessarily large for turbine blades. A special purpose workstation was therefore designed solely for the hard-surfacing of turbine blades. The main factors considered in the design were as follows.

A work-handling and blade fixturing system of low inertia to give a long operating life over tens of thousands of similar or identical operations per year. To this end aluminium was used extensively in the design. This did not involve a compromise with strength as laser hard-surfacing, being a non-contact process, imposes no loads on the workpiece or fixture.

Maximum utilisation of the laser beam, achieved by fully automating the operation of the process. With a total processing time of approximately seventy-five seconds, single button initiation of the cycle was incorporated freeing the operator to load a second workstation. With a simple single-clamp fixture, unloading a finished blade and loading a new one can be accomplished comfortably within a seventy-five second cycle. Maximum utilisation could therefore be achieved with a double workstation installation. The production double workstation, with the microprocessor controls is shown in Figure 5. The fixture with its simple clamping, and turbine blade in the process of being hard-surfaced, is shown in Figure 6.

Simple operation was ensured by designing the microprocessor control systems to monitor and perform the maximum possible number of functions. Separate units control each workstation but they are interlinked so that each may monitor the position of the changeover mirror which switches the beam from one station to the other and whether a programme is being run at the other workstation. When the operator loads a blade, closes the safety guard and initiates a cycle the microprocessor will monitor the other station and, if a programme is running, will wait until that cycle is completed before switching the beam to its own workstation and commencing the hard-surfacing operation. The operator can thus leave a workstation having initiated the cycle. Once the cycle is complete the safety guard is opened automatically by the microprocessor thereby indicating to the operator that the blade is ready for unloading.

In-process monitoring is carried out via the microprocessor. In addition to the change-over mirror it monitors laser operation, the hard-surfacing powder supply, carrier gas supply and also overchecks the interlocks on the safety guard and beam path. If any aspect is unsatisfactory the start of a hard-surfacing operation will be inhibited and the problem area indicated to the operator diagrammatically on the V.D.U. incorporated into the microprocessor control.

The microprocessor programme is recorded on a magnetic tape cartridge uniquely identified to the component and data card. The programme is initially entered on a keyboard on the control unit by which it can also be modified or edited. This facility enables programmes to be checked and proved out on the machine. To ensure the security of the programme, once approved, the machine keyboard is inoperative unless unlocked with the appropriate key. In ordinary day-to-day operation only four controls are available which enable the operator to move the blade to a datum point to check the laser beam position, return it from the datum to the normal programme start position, initiate a normal hard-surfacing cycle or run through the programme without opening the laser shutter, for programme checking.

SAVINGS

A direct cost saving arises from the greatly reduced processing time. Compared with the previous 'double-pass' T.I.G. welding technique the laser hard-surfacing operation reduces processing time from fourteen minutes to seventy-five seconds. Despite the higher operating costs of the laser this gives a reduction of 85% of the previous cost.

Due to the greater precision of the process and the elimination of the intermediate grinding operation material savings are considerable. It has been estimated that consumption of the expensive cobalt-based hard-surfacing material is cut by more than 50%.

Gains also accrue from the reduced number of separate operations which simplifies and shortens the manufacturing cycle, reducing inventory costs.

COMPARISON WITH CONVENTIONAL PROCESS

Laser hardsurfacing produces a higher quality hard-surfaced deposit.

Total heat input into the blade is greatly reduced. Indeed it is possible to handle a laser hard-surfaced blade immediately after processing despite raising a localised area to its melting point. This reduced heat input has led to the elimination of the cracking problem.

The accurate control of the power input possible with the laser enables melting of the substrate to be controlled more closely, and minimum dilution of the hardsurfacing alloy to be achieved using a single stage preparation. When using the manual T.I.G. welding process, deposits with minimum dilution and optimum wear performance can only be produced using a double-pass technique with an intermediate grinding operation.

The reproducibility of the power settings, the control of the powder delivery and the precise repeatability of the processing speed and programme movements give a consistency in the finished product which cannot be matched by a manual technique.

The combination of precise control giving an optimised result and the repeatability of programme and settings giving consistency enable the process to be automated. This in turn further assures process reproducibility and enables a high throughput capability to be achieved leading to the cost savings already outlined.

REFERENCES

1. W. M. Steen and C. G. H. Courtney "Hardfacing of Nimonic 75 Using 2kW Continuous Wave CO_2 Laser", Metals Technology, pp 232-237, June 1980.

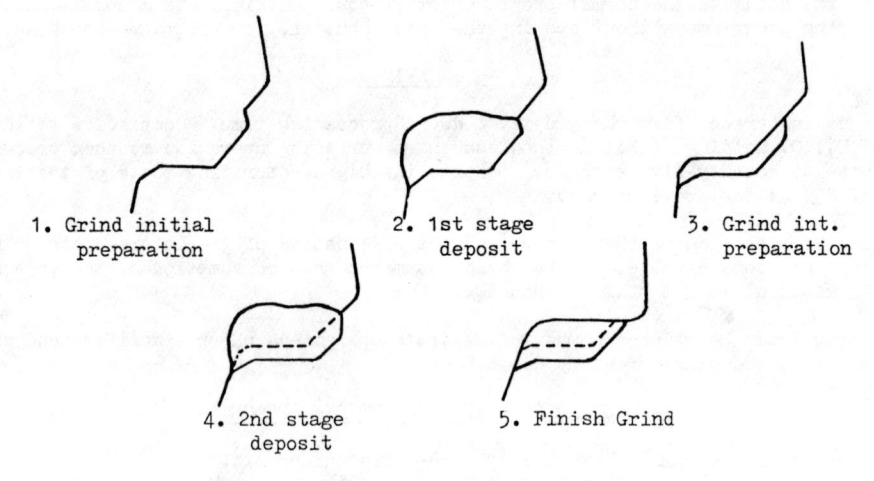

1. Grind initial preparation
2. 1st stage deposit
3. Grind int. preparation
4. 2nd stage deposit
5. Finish Grind

T.I.G. HARDSURFACING SEQUENCE OF OPERATIONS

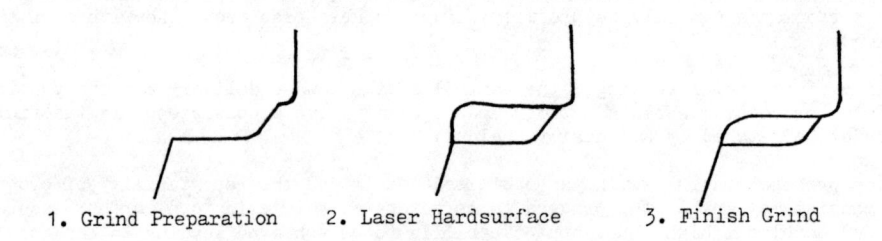

1. Grind Preparation
2. Laser Hardsurface
3. Finish Grind

LASER HARDSURFACING SEQUENCE OF OPERATIONS

FIGURE 1

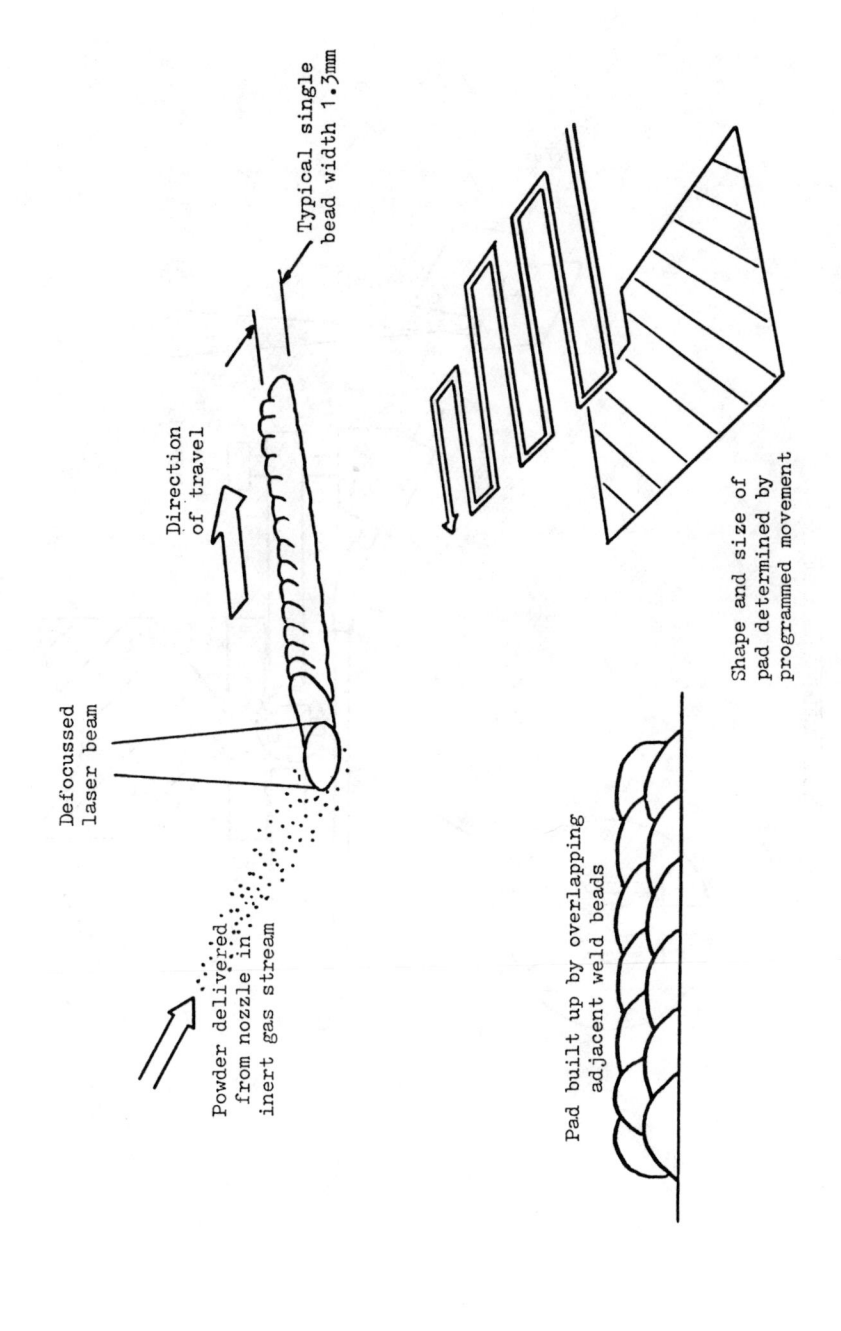

METHOD OF APPLYING SINGLE BEAD OF HARDSURFACING MATERIAL
AND BUILD UP OF LARGER PAD

Figure 2

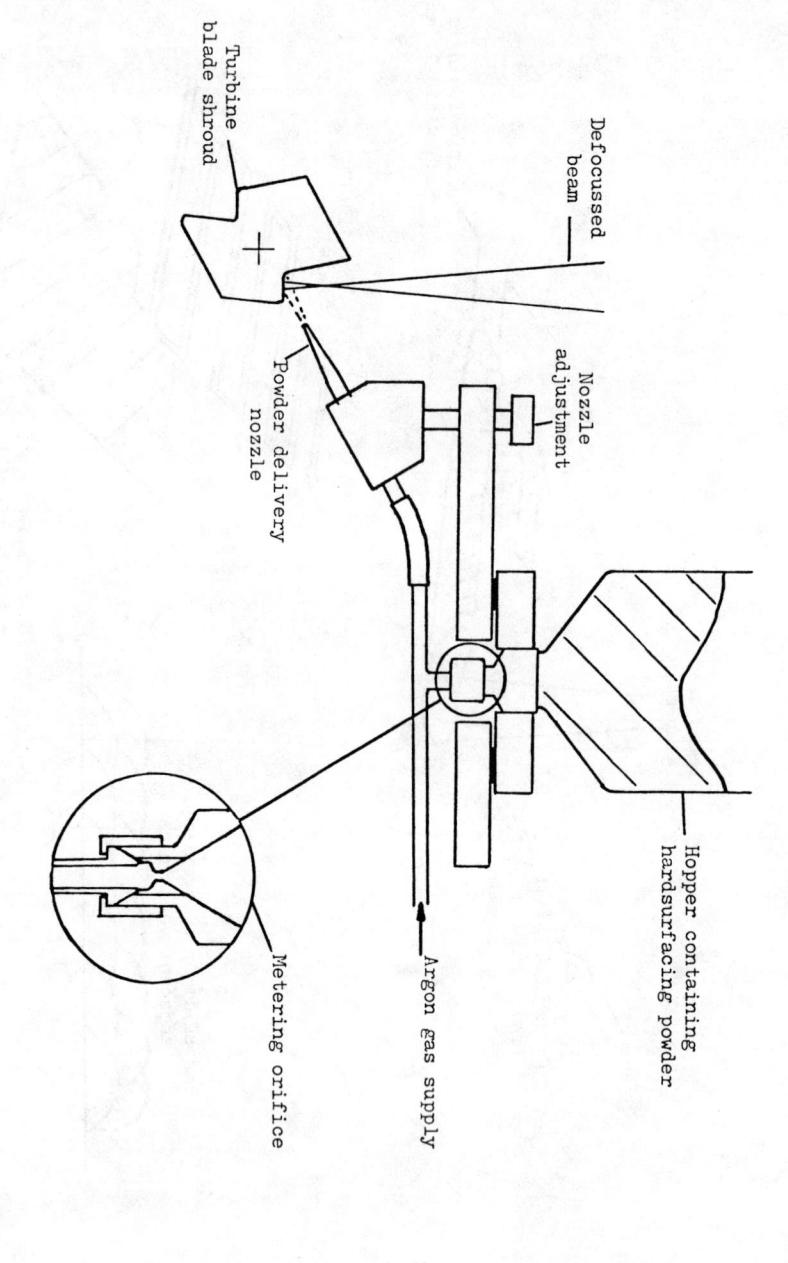

POWDER DELIVERY SYSTEM

Figure 3

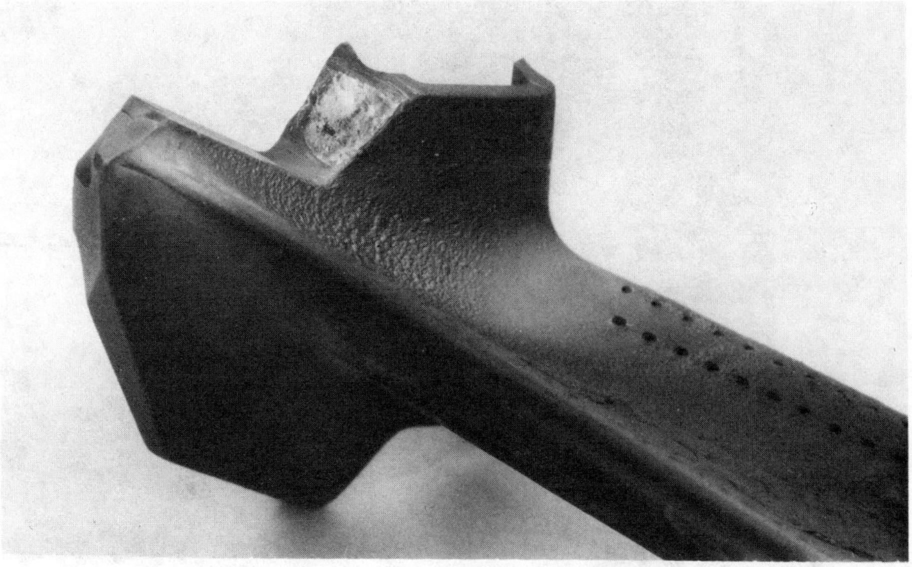

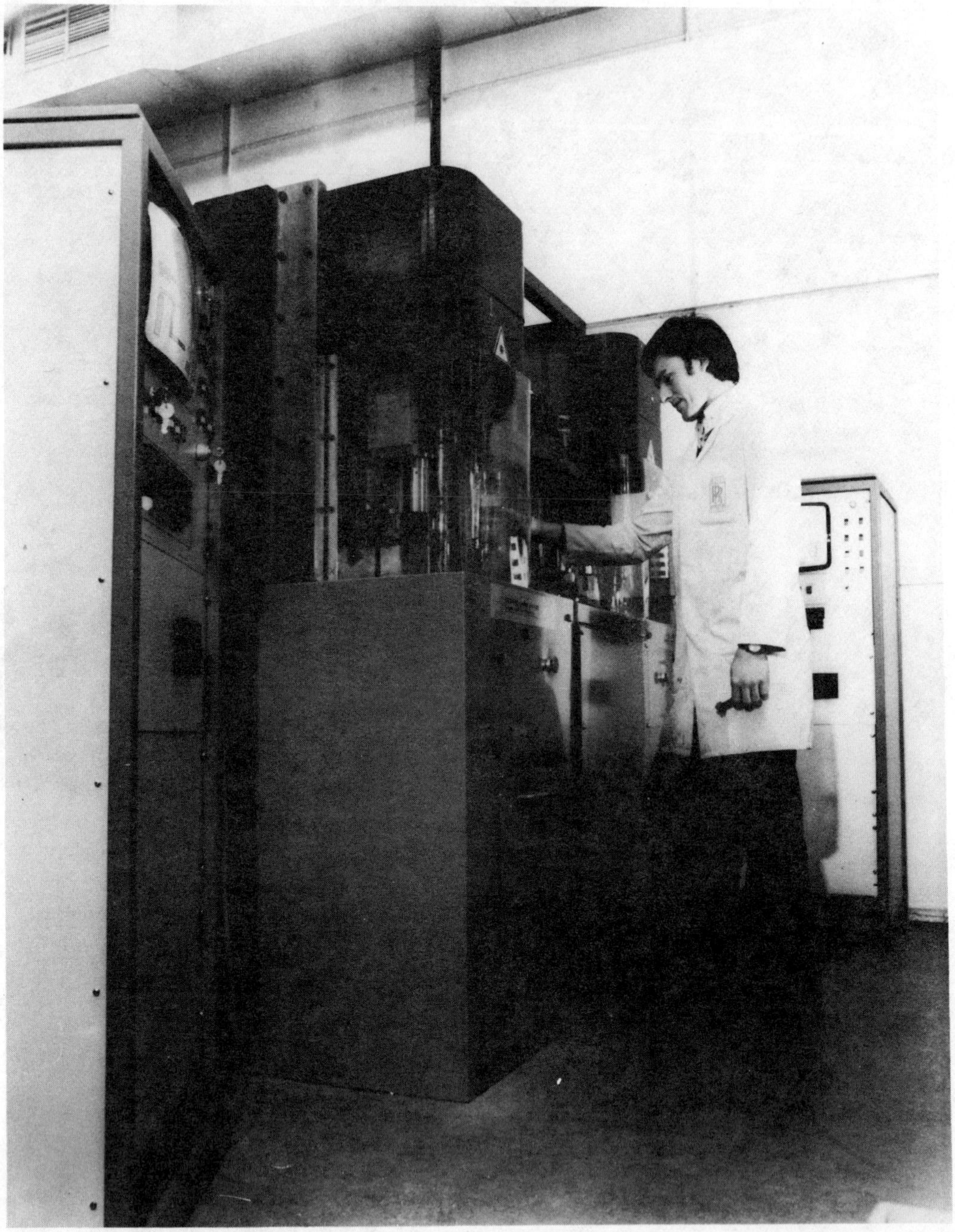

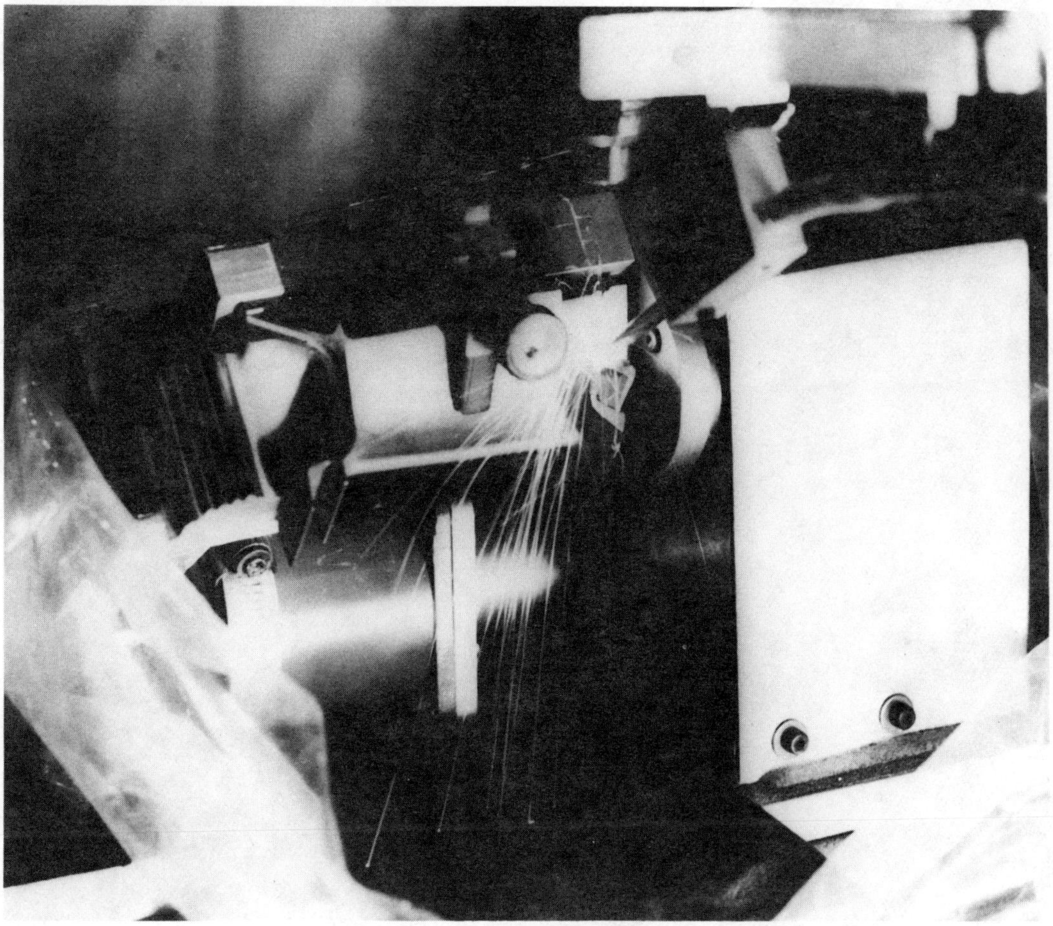

A CORRELATION BETWEEN DENDRITE-ARM-SPACING AND COOLING RATE FOR LASER-MELTED Ti-15V-3A1-3Sn-3Cr*

T. C. Peng, S. M. L. Sastry, and J. E. O'Neal
McDonnell Douglas Research Laboratories, St. Louis, MO. 63166

J. F. Tesson
St. Louis Community College at Forest Park, St. Louis, MO. 63110

INTRODUCTION

Rapid-solidification processes with cooling rates $> 10^3$ K/s can be used to obtain the strengthening of Ti-alloys.[1] However, different rapid-solidification processes have different cooling rates, depending on characteristics of the material as well as the heat-transfer mechanism of the specific process.[2] Thus, an evaluation of the cooling rates is needed to compare the merits of different rapid-solidification processes and to understand the metallurgical and physical changes for a specific process of rapid-solidification.

For most rapid-solidification processes, a direct measurement of cooling rates is difficult and not always reliable because of the extremely fast temperature changes and the substantial interference of the probe on the material under observation. The current practice is to establish correlations between the cooling rates and dendrite-arm-spacings within the material.[3] The correlations can then be further calibrated and interpreted through the use of heat-transfer modeling appropriate to the rapid-solidification process in question.

This paper describes the results of recent studies on rapid-solidification of Ti-15V-3Al-3Sn-3Cr alloy using laser surface-melting techniques. By matching the observed and the calculated melting front, a correlation between the observed dendrite-arm-spacings and the calculated cooling rates was established based on a constant thermal property, three-dimensional, heat-transfer model by Cline and Anthony.[4]

MATERIAL

A single-phase material is likely to produce clear patterns of dendrite-arm-spacing through rapid-solidification. Thus, Ti-15V-3Al-3Sn-3Cr (Ti-15-3) alloys with a stable β-phase that extends to room temperature was selected as the sample material. Rectangular 25 x 46 mm samples were cut from 2-mm thick Ti-15-3 sheets. The surface roughness of the sample was produced by a 320-grit grinding paper.

LASER SURFACE-MELTING

According to the schematic arrangement in Fig. 1, a 1.5-kW, continuous-wave (cw), CO$_2$ (10.6 μm) laser beam with a diameter of ~ 13 mm was focused to a 0.5-mm diameter spot in the sample surface plane by a 120-mm focal-length lens. The rectangular sample was mounted radially on a 15-cm diameter rotating disk located in a plane perpendicular to the laser beam. A slightly curved solid-liquid interface resulted as the sample passed beneath

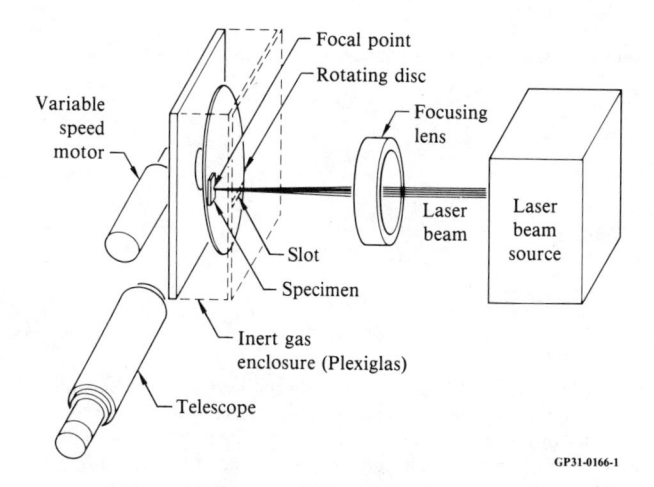

Fig. 1 Experimental arrangement for laser-melting titanium alloys.

the fixed laser beam. The laser scanning speed (0.5-40 cm/s) was determined by the disk rotational speed and the radial distance of the beam spot from the disk center. During the surface-melting, He gas flowed through the enclosure around the rotating disk and over the Ti-15-3 sample. The direction of the He flow was away from the surface through the slot for the incident laser beam (Fig. 1). The He flow served to 1) prevent undesirable oxidation at the melting surface, 2) eliminate formation of vapor plasma which can attenuate the laser energy before it reaches the sample surface, and 3) increase the heat-transfer rate at the melt surface.

The location of the focal point and the spot-size of the laser beam were determined with low-power beams on 3-mm thick Lucite targets. The beam power at the target (the incident power) was measured to be 58% of the beam power at the generator exit (the indicated power) as a result of transmission losses. At the target, a part of the incident beam was reflected or reradiated and the remainder was absorbed by the target material. The ratio of the absorbed to the incident laser power is the coupling efficiency. Generally, the coupling efficiency varies with roughness, temperature, and material state and composition at or near the sample surface and cannot be determined in advance. In this paper, evaluation of the coupling efficiency is a part of heat-transfer analysis.

HEAT-TRANSFER ANALYSIS

A mathematical model suggested by Cline and Anthony[4] was used to describe the heat conduction in laser surface-melting. Basic assumptions of this model are: 1) constant thermal properties throughout solid and liquid phases, 2) zero heat of fusion at

---

*This work was supported by the McDonnell Douglas Independent Research and Development program.

the melting temperature, 3) temperature limited to the boiling point, 4) constant beam-scanning speed, 5) a Gaussian distribution of power across the laser-beam diameter, and 6) a well-defined absorbed laser power at the target.

Assumptions 1 and 3 are necessary to obtain a closed-form solution, and their effects on temperature profiles need to be evaluated. Assumption 6 assumes a knowledge of the interaction of the laser beam and the material surface. Lack of this knowledge requires an alternative approach using an emperical coupling efficiency in the heat-transfer analysis.

According to Cline and Anthony, the temperature, T, at position (x,y,z) within the target material (Fig. 2) is given by

$$T(x,y,z) = \frac{\eta_c P_i}{C_p DR} \cdot \frac{1}{\sqrt{2\pi^3}} \int_0^\infty \frac{e^{-H(u)}}{1+u^2} du, \quad (1)$$

where η_c is the coupling efficiency, p_i is the incident power,

$$u^2 = \frac{2Dt'}{R^2},$$

$$H(u) = \frac{[x/R + \rho/2u^2]^2 + (y/R)^2}{2(1+u^2)} + \frac{(z/R)^2}{2u^2},$$

$$\rho = \frac{R}{D} v,$$

and C_p, D, R, v, and t' are respectively the heat capacity per unit volume, thermal diffusion, beam spot-size, laser scanning speed in the +x direction, and a time-variable of integration. The solid-liquid interface, or the melting front, is calculated by requiring $\partial T/\partial X = 0$ at $T(x,y,z) = T_m$, the melting temperature of the sample material. Thus, a melting front in the cross section of the sample (the yz plane in Figs. 3a–h) can be mapped for a given value of the coupling efficiency. Conversely, by requiring a match between the calculated and the observed melting fronts, the value of the coupling efficiency can be determined.

The cooling rate is estimated by

$$\frac{\partial T}{\partial t} = -v\left[\frac{x}{r^2} + \frac{V}{2D}\left(1 + \frac{x}{r}\right)\right] T, \quad (2)$$

where

$$r^2 = x^2 + y^2 + z^2.$$

Equation (2) is reasonably accurate if the distance from the beam, r, is greater than the beam spot-size, R. The maximum cooling rates for given values of y and z is obtained by requiring $\partial^2 T/\partial t^2 = 0$. For this paper, only maximum cooling rates at y = 0 are calculated and correlated with the measured dendrite-arm-spacings.

RESULTS

The operating conditions and the measured melt-depths and widths for Ti-15-3 alloy are shown in Table 1. Six photomicrographs of laser-melting cross-sections are shown in Fig. 4, where the melting fronts are evident from the changing features in the microstructure. The calculated melting fronts for different values of coupling efficiency are shown in Figs. 3a–h. Superimposed on the calculated lines are the observed melting fronts. For this study of laser melting of Ti-15-3 alloy, the coupling efficiency was determined to vary from 0.70 to 0.36.

With the determination of the coupling efficiency, the maximum cooling rates can be calculated according to Eq. (2), and the results are indicated in Figs. 3a–h. The calculated cooling rates were correlated with the dendrite-arm-spacings (DAS) measured from photomicrographs (Figs. 5–7 and Table 2). The best functional correlation between the cooling rates and the DAS is shown in Fig. 8 and is represented by

$$d = 80 \epsilon^{-0.34}, \quad (3)$$

where ϵ is the cooling rate in K/s and d is the dendrite-arm-spacing in μm.

Table 1. Dependence on specimen scanning rate of melt penetration and width.

Sample alloy	Laser-beam incident power (W)	Specimen scanning rate (cm/s)	Melted depth (cm)	Melt width (cm)
Ti-15-3	580	0.5	0.100	0.340
		2.0	0.062	0.175
		5.0	0.013	0.051
Ti-15-3	870	10	0.027	0.064
		20	0.013	0.047
		30	0.008	0.039
		40	0.006	0.022

Table 2. Dendrite spacing as a function of scanning rate for laser-melted Ti-15-3.

Melt Location	Power (incident) (1000 W) Scanning rate (cm/s)				Power (incident) (1500 W) Scanning rate (cm/s)		
	0.5	2	5	10	20	30	40
Top	8.37	3.22	1.29	2.11	1.07	0.57	0.74
Bottom	20.24	4.40	1.48	—	1.29	0.88	1.06
Average	14.31	3.81	1.39	—	1.19	0.73	0.90

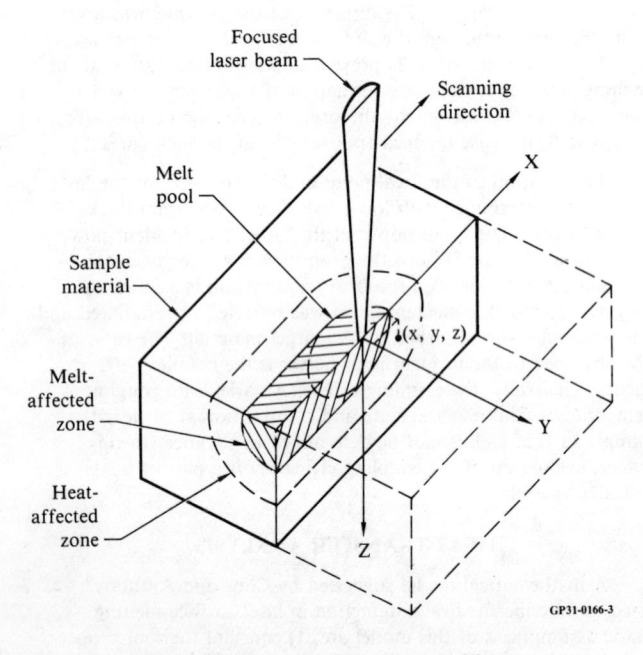

Fig. 2 The physical model of laser melting.

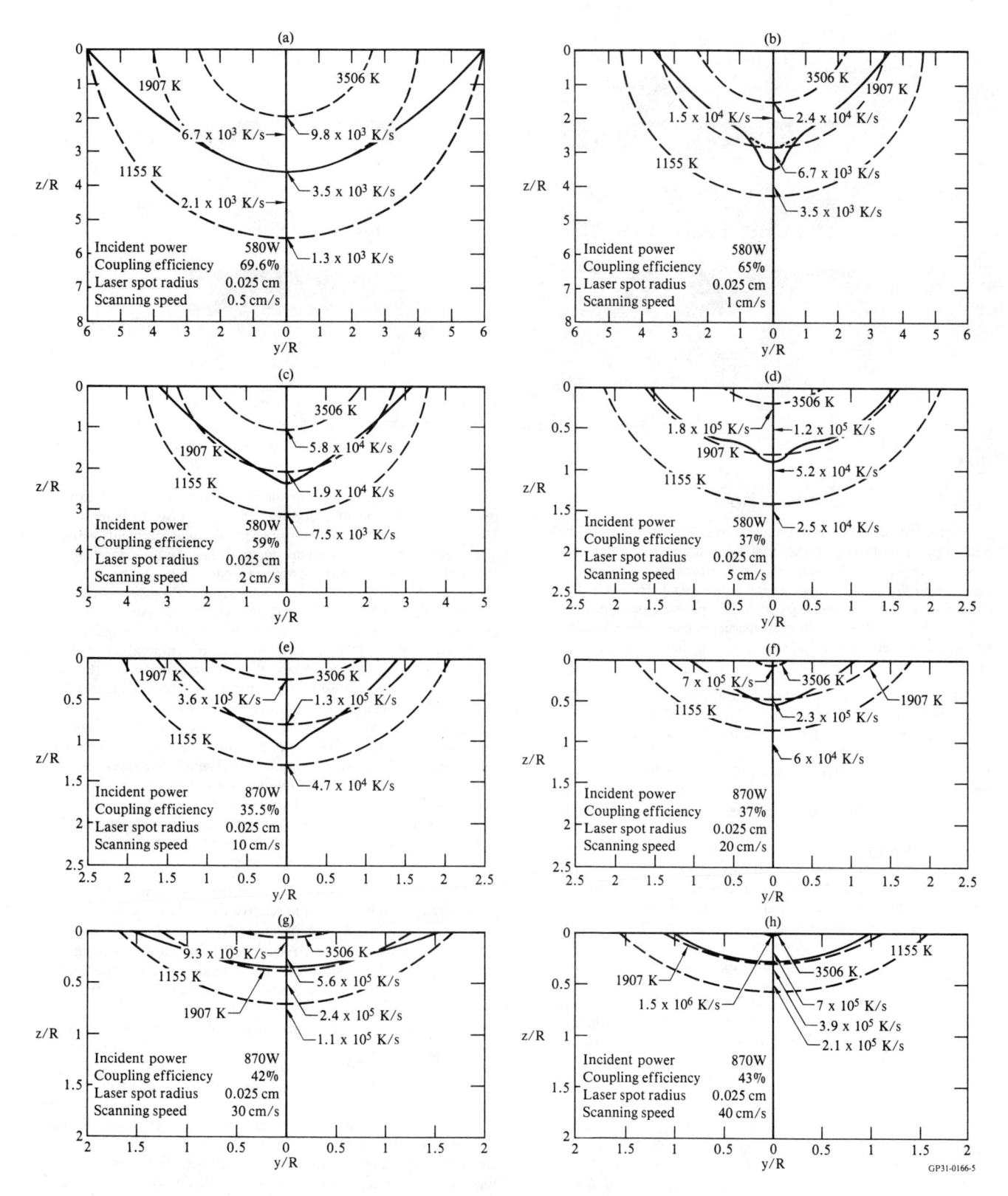

Fig. 3 Calculated and observed isotherms and maximum cooling rates for laser meltings of Ti-15V-3Al-3Sn-3Cr.

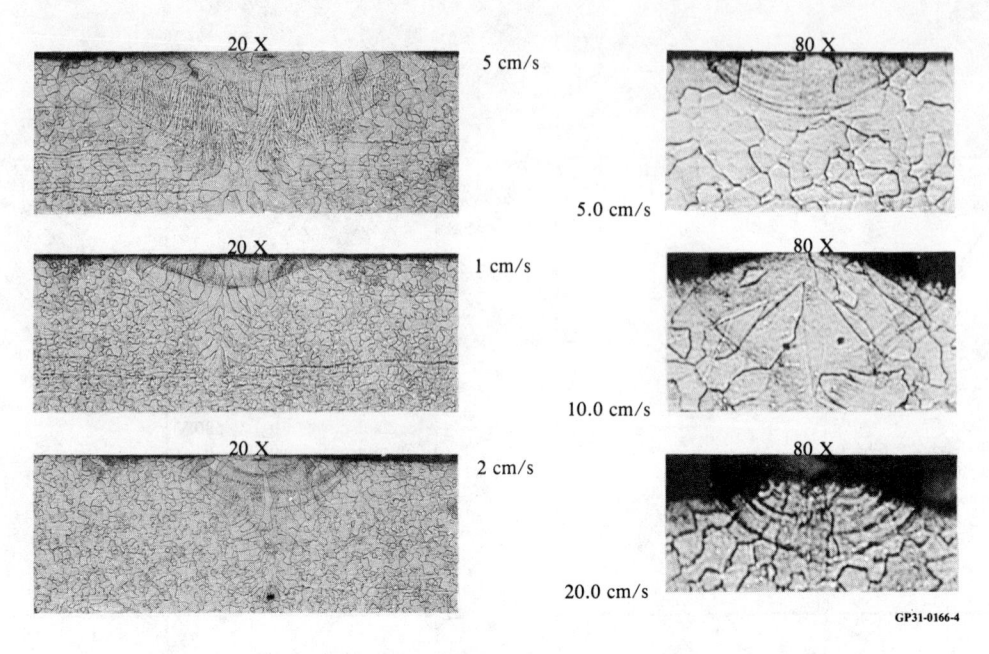

Fig. 4 Photomicrographs of laser-melted Ti-15-3-3-3.

DISCUSSION

Some calculated surface temperatures, well above the 3506 K boiling point of Ti-15-3, reveal that the Cline and Anthony model is not entirely applicable since the boiling process is not considered and the heat of vaporization is not small. However, the Cline and Anthony model does provide a physical description of laser melting below the boiling temperature and a closed-form solution that yields numerical values with a simple and fast computer program.

The Cline and Anthony model requires constant thermal property values. For this study, fixed values of 0.27 W/K·cm for thermal conductivity and 0.08 cm^2/s for thermal diffusion were used based on an estimated average temperature of 1304 K for the laser melting. Available data on thermal conductivity and specific heat for Ti-V-Σx alloys[5,6] up to 1000 K and linear extrapolations above 1000 K yielded thermal conductivity values from 0.086 W/K·cm at room temperature to 0.6 W/K·cm at the boiling point (3506 K), specific heat values from 0.52 J/K·g at room temperature to 0.84 J/K·g at the boiling point, and thermal diffusivity values from 0.034 cm^2/s at room temperature to 0.15 cm^2/s at the boiling point. The effect of variable thermal properties on the calculated temperature profiles in Fig. 3a–h was not evaluated because of the lack of reliable data above 1000 K and the incompleteness of the Cline and Anthony model.

Equation (1) does not include deep-well or key-hole punctuation effects in laser melting because the physical data in the liquid phase and an adequate model for the key-hole hydrodynamics are presently not available. The calculated and observed melting fronts were therefore compared only at the surface perimeters of the melting fronts and at the part of deep penetration (Fig. 3b–f).

Only a part of incident laser energy is absorbed by the sample because of reflection, reradiation, and convection losses at the surface of the laser-melt. All three losses are difficult to assess without reliable temperature data of the laser melt and high-temperature thermal and optical data for the sample. An alternative approach used in this study is to match the calculated melting-point (1907 K) isothermal contours to the observed melting fronts in the samples. For a given laser-beam spot-size and heat-transfer model (Eq. (1)), the depth of the melting front is directly related to the laser energy absorbed at the surface.

Thus, the absorbed energy, or the coupling efficiency (the ratio of absorbed-to-incident energy, η_c, in Eq. (1)) can be determined. Uncertainties involved in this approach are measurement of the laser-beam spot-size and inadequacies in heat-transfer modeling. Using the matching melting-front approach (Figs. 3a–h), the coupling efficiency was found to vary from 70% at 0.5 cm/s laser-scanning velocity to $\approx$ 40% at scanning velocities of 5–40 cm/s. For those cases where the coupling efficiency is significantly above 40%, there are large regions of temperature above the boiling point (3506 K). Thus, the heat-transfer model (Eq. (1)), which does not consider boiling and its associated losses, over-estimates the absorbed energy. The $\pm$ 4% variation of coupling efficiency around 40% for scanning velocities of 5–40 cm/s reflects experimental errors and the imperfect matching of calculated and observed melting fronts. From other investigations using the Cline and Anthony model for different materials, 16.3%[7] and 23%[8] coupling efficiencies have been reported. Aside from the material differences, the uncertainty in high-temperature material properties is a source of discrepancies among the reported values of the coupling efficiency.

Equation (2) is capable of predicting cooling rates at any point (x,y,z) within the sample relative to the laser beam if $r = (\sqrt{x^2 + y^2 + z^2}) > R$ (the beam radius). However, the actual heat-transfer rate at a fixed point in the target material changes as the laser beam passes by it. In terms of the laser-beam moving-coordinate frame (x,y,z in Eq. (2)), the time variation in the sample coordinate system is translated in the x-variation. Thus, by fixing the y- and z-axes, the cooling rates reach a maximum as the x values vary from a + x to a - x. This maximum cooling rate is correlated with the observed dendrite-arm-spacing in the sample after laser melting. For this study, the maximum cooling rates were evaluated only for the centerline region (y = 0) since most of the visible dendrite-arm-spacings were found in this region. The individual cooling-rate values on the z-axis are indicated in Figs. 3a–h, and the dendrite-arm-spacing/maximum-cooling-rates correlation is shown in Fig. 8. The curve that best fits all points has a curvature and appears to depart from the power function fit suggested by Mehrabian.[3] However, the first four data points are derived from cases where the boiling can be significant and hence are not suitable for the Cline and Anthony heat-transfer model used in this study. For the remaining data

0.5 cm/s 800 X

Top of melt
8.37 μm/dendrite

Bottom of melt
20.24 μm/dendrite

1 cm/s 2000 X

Top of melt
2.54 μm/dendrite

Bottom of melt
3.18 μm/dendrite

2 cm/s 2000 X

Top of melt
3.10 μm/dendrite

Bottom of melt
4.35 μm/dendrite

5 cm/s 2000 X

Top of melt
1.31 μm/dendrite

Bottom of melt
1.39 μm/dendrite

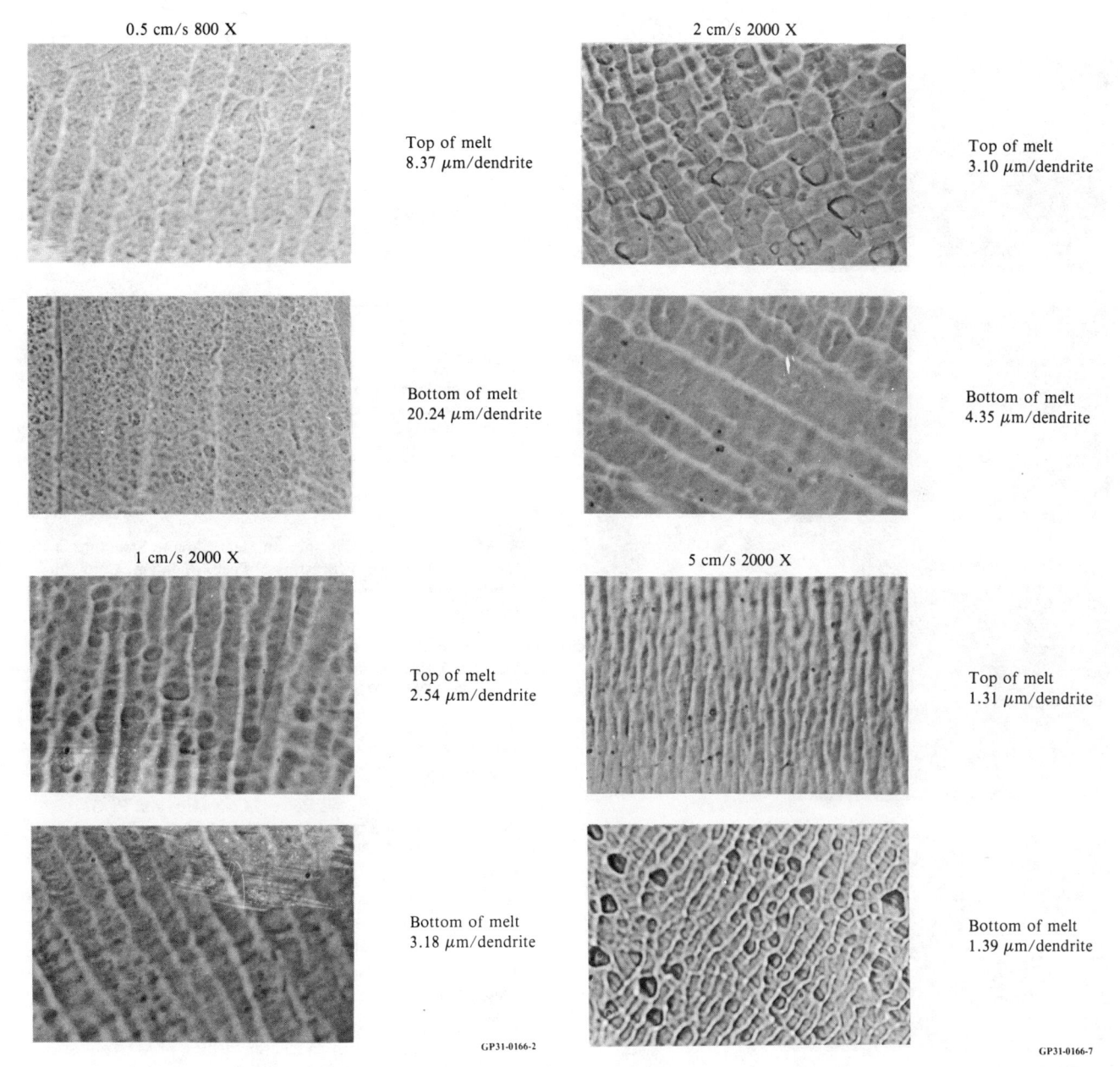

Fig. 5 Dependence of dendrite spacing on laser-beam scanning rate for laser-melted Ti-15-3-3-3.

Fig. 6 Dependence of dendrite spacing on laser-beam scanning rate for laser-melted Ti-15-3-3-3 (Continued).

between the maximum cooling rates of 8×10^4 to 1.5×10^6 K/s, the laser melting is described well by the Cline and Anthony model. For these data, a power function fit expressed as $d = 80 \epsilon^{-0.34}$ was established.

CONCLUSION

By matching the calculated and observed melting fronts, a coupling efficiency of ~ 40% is obtained for laser melting of Ti-15V-3Al-3Sn-3Cr alloy. In addition, a correlation between the dendrite-arm-spacing and the maximum cooling rate is established. This correlation is built on the Cline and Anthony model without considering the boiling and deep-well penetration phenomena. Further improvement of this correlation can be obtained by better heat-transfer modeling and additional thermal-properties data for sample material.

ACKNOWLEDGEMENT

The authors would like to express their appreciation to Mr. Gary Niemeyer for his assistance in the laser-melting experiment and to Mr. John Putnam for his help in computer programming.

REFERENCES

1. S. M. L. Sastry, T. C. Peng, J. E. O'Neal, and L. P. Beckerman, "Consolidation, Thermochemical Processing and Mechanical Properties of Rapidly Solidified, Dispersion-strengthened Titanium Alloys," Third Conf. on Rapid Solidification Processing, Gaithersberg, MD, 6–8 Dec. 1982.

2. H. Jones, "Some Principles of Solidification at High Cooling Rates," **Proc. of Intern. Conf. on Rapid Solidification Processing** (Claitor's Publishing Division, 1978), p. 28.

20 cm/s 3200 X

Top of melt
1.08 µm/dendrite

Bottom of melt
1.29 µm/dendrite

30 cm/s 3200 X

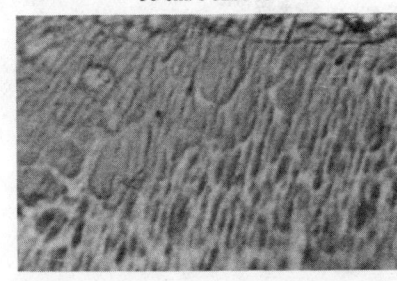

Top of melt
0.57 µm/dendrite

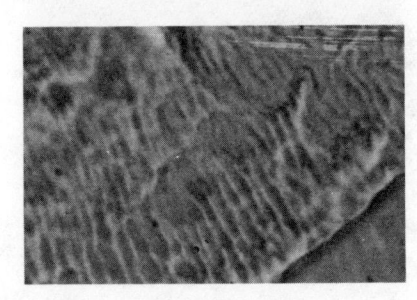

Bottom of melt
0.88 µm/dendrite

Fig. 7 Dependence of dendrite spacing on laser-beam scanning rate for laser-melted Ti-15-3-3-3 (Continued).

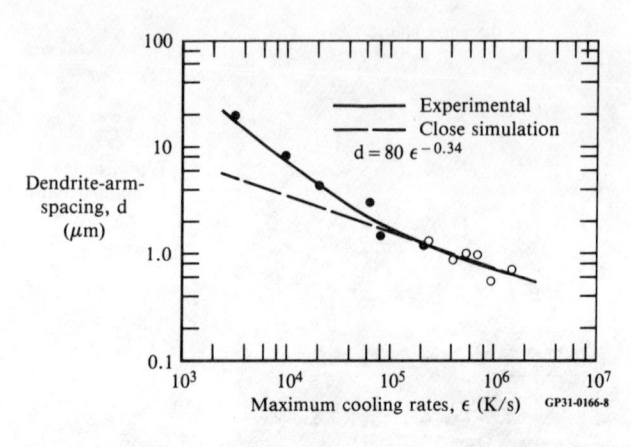

$$d = 80\,\epsilon^{-0.34}$$

Dendrite-arm-spacing, d (µm) vs Maximum cooling rates, ϵ (K/s)

Fig. 8 A correlation between the observed dendrite-arm-spacing and the calculated maximum cooling rates.

3. R. Mehrabian, "Relationship on Heat Flow to Structure in Rapid Solidification Processing," **Proc. of Intem. Conf. on Rapid Solidification Processing** (Claitor's Publishing Division 1978), p. 9.

4. H. E. Cline, and T. R. Anthony, "Heat Treating and Melting Material with a Scanning Laser or Electron Beam,: J. Appl. Phys. **48**, 3895 (1977).

5. Y. S. Touloukian, and E. H. Buyco, **Thermophysical Properties of Matter, Specific Heat**, (IFI/Plenum, New York-Washington, 1970), Vol. 4 p. 607.

6. Y. S. Touloukian, R. W. Powell, C. Y. Ho, and P. G. Klemens, **Thermophysical Properties of Matter, Thermal Conductivity** (IFI/Plenum, New York-Washington, 1970), Vol. 1 p. 1086.

7. O. Esquivel, J. Mazumder, M. Bass, and S. Copley, "Shape and Surface Relief of Continuous Laser-Melted Trails in Udimet 700" in **Rapid Solidification Processing: Principles and Technologies, II** (Claitor's Publ. Div. Baton Rouge, LA, 1980), p. 180.

8. P. R. Strutt, M. Kurup, and D. A. Gilbert, "Comparative Study of Electron Beam and Laser Melting of M2 Tool Steel" in **Rapid Solidification Processing: Principles and Technologies, II** (Claitor's Publ. Div. Baton Rouge, LA, 1980), p. 225.

THE EFFECTS OF INCONEL 600 ON THE TOUGHNESS OF HY-STEEL LASER WELDS

D. W. Moon and E. A. Metzbower
Naval Research Laboratory
Washington, DC 20375

ABSTRACT

Using a high energy CO_2 laser, 12mm (0.5 in) thick plates of HY-80 and HY-100 steels have been butt welded in a single pass, both with and without an insert of Inconel 600.

The microstructures of the welds, with and without the insert, were identified. Fusion boundaries of the heterogeneous welds were thoroughly examined with an electron beam microprobe to examine the mixing behavior between the parent materials (HY-80 and HY-100) and the Inconel 600.

The hardnesses of the weldments were measured. Impact properties of weldments were also measured from room temperature down to -50°C (-60°F) by Charpy V-notch test (CVN). Dynamic tear (DT) impact toughness test was also conducted at temperatures -1°C (30°F) and -29°C(-20°F).

The fracture toughness energies of autogenous welds of HY-80 and HY-100 failed to meet the Navy's requirement. On the other hand, those welds with inserts demonstrated excellent toughness values over the entire test temperature range and satisfy the Navy's criteria.

INTRODUCTION

High power lasers have been used since the early 1970's to weld thick structural materials including the HY-steels. Investigators have reported both the excellent properties and the advantages of laser welding. No experimenters, however, have successfully fabricated laser weldments which meet the Navy's impact fracture toughness criteria, particularly at low temperatures.

In an effort to improve the fracture toughness, various attempts such as pre-and post weld heat treatments, have been tried at NRL. Unfortunately, none of the above methods were successful. Typical problems found in the autogenous welds were solidification cracking and high hardness values of the fusion zones [1].

An attempt to introduce an inoculant to the fusion zone has been undertaken in order to improve the weld toughness. The inoculant was Inconel 600 sheet. The sheet was sandwiched between square butt-surfaces prior to welding. Inconel Ni-Cr-Fe alloy 600 is a standard engineering material for applications which require resistance to corrosion and heat, as well as good mechanical properties [2]. The versatility of Inconel alloy 600 has led to its use in a variety of applications involving temperatures from cryogenic to above 2000°F.

The Charpy V-notch and DT impact test results of these weldments revealed that the impact toughness values of the welds with Inconel 600 insert satisfy the values of the Navy's requirement.

MATERIALS AND PROCEDURE

Using a 15 kW, continuous wave, CO_2 laser, HY-80 and HY-100 steel plates of 12 mm (0.5 inch) thickness were square butt welded both with and without Inconel 600 sheet. The thickness of the Inconel 600 sheet was 0.12 mm (5 x 10^{-3} inch) thickness. The compositions of the steels and Inconel 600 are shown in Table 1. The welding conditions are summarized in Table 2.

TABLE 1. COMPOSITIONS, wt %

ELEMENT / MATERIALS	Ni	Cr	C	Mn	Si	Mo	Cu	Al	V	Ti	Sn	S	P	Fe
HY-80	2.40	1.63	0.17	.31	0.24	0.38	0.04	0.016	0.005	0.002	0.003	0.016	0.01	BAL
HY-100	2.89	1.60	0.15	0.30	0.26	0.30	0.15		0.003	0.003		0.011	0.005	BAL
INCONEL 600	75.68	15.14	0.02	0.22	0.18		0.14					0.002		8.62

TABLE 2
WELDING CONDITIONS

LASER POWER: 12 kW
TRAVEL SPEED: 1.27 cm/sec (30 ipm)
HEAT INPUT: 0.94 kJ/mm (24.0 kJ/in)
SHIELDING GAS (helium)
   PRESSURE: PLASMA - 60 CFH
                     LEAD - 30 CFH
                     TRAIL - 30 CFH
                     BOTTOM - 30 CFH

Pure helium gas was used to protect the reactive, hot weld and to control the plasma formed during laser welding action. Figure 1 depicts the joint configurations for the autogenous and the heterogeneous welds. All the weldments were given visual and radiographic examinations.

The microstructures of the laser beam weldments, with and without the Inconel 600 inserts, were determined by conventional metallographic techniques. The microhardness of the weldment was measured by using a diamond pyramid indenter. Compositional changes in the different zones of the weldment were determined using an electron beam microprobe. The energy absorbed by the weldment as a function of temperature was determined by the Charpy V-notch test and the dynamic tear test. The

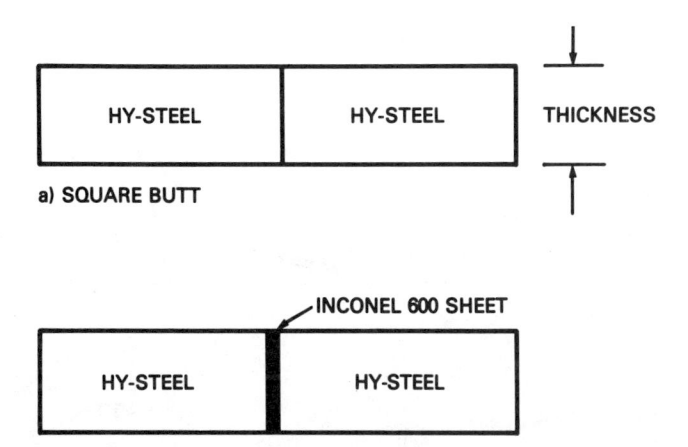

Fig. 1 - JOINT CONFIGURATIONS.

results of these tests are discussed and correlated in the following sections.

RESULTS

HY-80

Metallography. In order to examine the soundness of bonding and mixing behavior between the inoculant and the parent metal the microstructures of the fusion boundaries were closely inspected. Figure 2 is the solidification structure of the HY-80 laser beam weldment with the Inconel 600 insert. The right hand side is the fusion zone, whereas the left hand side is the base metal.

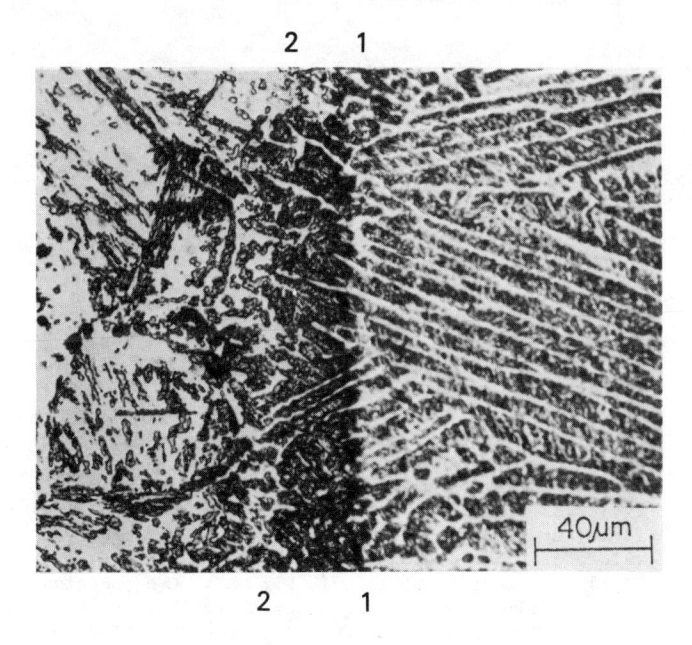

Fig. 2 - FUSION BOUNDARY OF HY-80/INCONEL WELD
1-1: APPARENT FUSION LINE,
2-2: TRUE FUSION LINE

Thorough metallographic examinations along the fusion boundaries revealed both an apparent fusion line and the true fusion line. Careful examination of Fig. 2 reveals that the origin of the epitaxial growth is not along the apparent fusion line (1-1) but along the true fusion line (2-2). No signs of cracks, porosities and other type of defects were found along the fusion boundaries.

The microstructure of the HY-80 base plate consisting of quenched and tempered martensite is shown in Fig. 3.

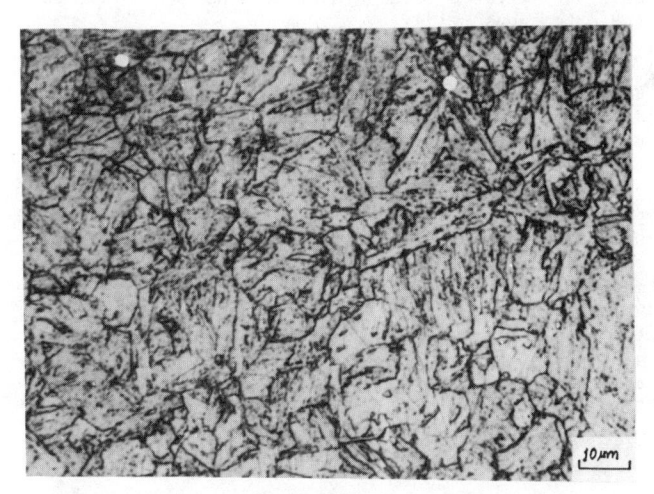

Fig. 3 - QUENCHED & TEMPERED MARTENSITE IN BASE METAL OF HY-80 STEEL. HARDNESS 21.0 Rc. ETCHED IN 1% NITAL.

a WITHOUT INSERT

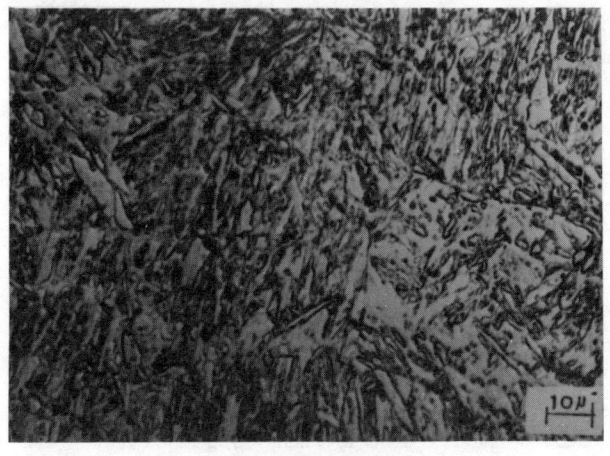

b WITH INCONEL 600

Fig. 4 - HY-80 FUSION ZONE MICROSTRUCTURES.

The fusion zone microstructure of autogenous weld comprises untempered martensite with some bainite as shown in Fig. 4a. The microstructure of the weld with the Inconel 600 insert is comprised of martensite with some bainite as shown in Fig. 4b. The fusion zone structure of the heterogeneous weld is a very refined structure in contrast with that of the autogenous welds.

Electron Beam Microprobe Analysis. Chemically different base metal and Inconel 600 insert will mix each other during welding. This mixing behavior can be examined by studying the redistributed alloying elements of both materials after welding. Thus we performed a point mode scan across the fusion boundaries. The concentrations of Ni and Cr as shown in Fig. 5 increased gradually from the base metal values 2.3 and 0.8 wt% to the fusion zone values 6.4 and 1.4 wt%. The scan from one side of the base metal, across the fusion zone to the other side of the base metal revealed that the compositions of the other alloying elements (Mo, Mn, Si and Cu) remained unchanged throughout the base metal, the HAZ and the fusion zone.

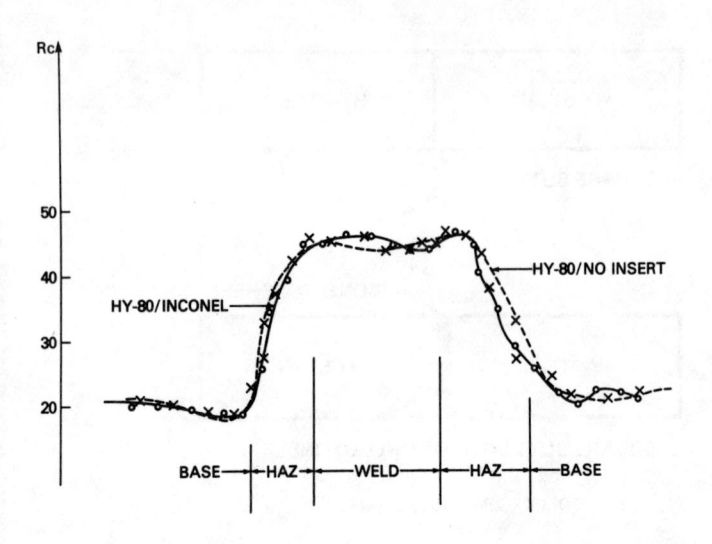

Fig. 6 - HARDNESS TRAVERSES OF HY-80 WELDS.

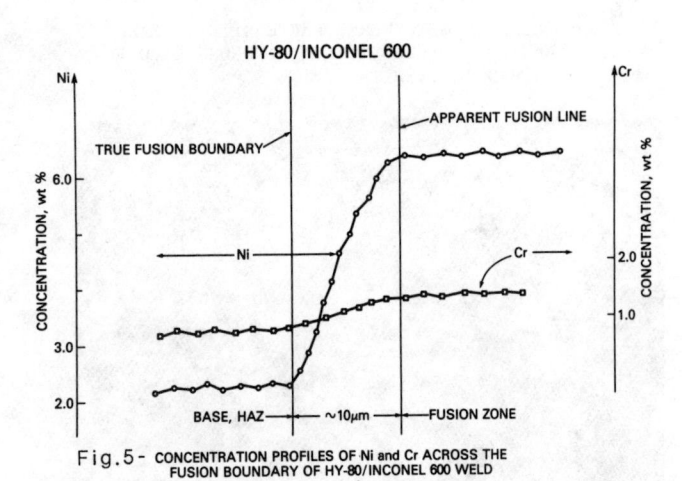

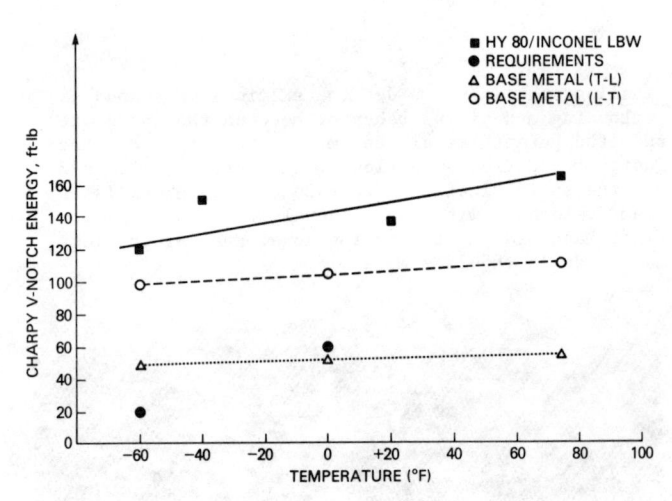

Fig. 5 - CONCENTRATION PROFILES OF Ni and Cr ACROSS THE FUSION BOUNDARY OF HY-80/INCONEL 600 WELD

Microhardness. Microhardness traverses across HAZ and fusion zones are shown in Fig 6. In the autogenous weld the average hardness value of the fusion zone is 45 R_c and the hardnesses of the HAZ varied from 25 R_c to 46 R_c. In the heterogeneous weld the average fusion zone hardness is 45 R_c and the HAZ hardnesses are from 25 R_c to 46 R_c. There is no difference in hardness between the autogenous weld and the heterogeneous weld. The extremely high hardness values of the HAZ and the fusion zones are characteristic of laser beam welded ferrous alloys at similar carbon levels. This is a result of the fast cooling rate inherent in the laser welding process.

Impact Toughness. Weld and base metal Charpy V-notch (CVN) impact specimens were tested at three temperatures: 25°C (77°F), -17.8°C (0°F) and -51.1°C (-60°F). Dynamic tear (DT) tests were also conducted at -1°C (30°F) and -29°C (-20°F). CVN impact energies of the laser beam weldments and the base metal as a function of temperature are shown in Fig. 7. The Navy's requirements are also shown for comparison.

Fig. 7 - CHARPY V-NOTCH ENERGY OF HY-80.

The fracture initiated in the fusion zone and then swiftly directed into the base plate at all temperatures as shown in Fig. 8.

The energies absorbed by the weldments are substantially higher than the required values. The average energy of the HY-80/Inconel 600 insert welds is 162.8J (120 ft-lb) at -50°C (-60°F) and 176J (130 ft-lb) at -18°C (0°F). The corresponding required values are 27.1J (20 ft-lb) at -50°C (-60°F) and 81.4J (60 ft-lb) at -18°C (0°F). The CVN energies of the weldments were also superior to that of base plate: the base metal values in the direction of L-T and T-L are 135.6J (100 ft-lb) and 67.8J (50 ft-lb) at -50°C (-60°F) and 141.1J (104 ft-lb) and 73.2J (54 ft-lb) at -18°C (0°F). The room temperature toughness of the welds is also higher than the base plate value. The average DT energies of HY-80 welds with Inconel 600 insert are 610.5J (450 ft-lb) at -1°C (30°F) and 530.5J (391 ft-lb) at -29°C (-20°F). The requirements are 576.6J (425 ft-lb) at -1°C (30°F) and 339.2J (250 ft-lb) at -29°C (-20°F).

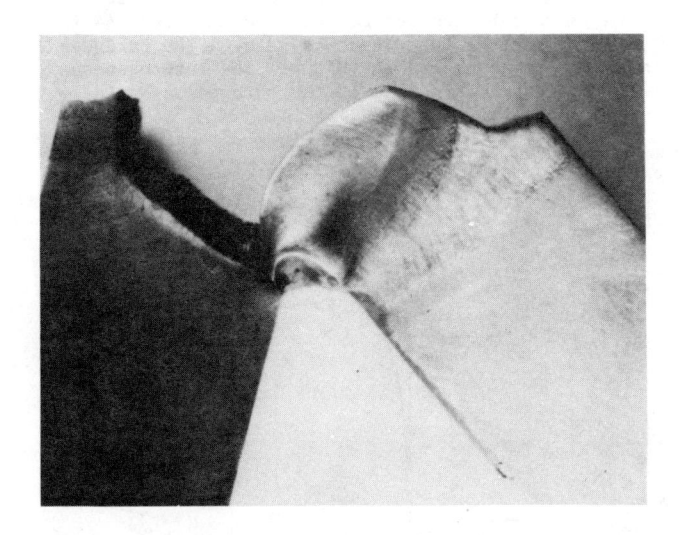

0°C

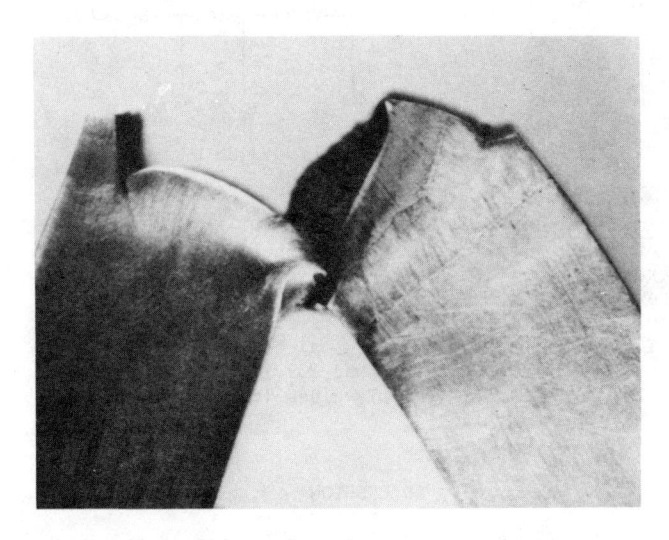

−50°C

Fig. 8 - CVN SPECIMENS OF HY-80/INCONEL LBW.

The fusion zone microstructure of the weld with Inconel insert is refined as shown in Fig. 11b, contrasting to the coarse and directional structure of the autogenous weld.

Fig. 9 - FUSION BOUNDARY OF HY-100/INCONEL WELD
1-1: APPARENT FUSION LINE,
2-2: TRUE FUSION LINE.

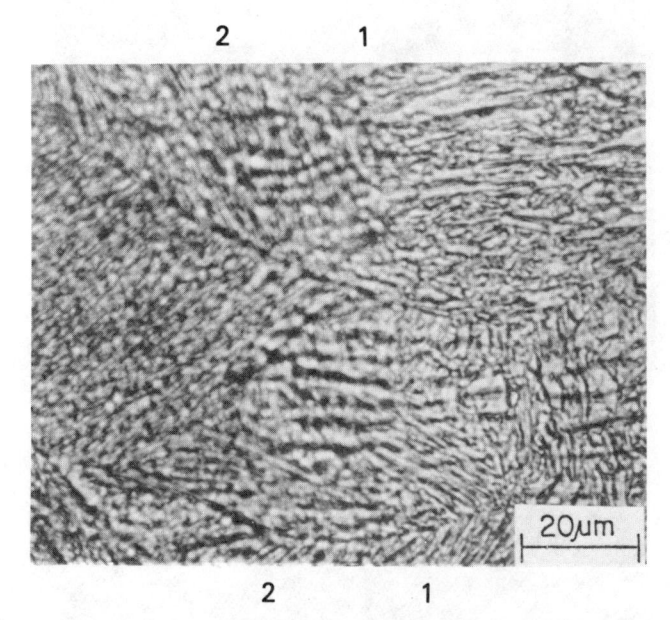

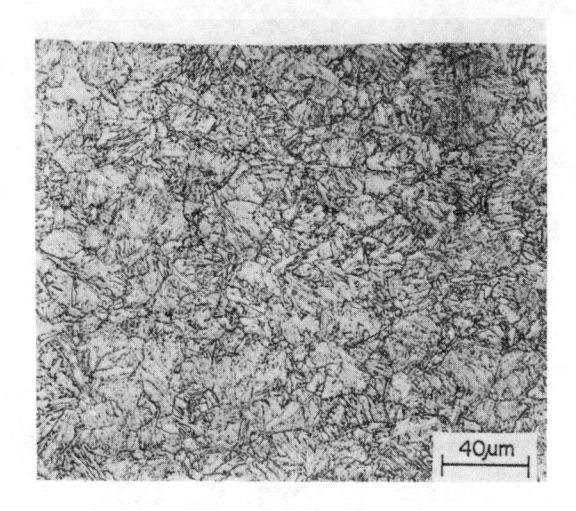

Fig. 10 - TEMPERED MARTENSITE IN BASE METAL OF HY-100; Rc 27.

HY-100

Metallography. Fig. 9 shows fusion boundary structures of HY-100 laser weld with Inconel 600 insert. Extensive studies on the fusion boundaries have been made to examine the bonding and the mixing behaviors between the parent metal (HY-100) and the insert (Inconel 600). The solidification structure of the fusion zone is on the right hand side which crossed over the apparent fusion line 1-1 and reached the true fusion line 2-2 as shown in Fig. 9. No signs of incomplete fusion, cracks, porosities were observed along the fusion boundaries.

The base plate microstructure is comprised of quenched and tempered martensite as shown in Fig. 10. The microstructure of the autogenous laser weld consists of untempered martensite with some bainite as shown in Fig. 11a. The fusion zone microstructure of the heterogeneous weld is martensite with some bainite as shown in Fig. 11b.

Electron Beam Microprobe Analysis. The mixing behavior of Inconel 600 with the base metal was examined by an electron microprobe scan across the fusion boundaries. Measured concentrations of Ni and Cr with base metal are 1.8 and 1.6 wt%. These values began to increase gradually at the true fusion line and reached the fusion zone values of 5.9 and 2.4 wt% at the apparent fusion line as shown in Fig. 12. These concentrations remained constant throughout the fusion zone.

All other alloying elements (Mo, Mn, Si and Cu) did not show compositional changes over the regions of base, HAZ and fusion zone.

a WITHOUT INSERT

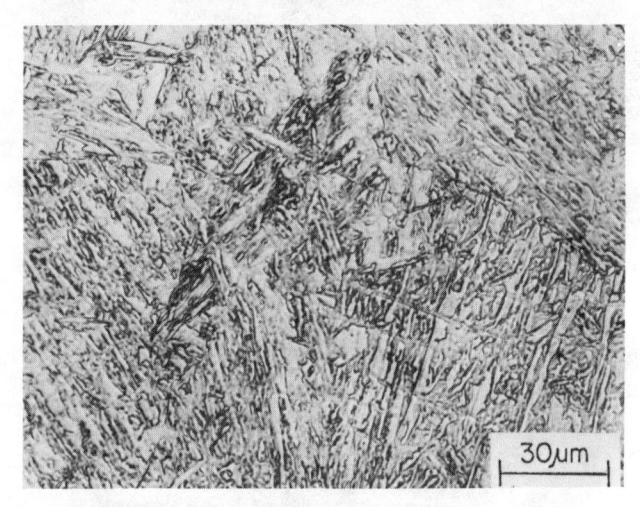

b WITH INCONEL 600

Fig. 11 - HY-100 FUSION ZONE MICROSTRUCTURES.

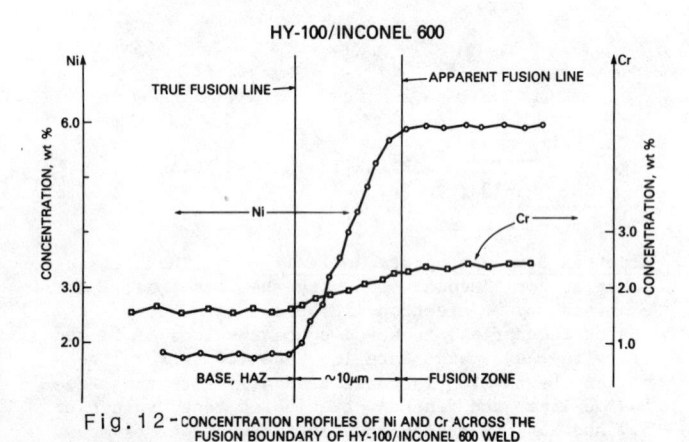

Fig. 12 - CONCENTRATION PROFILES OF Ni AND Cr ACROSS THE FUSION BOUNDARY OF HY-100/INCONEL 600 WELD

Hardness. Microhardness traverses across the HAZ and the fusion zone are shown in Fig. 13. The fusion zone hardness is approximately 47 R_C for the autogenous weld and 44 R_C for the weld with Inconel 600 insert. The HAZ hardness values range from 47 R_C to 29 R_C for the autogenous weld and from 44 R_C to 27 R_C for the heterogeneous weld. The hardness in the fusion zone and the HAZ of the heterogeneous weld is higher than that of the autogenous weld by approximately 3 R_C.

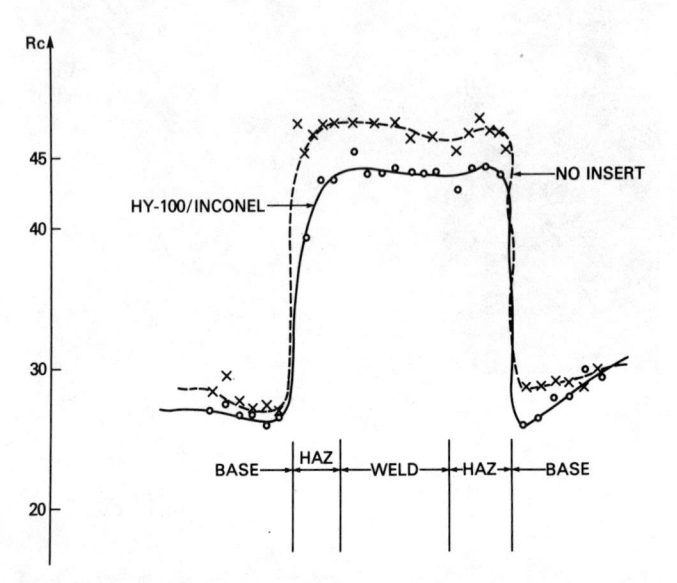

Fig. 13 - HARDNESS TRAVERSES OF HY-100 WELDS

Impact Toughness. Dynamic tear (DT) test specimens of HY-100 laser welds with Inconel 600 insert were tested at -1°C (30°F) and -29°C (-20°F). DT energies of the weldments are 873.8J (644 ft-lb) at -1°C (30°F) and 674.3J (497 ft-lb) at -29°C (-20°F). The corresponding required values are 576.6J (425 ft-lb) at -1°C (30°F) and 339.2J (250 ft-lb) at -29°C (-20°F). All the DT values for weldments meet the required values.

DICUSSION

Following welding, all the weldments were visually inspected. The weld beads of the heterogeneous weld were smoother than the beads of the autogenous weld. Particularly the bottom beads of the heterogeneous weld were more uniform and appeared more sound than the beads of the autogenous welds. More than 90% of the weldments passed radiography testing.

Radiography, metallography and electron beam microprobe studies indicated that no defects have been found in the fusion zones and particularly along the fusion boundaries, indicating good mixing between inoculant and parent metals. The electron beam microprobe study across the fusion boundaries showed gradual mixing over approximately 10 μm width at the fusion lines and then uniform mixing over the entire fusion zone.

At the fusion boundaries apparent and true fusion lines were observed, but no unmixed weld metal regions were found as Savage and Szekers[3] observed in a conventional welding with filler metal. One of the reasons could be that penetration mechanism in laser welding is not a slow melting process but a keyhole process of very short interaction time. A more exhaustive examination of the fusion zone boundaries is being undertaken.

Dynamic tear (DT) test values of HY-80 and HY-100 with Inconel 600 laser welds at temperatures -1°C (30°F) and -29°C (-20°F) satisfy the required values. Similarly Charpy V-notch tests demonstrated excellent fracture toughness properties of laser beam weldments in HY-80 with Inconel 600 insert.

The improved toughness of the welds is attributed to the increased Ni concentration approximately tripled in the fusion zone due to the addition of the Inconel 600 sheet. Ni is generally well recognized as the most effective alloying element in enhancing the fracture toughness of various steels[4], particularly at very low temperatures. The increased chromium concentration in the fusion zone provides an additional benefit. Chromium provides resistance to oxidizing conditions at high temperatures and in corrosive solutions.[2] Cr is often used in steel in combination with Ni. The general effect of the combination of alloying elements on the fracture transition temperature of welded steels is reported by Stout and Doty[5]. The report shows that the Cr-Cu-Ni steels had the lowest transition temperatures. However, there is no theory pertaining to toughness effect of solutes.[6]

CONCLUSIONS

- DT energy values of laser beam weldments of HY-80 and HY-100 with Inconel 600 insert meet the Navy's requirements.

- Charpy V-notch impact toughness values of laser beam weldments of HY-80 with Inconel 600 insert are superior to the base plate values, the values for autogenous welds and satisfy the Navy's requirements.

- A gradual increase in Ni and Cr concentrations from the true line to the apparent fusion line was found.

ACKNOWLEDGEMENT

The authors sincerely appreciate Mr. Edward R. Pierpoint for his excellent technical assistance in this experiment, particulary for his impact toughness measurements.

REFERENCES

1. E. A. Metzbower and D. W. Moon, "Fractography of Laser Welds," Fractography and Materials Science, p.131, ASTM STP 733, 1981.

2. Inconel Alloy 600, published by Huntington Alloys Inc., 1978.

3. W. F. Savage and E. S. Szekers, "Technical Note: A Mechanism for Crack Formation in HY-80 Steel Weldments," Welding Journal Research Supplement, February 1967.

4. G. E. Linnert, "Welding Metallurgy," Vol 2 p.425, 3rd edition published by AWS.

5. R. D. Stout and W. D. Doty, "Weldability of Steels," Welding Research Council, 1953 p.174.

6. William C. Leslie, "The Physical Metallurgy of Steels," p.123. McGraw-Hill Book Company.

POWDER-FEED LAYERGLAZE[SM]/ NARROW-GAP LASER WELDING OF TITANIUM-6A1-4V

Edward M. Breinan and David B. Snow
United Technologies Research Center
East Hartford, CT 06108

ABSTRACT

Laser welds of 1.27 and 2.54 cm thick Ti-6A1-4V (wt%) plate separated by a 4 mm straight-sided gap were achieved with the aid of continuous addition to the weld gap of prealloyed powder of the same composition at the point of beam impingement (the powder-feed LAYERGLAZE process). A continuous CO_2 laser was utilized in the unstable resonator mode with a set beam power of 5 kW, focused by a copper mirror of 45.7 cm focal length. Each successive layer of feedstock filled the gap width, and was overlaid by another until the gap was filled. Ambient temperature tests of cross-weld tensile and impact specimens revealed no fusion zone strength reduction but a somewhat lower toughness than the annealed base metal. The average fusion zone microhardness was 417 VHN vs 342 outside the HAZ, which reflected an increase in oxygen content of the fusion zone. The fusion zone grain structure was distinctively columnar, with grains traversing many successive deposited layers. No grain boundary α phase was detected; the grains consisted entirely of fine α' HCP martensite with a wide range of lath sizes and a high dislocation density.

INTRODUCTION

During the past few years, the perpetual need for better performance in advanced military aircraft and warships has led to the development of structural and engine materials with improved mechanical property combinations. In the case of nickel, titanium and aluminum base alloys, such improvements have become increasingly difficult to achieve through new alloy development alone. This situation has led to a greater awareness that significant advances in materials properties can be made through the development of advanced, precisely controlled processing techniques, which beneficially change the alloy microstructure during either consolidation or joining. One such technology is laser materials processing, which has been developed at the United Technologies Research Center as a result of its traditional interest in improved materials and continuous, high power carbon dioxide laser systems. A specific example is the LAYERGLAZE Process (1-3), in which wire or powder feedstock is introduced into the moving laser beam-metal interaction zone. By this technique it is possible to achieve the bulk consolidation of rapidly solidified alloys (10^4-10^5 °C/s) in the form of sequentially overlapping layers. The correct choice of beam intensity and energy distribution on the workpiece will reduce the amount of retained porosity to very low levels (1,4), thus eliminating the need for subsequent densification. A controlled amount of substrate melting occurs during the deposition of each layer, and this ensures complete interlayer mixing and consequent mechanical integrity between layers.

An important requirement for successful LAYERGLAZE processing is the capacity of an alloy to accommodate the high stresses and consequent high strain rates imposed by the rapid solid state cooling of each layer. Research at UTRC has demonstrated that titanium base alloys are particularly compatible with the LAYERGLAZE Process in this respect (4). In addition, laser welded and LAYERGLAZE processed titanium alloys have been reported to contain very low levels of entrapped porosity in most cases (4-6), with one exception (7). Further, the mechanical properties of as-laser welded, near-α and α-β titanium alloys are comparable to, and in some cases more ductile than, welds made by gas/tungsten arc or electron beam welding (6,8). These favorable structure/property characteristics of laser processed titanium are augmented by two factors which make LAYERGLAZE processing particularly applicable to the narrow gap welding of thick plates: (A) the deep penetration characteristics of high-power lasers achieved by the creation of a steady-state vapor column (the "keyhole" effect) and by sidewall reflection within the gap (9), and (B) the greater effectiveness of feedstock introduction into the gap in powder rather than wire form. It was the main objective of the first phase of the research program described herein to demonstrate that narrow-gap welding of Ti-6A1-4V plate could be achieved by this means without machining the weld gap into a "v" configuration or sacrificing the superior mechanical properties of autogeneous laser welds in this alloy. The data reported herein show that this objective was successfully achieved. Further, they suggested that it will be feasible to attempt to achieve additional ductility and strength enhancement of laser-processed titanium alloys through deliberate changes in the powder feedstock composition and the inherently rapid solidification of the fusion zone.

Experimental Procedure

The titanium alloy utilized for this research program was Ti-6Al-4V (wt%), purchased in the mill-annealed condition in thicknesses of 1.27 and 2.54 cm. The specimen sides were machined so that two pieces could be joined to form the narrow gap configurations shown in Fig. 1. The specimens were placed in a retaining fixture (Fig. 2) so as to form a 4 mm wide, straight-sided narrow gap. This fixture was equipped with retention bars to prevent warpage of the welded specimens due to solidification shrinkage as the gap was filled.

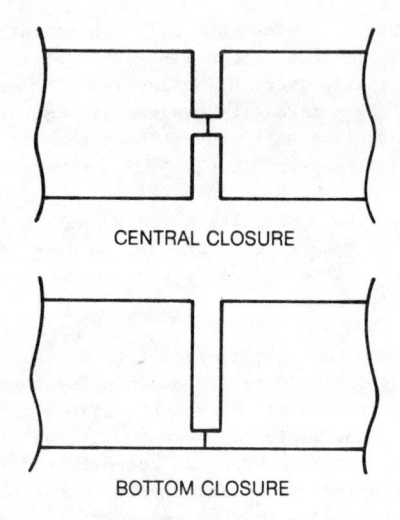

Fig. 1. Narrow gap weld specimen configurations

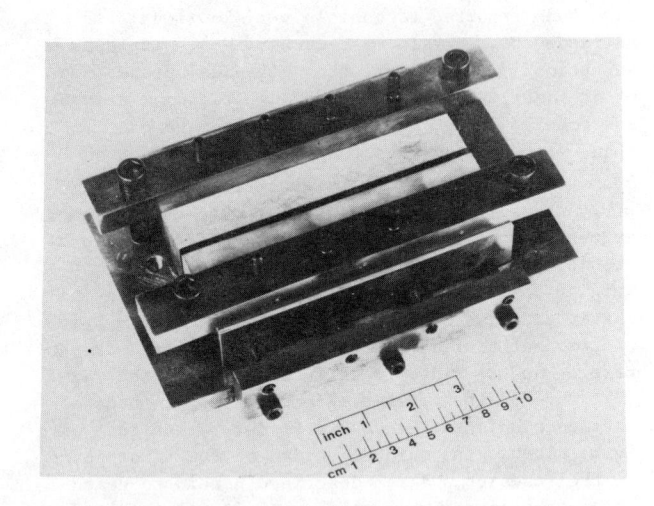

Fig. 2. Narrow gap positioning fixture

The protective atmosphere chamber and the work station utilized for all LAYERGLAZE/narrow-gap welds are shown in Fig. 3. The laser utilized for all welds was a continuous, cross-flow CO_2 laser operated in the unstable resonator mode (annular beam cross-section). The beam was focused to a 3.18 mm diameter spot at the bottom of the gap by means of a 47 cm focal length copper mirror positioned so that the bottom of the gap lay 2.86 cm below its focal plane.

Fig. 3. Work station, including protective atmosphere chamber for titanium/narrow-gap welding

The specimen traverse speed was 2.12 cm/s when continuous powder feed was used; 0.85 cm/s for preplaced powder. The laser power was set at 5 kW, so that ~3.9 kW was delivered to the workpiece with an intensity of ~47.5 kW/cm^2. The specimen height was lowered after every four layers were added to compensate for material accumulation at the bottom of the gap and thus maintain a relatively constant laser power input to the weld.

Transport of the powder feedstock into the weld gap was accomplished by gravity feed assisted by mechanical vibration of the powder container (1). An attempt was made to utilize both mechanically blended powder and prealloyed powder. However, an erratic flow rate was experienced with the elemental blend powder, and prealloyed powder was used for all subsequent welds. The specifications of both types powder feedstock are described in Table I.

Table I
Ti-6Al-4V Powder Feedstock Specifications

Powder Type	Mesh Size Range (as-received)	Source
Mechanically Blended	-170, +325	Amerimet
Prealloyed (Solution Blend)	-35, +325	Nuclear Metals

Prealloyed Powder Composition

Element:	Al	V	Fe	W	Cu
wt%	6.6	4.2	0.16	0.0007	0.0027

Element:	O	N	C	H_2	
wt%	0.173	0.012	0.012	0.0036	

Specimen contamination was minimized by chemically cleaning the Ti-6Al-4V plate minutes prior to welding in a solution of 30% nitric and 3% hydrofluoric acid, rinsed with water and dried with forced air. The protective chamber was filled with helium during welding, continuously introduced at a flow rate of 5.7 m^3/h.

An attempt was made to weld one specimen with the use of a 51-cm focal length oscillating mirror, which was driven by compressed air so as to rotate the focal spot in a circle at 2000 rpm. This innovation was unsuccessful and was replaced by the 47 cm fixed focusing mirror used for all other specimens. A shaker table was used underneath two specimens during welding, but it was subsequently removed (specimens 9 and 10) as it appeared to promote sidewall bridging by fusion of the powder above the bottom of the weld, thus causing macroscopic voids. Specimens 9 and 10 were welded with preplaced powder feedstock in an attempt to completely eliminate this problem, a procedure which also allowed the entire height of the gap to be filled in 14 passes. The height of these two specimens was lowered by 3.8 mm after placement of the first four layers, but was not lowered again as the beam was about to touch the sidewalls of the gap.

The macroscopic structure of all welds were examined in transverse section, but (2.54 cm gap) specimen No. 9 received the most extensive evaluation. This included a determination of the microhardness variation across the fusion zone at three different heights, and detailed microstructural observation by light microscopy and analytical electron microscopy using a Philips EM400T TEM/STEM.

Mechanical test specimens were prepared from sections cut transverse to the welding direction. A typical arrangement of test specimens within such a section is shown by Fig. 4. Both standard and one-half size Charpy bars were prepared, and were fractured at room temperature on an instrumented apparatus according to standard ASTM procedures. Tensile specimens were prepared with 2.61 gage length and a 3.18 mm gage diameter and were tested at room temperature at a strain rate of 0.01 min^{-1} (1.7 x 10^{-4} s^{-1}).

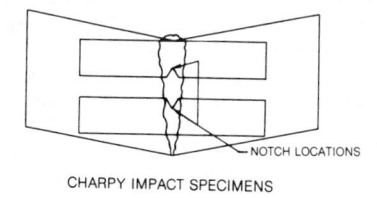

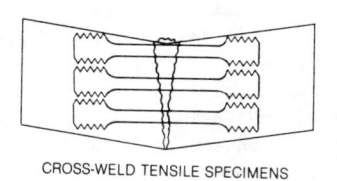

Fig. 4. Location of mechanical test specimens in LAYERGLAZE/narrow-gap welds.

RESULTS AND DISCUSSION

A. General Structure

The macrostructure and transverse microstructure of 2.54 cm specimens welded from one side (bottom gap closure) and from both sides (central gap closure) with continuous powder feed are shown in Figs. 5 and 6, respectively. These welds had reasonably good integrity, but exhibited occasional macroscopic porosity due to insufficient powder feed to the gap.

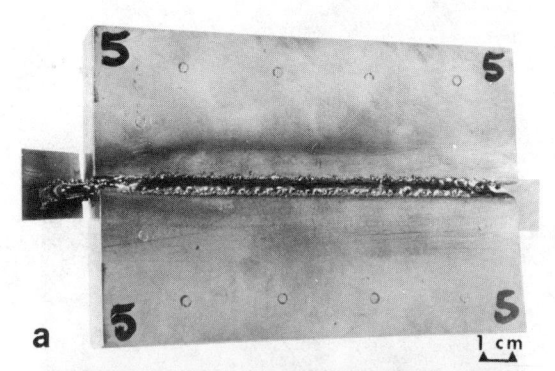

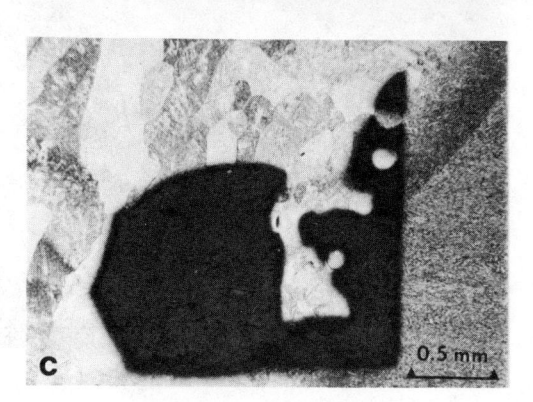

Fig. 5. Bottom gap closure, 2.54 cm thick LAYERGLAZE/narrow gap weld specimen. (a) Top surface. (b) Transverse microstructure. (c) Unfilled cavity at bottom.

257

Fig. 6. Central gap closure, 2.54 cm thick LAYERGLAZE/ narrow gap weld specimen. (a) Top surface. (b) Transverse microstructure. (c) Unfilled cavity at bottom.

The low magnification views of each cross section show the distinctive columnar grain structure generated by the LAYERGLAZE Process (1,2,4), which is formed by the sequential epitaxial nucleation of the grains as each layer is deposited. The orientation of the columnar grains tends to become more vertical toward the center of the fusion zone. This is in contrast to the columnar microstructure of autogeneous, deep-penetration laser welds in α-β titanium alloys reported in the literature (6-8), in which the grains are oriented in a direction more uniformly perpendicular to the plane of the weld center line. This difference in the Layerglazed microstructure reflects an additional heat flux toward the bottom of each layer during its solidification, in addition to that toward the sides of the fusion zone.

By way of contrast, the 2.54 cm-thick specimen welded from both sides, with a machined step bridging the gap at mid-thickness, displayed a fusion zone with generally better integrity since both exposed surfaces were now the top crown of the weld (Fig. 6a). The microstructure was generally free from macroporosity, although the gap was still not consistently filled in at the bottom (Figs. 6b & 6c). Again, the grain structure changed from vertically columnar at the center line to smaller and nearly horizontal at the edge of the heat-affected zone.

In order to ensure that subsequent specimens would have sufficient integrity for mechanical property evaluation, the powder feedstock was preplaced at the bottom of the weld gap prior to each laser pass. During the welding of specimens No. 9 and 10, each layer of powder was added to a depth of ∼2.5 mm, which allowed the 2.5 cm weld gap to be filled after ∼14 laser passes. The resulting weld had a smooth surface appearance, with no evidence of macroscopic porosity when sectioned. However, some instances of ≤0.2 mm interlayer porosity were observed (Fig. 7).

Specimen No. 10 was welded under the same conditions that were employed for specimen No. 9, but was double sided, with two gaps of 1.27 cm depth. However, when an attempt was made to weld the second side, the degree of warpage after the second laser pass closed the gap sufficiently to cause sidewall bridging and consequent macroscopic cavity formation. The fusion zone of the first side appeared to have the quality of weld No. 9, in which only small spherical voids (≤50μm) were infrequently observed at the interface between some of the layers.

B. Chemical Analysis and Microhardness

Samples of the fusion zone and of the surrounding unwelded plate were taken from weld specimen No. 9 for chemical analysis. The results (Table II) show that there was some increase in the average oxygen content of the fusion zone. Some degree of oxygen increase was anticipated because of the 0.17 wt% O content of the powder feedstock and its sequential deposition in layers. To examine this phenomenon in more detail, it was assumed that oxygen in solid solution would produce a measurable hardness increase.

Fig. 7. Bottom gap closure, preplaced powder, 2.54 cm thick LAYERGLAZE/narrow gap weld specimen No. 9. (a) Transverse microstructure. (b) Interlayer microporosity.

Table II

Chemical Analysis, Weld Specimen No. 9

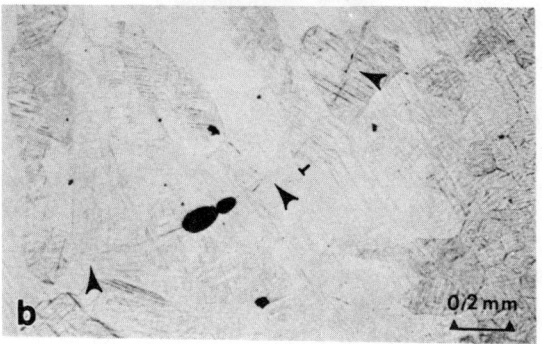

Element (Wt%):	Al	V	O
Base metal	6.1	4.01	0.11
Fusion zone	5.8	4.1	0.14

Horizontal microhardness profiles of one transverse cross section of weld specimen No. 9 were determined near the top surface, at mid-height and near the bottom surface across the welded region. As shown by Fig. 8, the microhardness of the fusion zone was significantly greater than the base metal outside the heat-affected zone. However, there was no significant variation in fusion microhardness from top to bottom. The overall microhardness increase in the heat-affected zone suggests that there was some oxygen pickup there, as well. A statistical analysis of the data is presented in Table III. A similar evaluation of laser-welded Ti-6Al-4V microhardness was reported (10) in which the base metal was determined to be 335-350 VHN, as observed in this study. However, the oxygen content of the fusion zone was reported to be 0.325% (presumably wt%), with a mean microhardness of ∼380 VHN in an identical microstructure of 100% α' martensite (10), vs. 0.14 wt% O and 417 VHN in the fusion zone of specimen No. 9. Nevertheless, the oxygen content observed here appears well within the range typical of Ti-6Al-4V (≲0.3 wt%).

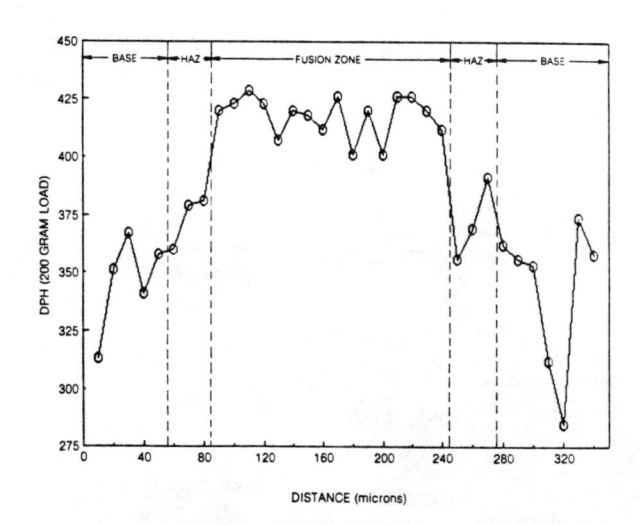

Fig. 8. Transverse microhardness scan, mid-height, specimen No. 9.

C. Mechanical Properties

The narrow fusion zones of the specimens welded during this investigation precluded the preparation of tensile specimens with a gage length composed entirely of as-welded material. However, tensile specimens oriented perpendicular to welding direction were prepared from weld specimens 9 and 10, with the fusion zone placed in the center of the gage length, and these were tested at ambient temperature. Additional data concerning the fusion zone yield strength was obtained by using two strain gages, one placed directly on the fusion zone and one on the material outside of the heat-affected zone. Thus the yield strength of the as-welded fusion zone, as well as that of the base metal, was obtained (Table IV). These data are presented in Table IV, and an example of the engineering stress-strain data from a dual strain-gaged tensile specimen are shown in Fig. The range of values of the fusion zone 0.2% yield stress was comparable to previous data from Ti-6Al-4V laser welds where the dissolved oxygen was reported to be higher (10); and significantly lower than the data of another investigation (% oxygen unspecified) (8). Overall, the tensile tests showed that these two narrow gap/LAYERGLAZE welds in specimens 9 and 10 were sound and were stronger than the unwelded plate. This strength differential is attributed both to the increased interstitial (oxygen) content of the fusion zone (11), and to its finer microstructure; and is typical of welds in alpha-beta titanium alloys (12). No evidence for fusion zone strengthening via martensite tempering (12) during sequential layer deposition was observed by transmission electron microscopy (subsequently discussed).

The fracture toughness of weld specimens No. 9 and 10 were evaluated by fracturing Charpy specimens according to standard ASTM procedures (13) (Table V). Apparently, the large prior-β grain size, which permitted the formation of a substantial number of long plates (long undeviated crack paths) in the acicular α' resulted in a significant reduction in fracture energy compared to the normal range for solution

Table III

Analysis of Microhardness Data, Weld Specimen No. 9

		Sample Size	Mean DPH	Unbiased Standard Deviation
UPPER SECTION	Base Metal	10	355	14
	Fusion Zone	17	414	13
MIDDLE SECTION	Base Metal	10	341	29
	Fusion Zone	16	418	9
LOWER SECTION	Base Metal	10	330	16
	Fusion Zone	16	419	20
OVERALL	Base Metal	30	342	23
	Fusion Zone	49	417	14

Table IV

Tensile Test Data, Weld Specimens No. 9 and 10

Weld Specimen	Identification	Weld Strain Gage 0.2% YS MPa(ksi)	Unwelded Zone Strain Gage 0.2% YS MPa(ksi)	UTS MPa(ksi)	% ε Fusion + Unwelded Zones
9	-*	-	841(122)	874(127)	12.3
9	9C-2**	>883(128)	800(116)	862(125)	9.9[1]
9	9F-2**	834(121)	807(117)	855(124)	10.7[1]
9	9H-1**	855(124)	800(116)	869(126)	10.8[1]
9	9C-1	-	-	862(125)	10.6[1]
9	9D-1	-	-	848(123)	9.1[1]
9	9D-2	-	-	855(124)	9.4[1]
9	9F-1	-	-	807(117)	1.1[2]
9	9H-2	-	-	862(125)	9.5[1]
10	10B-2**	821(119)	800(116)	876(127)	9.0[1]
10	10D-2**	>848(123)	793(115)	848(123)	9.7[1]
10	10B-2	-	-	862(125)	10.8[1]

*as-received only, no weld, average of four tests
**separate strain gages, fusion and unwelded zones

[1] failure in unwelded zone
[2] failure in fusion zone, macroscopic internal porosity

Table V

Ambient Temperature Fracture Toughness Data,
Weld Specimens, No. 9 and 10

Weld Specimen	Identification	Measured 1/2 - Size Charpy V-Notch Energy Joules(ft-lb)	Calculated Full Size Charpy V-Notch Energy Joules(ft-lb)
9	9A-1	5.55(4.09)	11.61(8.65)
9	9A-2	5.94(4.38)	12.19(8.99)
9	9B-1	5.02(3.70)	10.78(7.95)
9	9B-2	5.94(4.38)	12.38(9.13)
9	9B-3	4.83(3.56)	14.51(10.7)
9	9B-4	4.23(3.12)	8.65(6.38)
9	9E-1	11.81(8.71)	24.14(17.8)
9	9E-2	6.24(4.60)	12.95(9.55)
10	10A-2	4.01(2.96)	8.45(6.23)
10	10A-3	4.77(3.52)	9.89(7.29)
10	10C-1	4.46(3.29)	9.21(6.79)

annealed, or solution annealed and aged Ti-6Al-4V (14). This effect has been previously observed in similar microstructures formed both by β-quenching (15) and by electron beam welding (12). The force vs time curves recorded during these instrumented Charpy tests revealed considerable plastic strain occurred during the fracture process. A higher range of fracture energies were obtained from another investigation of laser welded Ti-6-4 (7), although the origin of this difference is not apparent as the microstructures are very similar.

D. Fusion Zone Microstructure

A detailed examination of the fusion zone microstructure shown in Figs. and was conducted by conventional metallography and analytical electron microscopy. As shown by Fig. 9, a grain structure typical of the LAYERGLAZE process was formed (1-4), in which the grains grew epitaxially as each successive layer was deposited. By this means the coarse, columnar grains were aligned approximately parallel to the heat flux during solification. Although this coarse grain structure might be interpreted as indicative of a relatively slow solidification rate, both a previously titanium-base alloys (16) and the fully martensitic microstructure strongly suggest that the solidification and solid state cooling rates were $\geq 10^4$°C/s. Consequently, the fusion zones consisted entirely of α' HCP martensite, while the heat affected zone displayed a sharp microstructural gradient, from all α' near the weld to finely mixed α+β near the base plate. The scale of the martensitic structure is somewhat deceptive, in that the very long α' plates which extend for hundreds of microns across the grains represent only the large end of the martensite plate size distribution. Examination of the fusion zone of specimen No. 9 by transmission electron microscopy (Fig. 10) revealed a much finer martensitic structure between these long plates, as the distance available for each new plate to form was successively reduced by those which had transformed earlier (geometric partitioning). Examination of relatively large amounts of this area, such as the ~ 480 μm^2 shown in Fig. 10, revealed no evidence of grain boundary α in the fusion zone. At higher magnification, most of the microstructural characteristics previously reported in β-quenched Ti-6Al-4V were observed (17), although no evidence of retained β was detected by selected area diffraction. Narrow stacking faults and longitudinally oriented $\{10\bar{1}2\}$ twins visible within the larger plates (Fig. 11), as well as the high dislocation density characteristic of as-LAYERGLAZED alloys (1,2). Qualitative EDS and microdiffraction data obtained from STEM observation of the same specimen revealed no evidence of discrete grain boundary phases or chemical segregation (Fig. 12). However, the electron microscope was operated with a tungsten rather than a LaB$_6$ electron source, so that the spot intensity was not sufficient to reveal grain boundary segregation at the level of a monolayer or less.

Overall, the examination of narrow gap/LAYERGLAZE weld microstructure at high magnification revealed no unexpected features. Although the grain structure was large and columnar, the martensitic structure within the grains was homogeneous and contained a very wide range of α' HCP martensite plate lengths (<1μm to >100μm). No phase distribution or chemical segregation unique to the grain boundaries could be observed, and the dislocation density appeared to be high enough to permit recrystallization of the fusion zone, should this prove desirable.

Fig. 9. Transverse microstructure, fusion zone of specimen No. 9.

Fig. 10. α' HCP martensitic structure of fusion zone, specimen No. 9.

Fig. 11. {10$\bar{1}$2} twins (A) and stacking faults (B) in large martensite plate, fusion zone, specimen No. 9.

Fig. 12. Grain boundary in fusion zone, specimen No. 9 (a) STEM image, area of EDS analysis containing grain boundary (arrows). ⌖ marks position of beam at grain boundary (arrows). (b) EDS data; no significant difference between (line) spectrum from entire area shown in (a) and that from a point on the grain boundary (dots), accelerating potential 120 kV; beam spot size 50 Å.

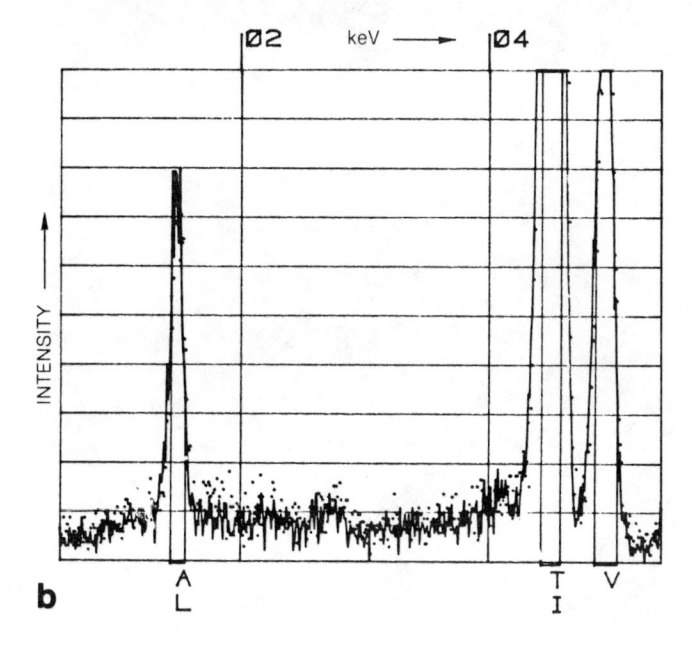

CONCLUSIONS

1. The powder-feed LAYERGLAZE Process can be used to make narrow, 4 mm straight-sided gap welds in 1.7 and 2.5 cm thick Ti-6Al-4V plate.

2. Solution-blend Ti-6Al-4V powder can be used to form narrow-gap welds with mechanical properties comparable to close-fitting laser butt welds in Ti-6Al-4V plate.

3. Narrow gap welds in Ti-6Al-4V plate can be made by LAYERGLAZING with continuous powder feed, but improved beam control to eliminate premature sidewall fusion above the gap floor will be required to avoid sidewall bridging and consequent macroscopic cavities.

4. The fusion zone cooling rate was sufficiently rapid ($\geq 10^4$°C/s) so that the resulting microstructure was entirely composed of acicular α' HCP martensite, with no formation of grain boundary α.

5. The high dislocation density and phase metastability inherent in LAYERGLAZED Ti-6Al-4V welds would allow grain refinement by post-weld recrystallization anneals.

ACKNOWLEDGMENTS

The authors gratefully acknowledge the financial support provided by the Office of Naval Research under contract N00014-79-C-0649. The assistance with the LAYERGLAZE processing by C. O. Brown of the Industrial Lasers Department of UTRC is also gratefully acknowledged.

REFERENCES

1. E. M. Breinan, D. B. Snow and C. O. Brown: "Program to Investigate Advanced Laser Processing of Materials", UTRC Final Report R81-914346-8, Contract N00014-78-0387, ARPA Order No. 3542, Jan. 1981.

2. E. M. Breinan, D. B. Snow, C. O. Brown and B. H. Kear: "New Developments in Laser Surface Melting Using Continuous Prealloyed Powder Feed", Rapid Solidification Processing, Principles and Technologies, II, R. Mehrabian, B. H. Kear and M. Cohen, eds., Claitor's Publishing Div., Baton Rouge, 1980, p 440.

3. B. H. Kear, J. W. Mayer, J. M. Poate, and P. R. Strutt: "Surface Treatments Using Laser, Electron and Ion Beam Processing Methods", Metallurgical Treatises, J. K. Tien and J. F. Elliott, eds., TMS-AIME, Warrendale, PA, 1981, p 321.

4. L. M. Masur, E. M. Breinan and C. O. Brown: "Application of the LAYERGLAZE Process to JT-8D Turbine Spacer Knife Edge Seal Repair", UTRC Report R79-210405-1, August 1979.

5. E. M. Breinan and C. M. Banas: "Fatigue of Laser-Welded Ti-6Al-4V", UTRC Report R75-412260-1, July 1975.

6. W. A. Baeslack III and C. M. Banas: "A Comparative Evaluation of Laser and Gas Tungsten Arc Weldments in High Temperature Titanium Alloys", J. Welding 60, 121s-130s (1981).

7. F. W. Fraser and E. A. Metzbower: "Laser Welding of a Titanium Alloy", *Advanced Processing Methods for Titanium*, D. F. Hasson and C. H. Hamilton, eds., TMS-AIME, Warrendale, PA, 1982, p 175.

8. T. C. Peng, S. M. L. Sastry, J. E. O'Neal and J. F. Tesson: "Microstructures and Mechanical Properties of Laser-Welded Titanium Alloys", *ibid.*, p 189.

9. Y. Arata: "Narrow Gap High Energy Beam Welding (1) - Principle", *Trans. Japan Welding Institute*, 2, 119-120 (1973).

10. J. Mazumder and W. M. Steen: "Microstructure and Mechanical Properties of Laser Welded Titanium 6Al-4V", *Met. Trans. A*, 13A, 865-71 (1982).

11. K. Shimasaki, K. Ono, and T. Tsuruno: "Relation Between Brinell Hardness of Titanium and Impurities (O_2, Fe, N, and C)", *Titanium '80, Science and Technology*, H. Kimura and O. Izumi, eds., TMS-AIME, Warrendale, PA, 1980, p 1131.

12. D. W. Becker, R. W. Messler, Jr. and W. A. Baeslack III: "Titanium Welding A Critical Review", *ibid.*, p 256.

13. *1981 Annual Book of ASTM Standards, Part 10*, Section E23-81, ASTM, Philadelphia, 1981.

14. H. Margolin, J. C. Williams, J. C. Chesnutt and G. Luetjering: "A Review of the Fracture and Fatigue Behavior of Ti Alloys", *Titanium '80, Science and Technology*, H. Kimura and O. Izumi, eds., TMS-AIME, Warrendale, PA, 1982, p 169.

15. M. Ashraf Imam and C. M. Gilmore: "New Observations of the Transformations in Ti-6Al-4V, *ibid.*, p 1534.

16. T. C. Peng, S. M. L. Sastry and J. E. O'Neal: "Exploratory Study of Laser Processing of Titanium Alloys", *Lasers in Metallurgy*, K. Mukerjee and J. Mazumder, eds., TMS-AIME, Warrendale, PA, 1981, p 279.

17. J. C. Williams and M. J. Blackburn: "A Comparison of Phase Transformations in Three Commercial Titanium Alloys", *Trans. ASM*, 60, 373-83 (1967).

LASER BEAM WELDING AT NIROP, A NAVY MANUFACTURING TECHNOLOGY PROGRAM

E. A. Metzbower
Naval Research Laboratory

R. A. Hella
AVCO Everett Metalworking Lasers

G. Theodorski
Northern Ordnance Division
FMC Corporation

Abstract

A high power laser beam welding system has been purchased by the Naval Research Laboratory (NRL) and installed at the Naval Industrial Reserve Ordnance Plant (NIROP), Minneapolis. The purchase was funded by the Manufacturing Technology Division of the Naval Material Command. NRL acted as project manager on the purchase and installation and also provided technical expertise on laser beam welding. The NIROP, Minneapolis, is an ordnance facility of the Naval Sea Systems Command, operated by the Northern Ordnance Division of the FMC Corporation. The plant manufactures guided missile launch systems and gun mounts for the U.S. Navy.

The laser beam welding system was procured from AVCO Everett Metalworking Lasers (AEML). The laser is a high power, continuous wave, carbon dioxide laser and can deliver 15 kilowatts to the workpiece. The linear workstation has the capability of producing either horizontal or vertical welds up to 10 feet (3.0 m) long. The beam director can place the focused beam at any position within a volume 10 ft x 10 ft x 6 ft (3.0 m x 3.0 m x 1.9 m). The workstation has a variable traverse speed of 1.7 to 100 inches per minutes (43.2 to 2540 mm per minute).

The comparison of the time required to weld a component with the present process and of the anticipated time necessary to weld the same component with the laser beam has resulted in an economic analysis. This economic analysis indicates that a cost savings of over $500,000 per year can be realized. Three illustrations of different weld joints that will be fabricated with the laser beam are given. An example of how the laser can possibly reduce costs even more by replacing extrusions with weldments is also shown.

The mechanical properties of laser beam weldments of A36 structural steel have been measured by transverse tension specimens. The laser beam weld is stronger than the base plate. Hardness traverses across the base plate, heat-affected and fusion zone indicate an increased hardness in the heat-affected and fusion zones. This increased hardness can be modified by a stress relief heat treatment. Charpy V-notch specimens have been used to measure the amount of energy absorbed as a function of temperature. The microstructure of the base plate is ferrite and pearlite. The heat-affected zone is refined ferrite/pearlite, whereas the fusion zone is bainite.

INTRODUCTION

The primary concern of the Navy Manufacturing Technology Program is to address problems of the production process and to increase manufacturing productivity for systems and equipment being procured by the Navy. More specifically, program objectives include the promotion and establishment of improved processes, methods, techniques, and equipment for the most efficient and economical production of defense material and providing the technology required to advance manufacturing capability. In order to achieve these objectives, the Naval Sea Systems Command (NAVSEA) sponsored a manufacturing technology program, "Laser Welding of Mild Steel at NIROP, Minneapolis." This program was administered by the Naval Research Laboratory (NRL) which also provided the technical management. The NIROP, Minneapolis, is the Naval Industrial Reserve Ordnance Plant, Minneapolis, a government-owned, company-operated facility which fabricates guided missile launching systems and gun mounts for the Navy. The plant is operated by the Northern Ordnance Division of the FMC Corporation.

This manufacturing technology program has purchased and installed a laser beam welding system at NIROP. Through the technology transfer phase of the program, welding engineers of FMC have been instructed in the use of high power lasers for welding. The laser system offers the capability of welding mild steel efficiently, soundly and at a much faster welding speed than the present submerged arc welding process. The laser and the workstation are described in this article. A brief description of the welding process is followed by a more detailed exposition of the properties of the weldments. An economic analysis on several different weldments is given and finally a preview of future work is stated.

LASER

The laser beam welding system was purchased by NRL as a result of a competitive procurement from AVCO Everett Metalworking Lasers (AEML). The laser

(Fig. 1) is a continuous-wave, closed cycle, electric discharge carbon dioxide laser with a wavelength of 10.6 micrometers (µm). The characteristics of the laser are stated in Table 1.

Fig. 1. Continuous wave, CO$_2$ Laser

Table 1

Characteristics of NIROP Laser

Power Output	1-15kW (at work piece)
Repeatability and Accuracy	+3%
Duty Cycle	Continuous (CW)
Lasing Gas Mix	He:H$_2$:CO$_2$:CO 6:4:1:1
Wavelength	10.6 µm
Operating Pressure	0.1 Atm (nominal)
Welding Beam	
Focal Diameter	0.040 in. (1.0mm) (f/7)
Power in Focal Spot	85% of Laser Output

An electron-beam stablized discharge provides the required pumping of the lasing mixture to generate the coherent 10.6 µm radiation. Lasing occurs in an "unstable resonator" optical cavity of the discharge region. The lasing gas is a mixture of helium, nitrogen, carbon dioxide and carbon monoxide. A high energy e-beam ionizes the gas, raising its electrical conductivity sufficiently to allow the main sustainer current to flow in the laser gas between the sustainer electrodes volumetrically.

The main chamber is a closed-circuit wind tunnel through which the lasing gas mixture circulates. Heat exchangers, cooled by the primary water supply, remove waste heat. Axial flow compressors or blowers drive the lasing gas mixture through the channel where lasing occurs.

The cathode and anode are attached to opposite walls of the channel. Sustainer current flows across the gas stream between the electrodes. The ionizer injects a broad, uniformly accelerated beam of high-energy electrons into the flowing laser gas, which becomes ionized, in order to provide a stable sustained discharge in the laser gas between the sustainer electrodes. In a closed-loop mode, a power monitor senses the laser output and adjusts the ionizing current and hence sustainer current to maintain programmed output power.

Water-cooled copper mirrors at each end of the channel form the resonator cavity. The transfer mirror is tilted to direct the output beam at a slight off-axis slope out of the aerodynamic window. The aerodynamic window permits the laser beam to exit from the low-pressure main chamber without allowing outside air to enter the chamber.

The laser output power is controlled by the e-beam current. The basis of the output is the set point on the operating panel and the maintainence of this power setting by the power monitoring subsystem. The time duration of the laser output is controlled by a dual shutter subsystem sequenced by the programmable controller. The output power at the cone power meter is variable from a threshold of 1 kilowatt (kW) to at least 20 kW. The maximum output power does not vary more that +3% within a 30 minute period or more than +5% during an 8 hour period.

The expanding beam emerging from the aerodynamic window encounters a transmissive chopper blade which directs part of the energy to the fast detector system used for monitoring laser power stability. The beam can then encounter a sliding mirror which directs the beam to a cone power meter. The expanding beam is collimated and directed to a rotating mirror which is used to direct the beam to the desired workstation.

WORKSTATION

The linear weld station is arranged with the work motion car moving normal to the optical axis. The track length is sized to accommodate a 10 ft. (3 m) long weld seam and the variable speed traverse car can move at speeds of 1.7 to 100 ipm (.7 to 42 mm/s). The characteristics of the workstation are stated in Table 2.

Table 2: Characteristics of Laser Weld Workstation

Linear Weld Workstation	
Work Piece Dimensions	10'x10'x6' (3X3X2m)
Work Piece Load	20,000 lbs. (9100 kg)
Work Motion	1-100 ipm (0.7 - 92mm/5)
Seam Tracking	+ 4 in. (I 10cm) (in two axis orthogonal to weld direction)
Wire Feed	Variable
Inert Gas Shield	

The focusing optics (Fig. 2) are mounted on an arm on the optical support, all adjustable along the optical axis with the arm adjustable in the elevation axis. For down hand welding the arm is raised to the proper height in order to place the telescope focal point approximately at the weld elevation. The support is moved towards the work until the focal point is over the weld location. Thus the focal point can be placed anywhere within a

10x10x6 ft. (3x3x2 m) volume at the work station (Fig. 3).

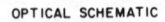

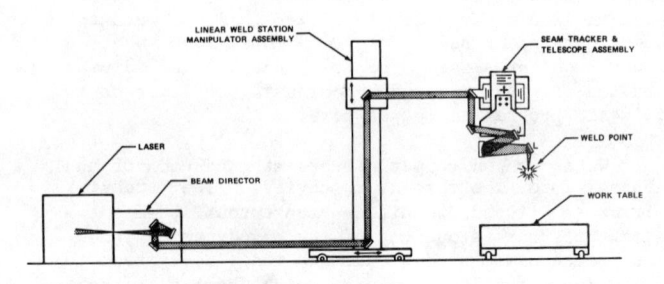

Fig. 2. Beam Delivery System

Fig. 3. Laser Welding Workstation

Linear welds can equally well be performed in the vertical plane by removing the down hand mirror and adjusting the optical support away from the workcar to a position where the focal point is at the weld plane. A seam tracker, wire feeder, gas shielding and plasma suppression devices are incorporated into the end of the beam director at the focal point of the laser beam.

LASER BEAM WELDING

Thick section laser beam welding is achieved by focusing a high power laser beam onto the surface of an alloy. This focused, highly collimated source of heat evaporates the alloy and creates a "keyhole", or metal-vapor-filled cavity, through the thickness of the workpiece. Surrounding this keyhole is the weld pool, a volume of molten metal in dynamic equilibrium with the metal vapor. As the workpiece is moved, melting takes place at the leading edge of the weld pool and solidification occurs to the rear. The depth of penetration of the keyhole and the shape of the weld pool are governed by a variety of factors such as: the amount of energy absorbed and the amount reflected at various temperatures, laser beam power at the focused spot, the size of the focused spot, travel speed, the gas used to shield the weld pool, the method of gas shielding, the thermal properties of the alloy being welded, etc.

At the top of the keyhole, above the workpiece, the evaporated alloy forms an ionized gas called a plasma. If this plasma is allowed to remain on top of the keyhole, it will absorb some of the laser beam and reradiate it in all directions, thus decreasing the depth of penetration of the weld. The plasma can easily be displaced by a stream of inert gas across the top of the keyhole. Helium, because of its ionization energy, is usually used for plasma control, although a helium-argon mixture may be used effectively (1). Thus, plasma formation is not considered to be a limiting factor in laser beam welding.

The relative positions of the top of the plate and the focal point of the laser beam have been shown (2) to affect the penetration capability for a fixed power and speed. In order to fabricate the deepest welds with a high depth-to-width ratio, the focal point of the laser is usually positioned about 2.5mm (0.1 in) below the top surface of the plate.

EXPERIMENTAL METHOD

Process

The most practical carbon steel commercially available for welded structures is ASTM A36, which is a steel with optimum composition for strength and weldability (3). This steel is widely used in many thicknesses and has excellent mechanical properties when welded. Welding of this alloy can be achieved by many processes: shielded metal arc (SMA), gas metal arc (GMA), submerged arc (SA), electron beam (EB), etc. The resulting weldments have good mechanical properties and ductility. Although not normally used in structures subjected to low temperatures or impact loading, the Charpy V-notch (CVN) energy of the welds is equal to or greater than that of the plate. In large structures, the weldments are often given a stress-relief heat treatment in order to minimize distortion and residual stress, and to enhance fit-up.

The laser processing parameters for the A36 are given in Fig. 4. These parameters have resulted in weldments which have yielded satisfactory mechanical properties.

Properties

The mechanical properties (yield and ultimate strengths, and elongation) were measured by transverse tension specimens, either round or reduced section plate specimens. The mechanical property data reflect the base plate values, since the specimens broke in the base plate and not the fusion or heat-affected zones. For all of the thicknesses of plates welded, transverse bend specimens were machined from the A36 weldments as well as from the base plate (L-T and T-L orientations) (5). These specimens were tested as a function of temperature. Additional bend, mechanical property, and Charpy V-notch testing was performed on weldments of A36 that had been given a stress-relief heat treatment at 635°C (1175°F) for one hour. The values of the Charpy V-notch energy as a function of temperature are shown in Fig. 5.

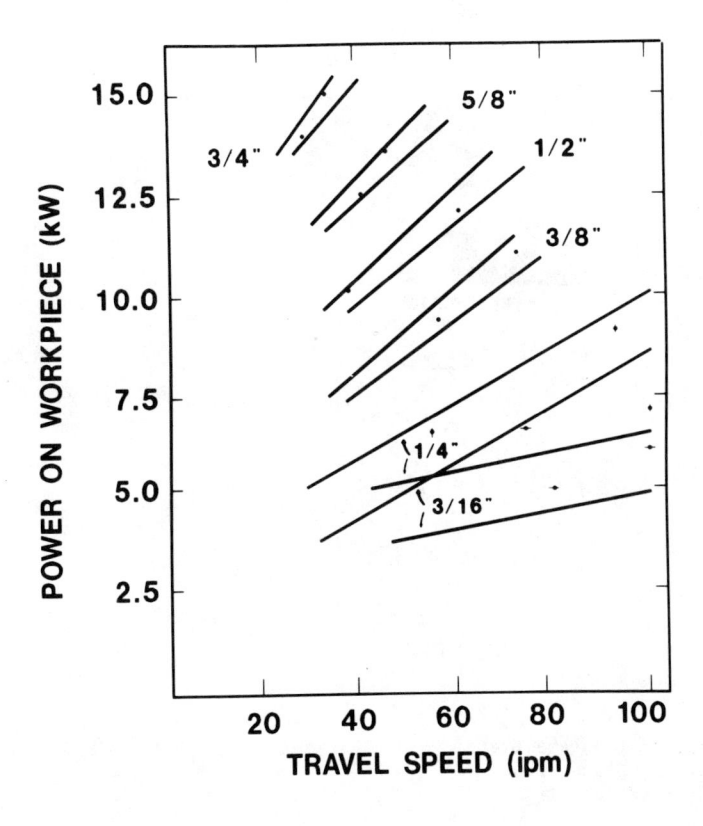

Fig. 4. Laser Power/Welding Speed Relationships for Different Thicknesses of Plate

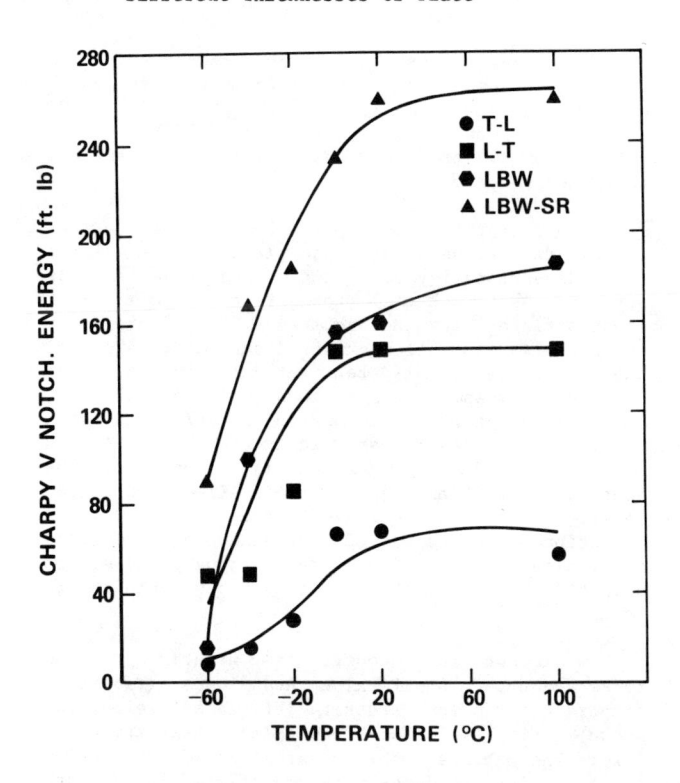

Fig. 5. Charpy V-Notch Energy as a Function of Temperature for the Base Plate (T-L, L-T) and Laser welds (as welded and stress-relief).

Microstructures

A36 steel has a ferrite/pearlite structure. When this steel is laser beam welded, the heat-affected zone (HAZ) has a refined, smaller grain size ferrite/pearlite structure, whereas the fusion zone has a bainitic structure. A cross section of the laser beam weldment is shown in Fig. 6. These microstructures are shown in Fig. 7. Hardness traverses across the HAZ and the fusion zone, for both the as-welded and stress-relief laser beam welds, are shown in Fig. 8.

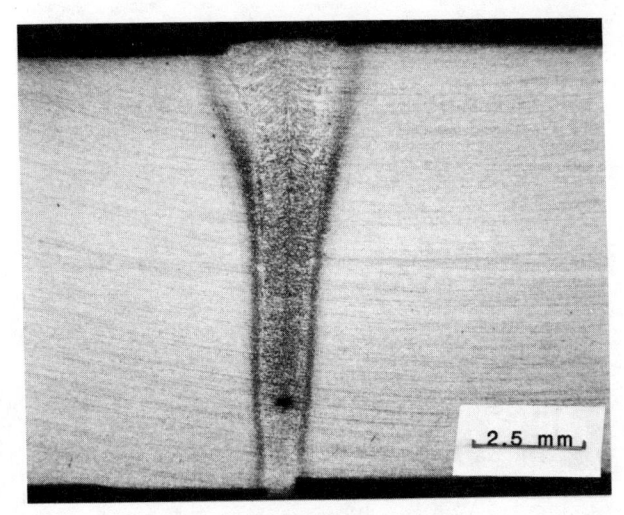

Fig. 6. Cross Section of Laser Beam Weld

ANALYSIS OF RESULTS

The mechanical property data gathered in these tests are a reflection of the base plate properties. But they also imply that the HAZ and fusion zone are stronger than the base plate. The bend specimens, which typically expose defects in the weldment such as undercut and interior defects (4), satisfactorily passed their tests.

Considering the chemical compositon of the A36 base plate, the microstructures observed in the HAZ and fusion zone are what is to be expected from a high energy density, low heat input process. Solidification takes place very rapidly, and the heat is conducted away from the fusion zone quickly, resulting in a very rapid quench from a high temperature. The increase in hardness values in the A36 laser beam weldments is moderate and indicates that no appreciable change in properties should occur.

The results of testing the Charpy V-notch specimens as a function of temperature are quite interesting. The base plate specimens indicated two different levels of upper shelf values depending

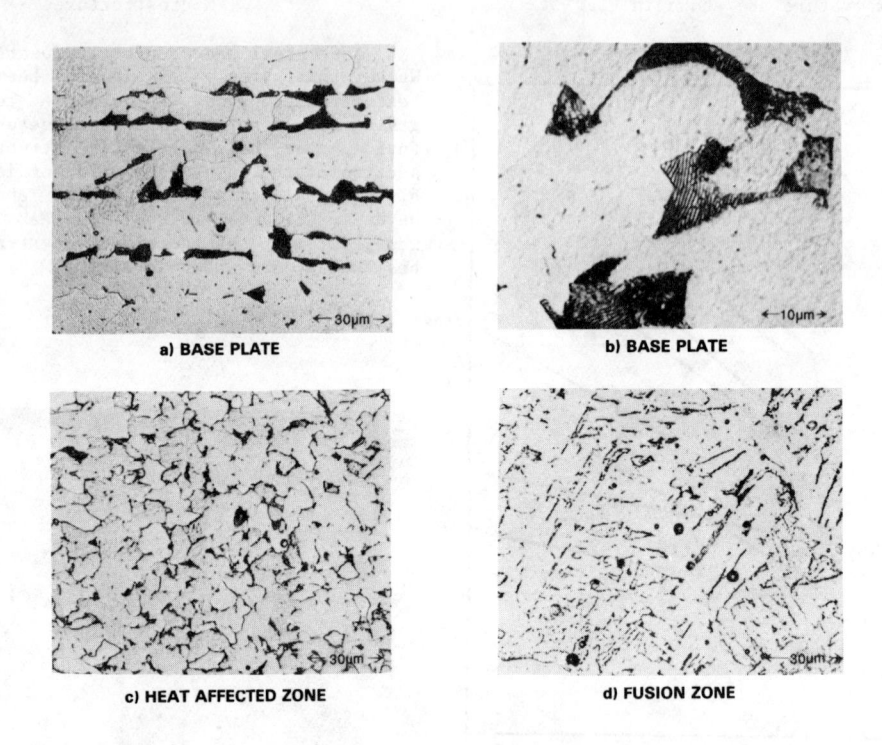

Fig. 7. Microstructures of (a) base plate (ferrite/pearlite), (b) base plate (resolved pearlite), (c) heat-affected zone (refined ferrite/pearlite), and (d) fusion zone (bainite)

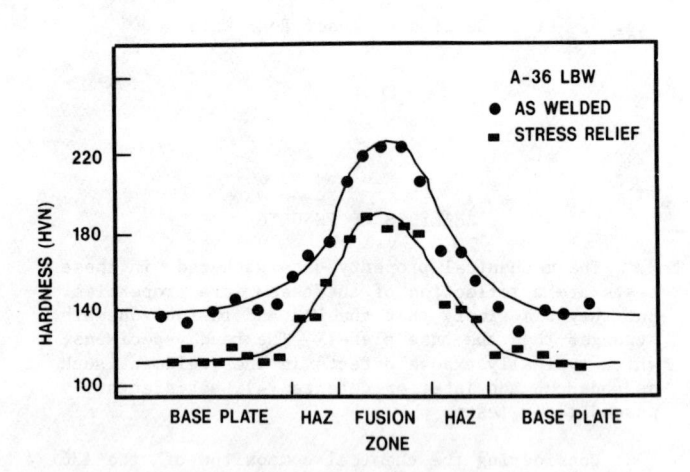

Fig. 8. Hardness Traverses Across A36 Laser Beam Weldments (as-welded and stress relief)

upon the specimen orientation (L-T or T-L). The as-welded laser beam weldments have a higher upper shelf value than those of the base plate. The stress-relief laser beam specimens has an even higher upper shelf value (most of the specimens stopped the machine at impact) and also a higher transition temperature. The values plotted in Fig. 5, for the absorbed energy at the various temperatures, are an average of at least three, and as many as five, tests. The scatter in the energy values decreased with decreasing temperature. The Charpy requirement for an E70 type weld wire, which would normally be used to weld the A36 steel, is 27 joules at -30°C (20 ft-lb at -20°F). The autogeneous laser beam welds of A36 easily surpass that requirement.

The difference in upper shelf values between the two directions in the base plate is a result of the directionality of the manganese sulfide stringers. A crack will propagate more easily along these stringers than across them. The difference in upper shelf values between the laser beam welds and the base plate can also be related to the inclusion content and shape. Fusion zone purification, i.e., the evaporation of inclusions in the weld zone as a result of the laser beam welding process, has been shown to reduce, significantly, the inclusion content, as well as changing their sizes and shapes (6,7). When an equivalent treatment, i.e., a reduction in the number of inclusions in a steel, is carried out as part of the steel-making process, the properties of the steel become less anisotropic (8).

A factor that contributes greatly to the increased upper shelf value and lower transition temperature of the stress-relief laser welds was shown in Fig. 8. The stress-relief heat treatment lowers the hardness of the heat-affected and fusion zones. Since hardness can be related to strength and as the strength increases, absorbed energy decreases (8), the lower hardness contributes to the higher upper shelf values.

This type of phenomenon, increased absorbed energy after stress-relief, has been shown to exist in arc and electron beam welds (11, 12). Davies and Garland attribute the increase in the impact strength to the degeneration of the bainitic carbides, formed along the prior solidification boundaries, into ferrite-carbide aggregates. Russell, et al, report a decrease in hardness as a result of the stress-relief heat treatment and do not report in any microstructural changes. They also report a significant increase in the energy absorbed as a result of stress-relief heat treating.

ECONOMICS

A comprehensive feasability study (13) sponsored by NAVSEA at the NIROP, Minneapolis examined the cost savings of laser beam welding over conventional welding fabrication. As part of this study, a model was constructed of the weldshop procedures. This model identified both on-station and off-station operations associated with various weld-joint configurations. Actual production-time standards were then utilized to estimate the time savings resulting from laser beam welding for various generic weld joints. As anticipated, the major time savings were derived from a reduction in burn-time, i.e. the time during which the arc is struck in conventional welding or the beam-on time in laser welding. Significant time savings also occurred from reduced straightening (less distortion) associated with the adoption of laser beam welding.

The FMC Corporation under contract with the Navy developed an integration plan (14) for the laser beam welding system. A detailed analysis of weldments to be fabricated by the system was completed.

An example of laser beam welding is the ready service ring components on the Mark 26 Guided Missile Launching System. The ready service rings store missiles and transfer them into position for hoisting. The laser beam welding application is a linear 0.875-inch (22.5 mm) thick butt weld on the 6-and 10-section ready service rings. The joints to be laser welded are the flange-to-web joints in 1015 steel (Fig. 9). The present method is submerged arc welding, requiring seven passes and a backgouge operation. Laser beam welding will require only one pass with no filler wire addition and no backgouge operation. Also grinding and straightening will be reduced due to the noncontact, low distortion properties of laser welding. The cost savings resulting from laser beam welding are shown in Fig. 10.

FUTURE WORK

This implementation of the laser beam welding system requires an analysis of the factors influencing the physical and functional integration of the system into the production line. Assuming that the laser beam welding system and the operator are qualified, the successful execution of the following events are required for each candidate laser beam weld before that joint can be made with the laser:

- Review of necessary weld joint design changes;
- Preparation of engineering drawings;
- Preparation of alternate process sheets and sketches;
- Selection and design of laser beam welding, tooling and fixturing;
- Demonstration of laser beam welding on prototype production model; and
- Production and scheduling of laser beam welding.

FLANGE-TO-WEB
105.5" LONG, 7/8" BUTT WELD
1015 STEEL

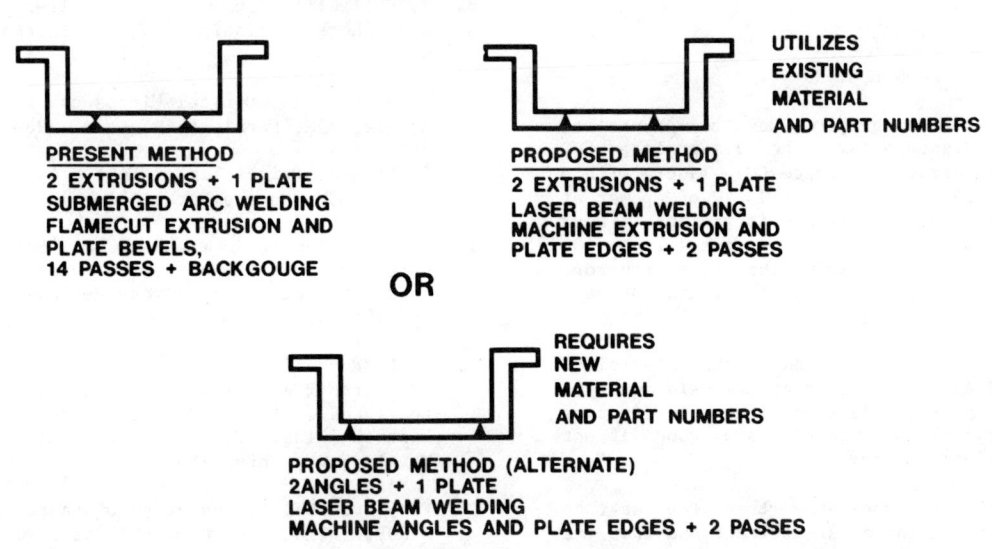

PRESENT METHOD
2 EXTRUSIONS + 1 PLATE
SUBMERGED ARC WELDING
FLAMECUT EXTRUSION AND
PLATE BEVELS,
14 PASSES + BACKGOUGE

OR

PROPOSED METHOD
2 EXTRUSIONS + 1 PLATE
LASER BEAM WELDING
MACHINE EXTRUSION AND
PLATE EDGES + 2 PASSES
UTILIZES EXISTING MATERIAL AND PART NUMBERS

PROPOSED METHOD (ALTERNATE)
2 ANGLES + 1 PLATE
LASER BEAM WELDING
MACHINE ANGLES AND PLATE EDGES + 2 PASSES
REQUIRES NEW MATERIAL AND PART NUMBERS

Fig. 9. Mark 26 Ready Service Rings, Flange-to-Flange and Flange-to-Web Butt Welds

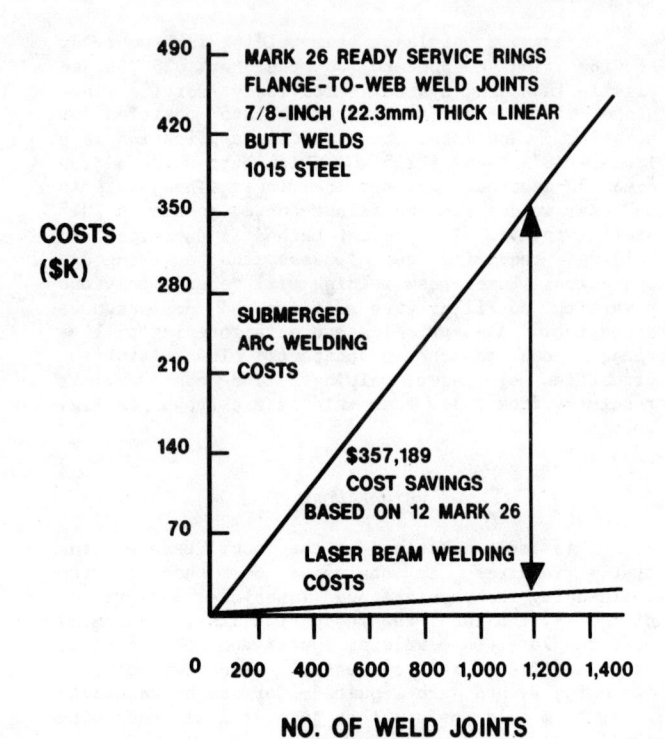

Fig. 10. Cost Savings for the Mark 26 Ready Service Rings

The original economic analysis (13) of laser beam welding at NIROP, Minneapolis, indicated that whereas the workstation usage would be high, i.e., about 2000 hours, the actual laser burn time is low, about 60 hours. The implication of this is straightforward: have many workstations. However, the practical implications require a finite number of workstations for welding. Since the laser has been purchased and installed, a demand to utilize its unique capabilities as a heat source arises in the areas of transformation hardening, alloying, cladding and cutting. These areas are being examined for potential manufacturing technology programs.

CONCLUSIONS

A laser beam welding system has been installed at the NIROP, Minneapolis. The laser is a CW, closed-cycle, electric discharge CO_2 laser with a wavelength of 10.6 μm, capable of 15 kW of power at the workstation. The workstation is a simple linear tranverse table, capable of speeds from 1.7 to 100 ipm (.7 to 42 mm/s). Utilizing the beam director, the focal point of the laser can be placed anywhere within a volume 10x10x6 ft. (3x3x2 m).

Preliminary laser beam welds on the material to be welded, ASTM A36, indicate that the weld joint is stronger than the base plate and that the fracture toughness (Charpy V-notch test) is as good if not better than the base plate.

The economic analyses show that the savings accrued by laser welding is significant and that the cost of the laser beam welding system will be recovered within three years. The economic analyses also indicate the potential for time sharing the laser between several workstations for doing welding, heat treating, alloying, cutting and cladding.

ACKNOWLEDGMENTS

This program required the cooperation and work of many people. Some of these people are: W. Garth Brown, NAVSEA; J. W. McInnis, NAVMAT; L. R. Hettche, D. W. Moon, E. R. Pierpoint at NRL; M. Sirchis at AEML; D. Brazys and E. Wigand at NOD/FMC. We are grateful for their support and cooperation.

REFERENCES

1. F. D. Seaman, "The Role of Shielding Gas in High Power CO_2 (CW) Laser Welding," Technical Paper MR77-982, Society of Manufacturing Engineers, Dearborn, Michigan, 1977.

2. F. D. Seaman, "The Role of Focus in Heavy-Plate Laser Welding," Technical Paper MR78-345, Society of Manufacturing Engineers, Dearborn, Michigan, 1978.

3. G. Linnert, Welding Metallurgy, 3rd Edition, Vol. 2, American Welding Society, Miami, Florida, 1967, p. 350.

4. "Standard Methods for Mechanical Testing of Welds," ANSI/AWS B4.0-77, American Welding Society, Miami, Florida, 1979.

5. ASTM Standard E399, 1980 Annual Book of ASTM Standards, Part 10, ASTM, Philadelphia, PA, 1980, p. 596.

6. E. M. Breinan and C. M. Banas, "Fusion Zone Purification During Welding with High Power CO_2 Lasers," Proceedings of the Second International Symposium of the Japan Welding Society, Osaka, Japan, Aug. 24-28, 1975.

7. E. A. Metzbower and D. W. Moon, "Fractography of Laser Welds," in Fractography and Materials Science, ASTM STP 733, L. N. Gilbertson and R. D. Zipp, Eds., ASTM, Philadelphia, PA, 1981.

8. W. C. Leslie, The Physical Metallurgy of Steels, Hemisphere Publishing Co., Washington, DC, 1981, p. 207.

9. L. E. Samuels, Optical Microscopy of Carbon Steels, ASM, Metals Park, Ohio, 1980, p. 121.

10. G. Krauss, Principles of Heat Treatment of Steel, ASM, Metals Park, Ohio, 1980, p. 175.

11. G. J. Davies and J. G. Garland, "Solidification Structures and Properties of Fusion Welds," International Metallurgical Reviews, Vol. 20, No. 196, June, 1975.

12. J. D. Russell, A. J. Rodgers, and R. J. Stearn, "Electron Beam Welding of Structural Steel," reprinted in Source Book on Electron Beam and Laser Welding," M. M. Schwartz, Editor, ASM, Metals Park, Ohio, 1981.

13. F. W. Gobetz, "Feasability of Laser Welding at NIROP," UTRC, Report R79-914782-1, 1979.

14. D. Brazys, "Laser Welding Integration Plan for Naval Research Laboratory," FMC Corporation Report E-2050, March 1982.

		Al
Q 621.366 L343	Lasers in materials processing	
	83072954	

		Al
Q 621.366 L343	Lasers in materials processing	
	83072954	

FEB 8 7

RODMAN PUBLIC LIBRARY
215 East Broadway
Alliance, OH 44601